Cooperative & Commune

Proceedings of a Conference on Group Farming
Madison, Wisconsin, 1975
Jointly sponsored by the Research Training Network
of The Agricultural Development Council, New York,
and the Land Tenure Center, University of
Wisconsin-Madison

Cooperative and Commune

Group Farming in the Economic Development of Agriculture

Edited by Peter Dorner

The University of Wisconsin Press

Published 1977
The University of Wisconsin Press
Box 1379, Madison, Wisconsin 53701

The University of Wisconsin Press, Ltd.
70 Great Russell Street, London

First printing

Printed in the United States of America

For LC CIP information see the colophon

ISBN 0-299-07380-7

List of Authors

18 Bodoni / 22 c & lc

John W. Bennett
Professor of Anthropology
Washington University
St. Louis, Missouri

Jacques Chonchol
Minister of Agriculture under the Allende Government in Chile
Presently Associate Professor
Institut de Hautes Etudes de l'Amerique Latine
Université de Paris, France

Yehuda Don
Professor of Economics
Bar Ilan University
Ramat Gan, Israel

Peter Dorner
Professor of Agricultural Economics and the Land Tenure Center
University of Wisconsin-Madison

Antony Ellman
FAO Advisor
Agrarian Research and Training Institute
Colombo, Sri Lanka

Gilbert Etienne
Professor of Development Economics
Graduate Institute of International Studies and Institute of Development Studies
Geneva, Switzerland

Boguslaw Galeski
Professor of Sociology
Institute of Philosophy and Sociology
Warsaw, Poland

Douglas Horton
Agricultural Economist
The International Potato Center
Lima, Peru

Natsuki Kanazawa
Professor of Agricultural Economics
Tokyo University, Japan

Don Kanel
Director of the Land Tenure Center and Professor of Agricultural Economics
University of Wisconsin-Madison

Moise C. Mensah
FAO Regional Representative for Africa
Accra, Ghana

Robert F. Miller
Senior Fellow in Political Science
The Australian National University, Canberra

Philip M. Raup
Professor of Agricultural and Applied Economics
University of Minnesota, St. Paul

Edward P. Reed
Graduate Student-Project Assistant in the Department of Agricultural Economics and the Land Tenure Center
University of Wisconsin-Madison

Nimal Sanderatne
Agricultural Economist
Economic Research Department
Central Bank of Ceylon
Colombo, Sri Lanka

Robert C. Stuart
Professor of Economics
Rutgers University
New Brunswick, New Jersey

Karl-Eugen Wädekin
Professor of International and East European/Soviet Agricultural Policy Economics
Justus-Liebig University
Giessen, Federal Republic of Germany

John Wong
Senior Lecturer in Economics
University of Singapore

Contents

18/22 Bodoni

Preface

The present volume is based on papers presented at a conference jointly sponsored by the Research and Training Network of The Agricultural Development Council, New York, and the Land Tenure Center of the University of Wisconsin-Madison. This Group Farming Conference was held at Madison June 10-12, 1975. It was the culmination of a three-year effort by a small planning group that met periodically, exchanged correspondence on review of literature, located potential authors, developed guidelines for invited papers, etc. Authors were selected on the basis of their past research on group farming in specific countries or their personal and professional involvement in establishing group farming systems. The intent was to bring together in a systematic way present knowledge regarding experiences with group farming.

"Group Farming" was defined to include agricultural production cooperatives, with varying degrees and levels of integration, as well as other forms of group or collective farming. In many countries of the less developed world there is a growing interest in some form of group farming. At times this may be advocated on the basis of ideological-political concerns. However, in most cases governments are seeking ways and means of reorganizing their agriculture in order to deal with the fundamental economic and social issues in its development.

In many cases there is little or no previous experience with cooperative farming. Consequently, countries may look for models of these forms established elsewhere. Ideas about production cooperatives and other forms of group farming are sometimes imported and adopted without critical evaluation of the conditions necessary to make them function effectively (i.e., to achieve the purposes for which they were intended). The objective of the conference, then, was to mobilize international knowledge on the experience with cooperative-collective farming. The hope and the expectation is that such information, widely disseminated, will serve as a useful input for policy-makers in their attempts to formulate programs for resolving the agrarian issues in their countries.

In the "Guide for the Preparation of Papers for the Group Farming Conference" sent to authors, four major categories were suggested as an organizational framework around which to discuss a specific country experience:

1. External influences and conditions affecting the establishment of group/collective farming enterprises;
2. Internal operating mechanisms and procedures which characterize such enterprises;
3. Measures of socio-economic performance of these enterprises (relative both to external criteria as well as to objectives set for such enterprises in the legislation or decree under which they were established);
4. Nature of change, transition, and evolution over time, and how changes in 1, 2, and 3 above may have influenced such evolution.

Obviously all experiences analyzed in the formal papers did not lend themselves to a rigid adherence to these categories and the numerous sub-topics suggested under each one. Nevertheless, this classification did provide somewhat of a unified focus and was useful in facilitating communication among conference participants.

As editor of this volume, I wish to express my deep appreciation to the many people who devoted their time and skillful assistance in organizing the conference and in getting this volume into print. Project Assistants Ira J. Cohen and Edward P. Reed coordinated literature, assembled references, and wrote brief summaries of many of them in the year preceding the conference. Their task was facilitated by the suggestions received from many professionals the world over who responded to our earlier inquiries seeking advice on literature, countries to be included, potential authors, etc. I am grateful for their assistance.

The numerous mailings to authors, discussants, and other conference participants, manuscript reproductions, and general supervision were under the skillful management of the Land Tenure Center's Program Coordinator, Julia Schwenn. Others at the Center, especially Head Librarian Teresa Anderson and Administrative Assistant Don Esser, were most helpful in providing assistance both before and during the conference, as was Technical Typist Helaine Kriegel. I wish to express my sincere thanks and appreciation for their help.

Final editing, correspondence with authors, liaison with the Press, and the over-all management involved in getting the manuscript to the Press were under the direction of Jane Knowles, Editor and Executive Assistant at the Land Tenure Center. Her expert knowledge and good judgment were indispensable and I am most grateful for her valuable assistance. For the very skillful typing and re-typing of the final manuscript, I sincerely thank Jane Dennis, Technical Typist at the Center.

The Conference and thus the present volume were made possible through the financial and other support provided by the co-sponsoring organizations--the Research and Training Network of The Agricultural Development Council and the Land Tenure Center. Both RTN and LTC receive part of their financial support from the U.S. Agency for International Development. I wish to express my appreciation to these organizations, and to a number of other international organizations who sent participants to the Conference. Finally, my thanks to the Department of Agricultural Economics and administrators in the College of Agricultural and Life Sciences and the University of Wisconsin-Madison for their interest and support in providing a favorable environment for work on these issues of international agricultural development.

To the above and to all participants at the conference, my sincere thanks and appreciation.

Peter Dorner

Madison, Wisconsin
September 1976

Cooperative & Commune

12 Bodoni
c & lc

Introduction: Some Economic and Administrative Issues in Group Farming

Peter Dorner and Don Kanel

The kind of economic organization to establish in farming is not likely to be a purely economic issue. Although this issue inevitably arises if a country embarks on a large-scale land reform, political and ideological factors will probably play a key role in determining the outcome. And it is of little value to look only superficially at present circumstances and then project a desired end product without due consideration to the means by which the present situation can and must be used painfully to reconstruct that future. Nations must deal with the agricultural system as it is and as it might reasonably be modified, not as it might be if there were a clean slate upon which to design such a system. The rural people with their specific skills or lack of them, the physical resources with peculiar capacities, obstacles and locations, the attitudes, beliefs, and institutions generated by unique historical antecedents--all these make up the real world situation to be dealt with. Any basic transformation of the economic and social organization in farming (of the land tenure system) is likely to involve a slow, painful process of experimentation pursued over a number of years, with built-in mechanisms of change for continuing evolution.

But all countries do not have the option, nor perhaps the desire, for massive transformation of the organizational patterns in the entire agricultural sector. Indeed all the country cases discussed in this volume, with the exception of the Soviet Union and the Peoples' Republic of China, have mixed systems--partially collectivized and partially organized on the basis of private property and individual farming. Such mixed systems create special and unique problems both for governments and for people in farming. Much of our discussion refers to such mixed systems.

In this paper we leave aside the political and ideological considerations leading to a decision to collectivize, or to establish group farming in, parts of the agricultural sector. We focus instead on some of the economic and administrative issues that arise after such a decision has been made. Also, except for an occasional reference, we do not deal with state farms. Since they are operated by state-appointed managers and hired laborers (all state employees), such enterprises are somewhat simpler in their organization and decision-making than are production cooperatives.

We begin by listing some of the fundamental issues in the development of agriculture. In order to gain perspective with respect to some of them, we have included brief comments indicating the manner in which such issues are likely to be resolved (or accentuated) under a dualistic farm-size structure in agriculture and under a family farm system. The main body of the paper is then devoted to a discussion of how group farming attempts to deal with these issues, and some of the major economic and administrative problems that are likely to arise in this process.

Policy-makers in many countries are struggling with many difficult economic issues in the development of agriculture. These include:

a) increasing rural (farm and nonfarm) employment opportunities;
b) expanding food and agricultural output;
c) achieving a more equal distribution of income;
d) obtaining greater governmental control over agricultural output;
e) preserving the infrastructure of expropriated large farms;
f) including the landless among land reform beneficiaries;
g) improving rural infrastructural and service systems.

It is in trying to resolve these and other basic issues that the form of economic organization in agriculture is brought into question.

The evidence is quite clear that a dualistic farm-size structure in agriculture does not provide adequate answers to the issues raised above. Incomes generated within such a system go mainly to the operators of large farms. Further, once a dualistic system begins to modernize it is highly inclined to mechanize to avoid, among other things, labor problems in the management of a large work force. Typically, small farmers and many of the landless become marginal to or nonparticipants in the process of economic growth within such a dual structure.

A family farm system is better able to provide some answers to the issues raised. We consider such a system as one in which the majority of the farms--i.e., operating the bulk of the land and providing the bulk of the farm output--employ family labor with only minor hiring or labor exchanges with other farmers at peak seasons. A family farm system requires an integration with private or cooperative (and perhaps governmental) marketing and supply firms, and a set of government services designed to serve it--agricultural research, extension, credit and transport systems, market facilities, regulations and quality controls, etc. There may be tenancy and/or sharecropping in addition to owner operatorship of family farms, but this would usually not be the predominant form of tenure.

There is sufficient evidence in many countries that a family farm system can be highly productive. Japan and Taiwan are the obvious but not the only examples; small farms in most countries utilize land very intensively through application of family labor, the use of which is not limited by its factor market cost, for any labor use which raises production and/or income benefits the family. Moreover, these farms are highly absorptive of labor and can provide at least some productive opportunities for an expanding rural labor force. Recent experience indicates that family farms are capable of adopting new technology such as the high-yielding varieties of seed and the related inputs provided they have access to such strategic variables as credit and water.

Despite its many efficiencies, a family farm system does present at least two major drawbacks to planners trying to cope with the development issues raised above. First, a highly productive family farm system requires an elaborate service structure which is both expensive and time-consuming to develop; a government cannot deal effectively with such a system until all the necessary infrastructure is in place and markets have begun to function more or less competitively. Second, a family farm system can allow great inequalities to develop. Such inequalities may be a function of variations in individual entrepreneurial abilities. Whatever their cause, they can accumulate over time and present serious obstacles to achieving a resolution of some of the issues noted above.

A number of countries are experimenting with group farming or production cooperatives in an effort to overcome the economic problems inherent in either a dualistic or a family farm system. We consider group farming as an organization in which the bulk of the land is held and farmed in common by the group (it may actually be owned by either the group or the state). Although the organization of group farms is committed in principle to participation by the members in management and policy decisions, production and marketing decisions are usually made by managers. However, whether management is effective and whether membership participation is meaningful are among the most difficult problems in group farming. Payments to participants often are in the form of advances per day worked (with different rates of payment for different types of work), plus a distribution of net earnings at year's end after deduction of all operating expenses, depreciation, funds for construction of communal buildings, other investments, welfare reserves, etc. Unless there is clear understanding and participation on the part of the members in the accounting processes, members may interpret the advances as wages rather than a partial payment for returns based on decisions--good or bad--made throughout the year for which final accounting can be made only at year's end. To develop such a view, it is imperative that members indeed participate in making these decisions.

A common feature in group farming is a private or individual plot of land for each member family. An alternative, though relatively infrequent, arrangement is the commune. Here the group provides the main consumption needs, including communal food preparation, with cash payments reduced to very nominal amounts.

The output of group farms may be marketed in a variety of ways. In socialist countries a large part of the marketed portion of the collective output is delivered to state agencies; a portion of the output may be distributed in kind to members, while some may go on the open market. The production of private plots of members may be allowed to be sold in private markets directly to consumers. In mixed economies group farms may deal with specialized agencies organized to assist them and market their produce, or they may have to deal with marketing firms which serve all kinds of agricultural producers; in the former case the assistance may include assured markets, guaranteed prices, and technical assistance.

As we have seen, economies of scale are not a crucial issue in that small family farms can be highly productive. Large-scale group enterprises may have some advantages, however, in certain circumstances. For example, if the prior system was one of large estates, the existing infrastructure will be geared to this scale of operation and major capital losses would result from any subdivision into smaller units. Also, where there is a great diversity in land quality, a large collective unit can be more flexible than a small one in adapting its land-use patterns to fit variations in soil or topography. This may become an increasingly important factor as population pressures on the land shorten the cultivation-fallow cycle in tropical areas, and as more farmers are forced to seek a subsistence living through cultivation of steep slopes subject to soil erosion. Likewise, a large unit may have an advantage over a large number of small ones in certain types of infrastructural investments, e.g., a drainage system.

When individual holdings are abolished and practically all peasants are forced into collective farming, as has happened in many socialist countries, then previous inequalities disappear. A rather different situation is where the state expropriates large private farms which formerly employed permanent resident workers and seasonal nonresident workers. In such cases the main beneficiaries tend to be the resident workers and it has proved to be very difficult to extend land reform benefits to the seasonal nonresident workers or to independent small peasant farmers. Resident workers resist the inclusion of outsiders in the membership of group farms, unless farm production can be significantly intensified. But even this kind of land reform reduces the cumulative inequalities often associated with early stages of economic development.

The continued existence of large private farms creates the danger of development based on labor-saving technology; with group farming, farm workers gain greater control over the type of technology that will be used, and they are more likely to preserve employment opportunities for themselves and their children. Thus group farming is likely to increase the capacity to provide equalitarian income opportunities for the succeeding generations of the original group of land reform beneficiaries. Still another situation is one in which individual peasant property has not yet emerged, as in the African case. Here group farming may serve to help farm people advance as a group, whereas the emergence of individual property and family farming may foster economic inequality and social differentiation.

Two other considerations should be mentioned under the general topic of labor absorptive capacity of farm production cooperatives. First is the development of entrepreneurial talents. As we have seen, in a family farm system opportunities exist for each farmer to test his entrepreneurial capacities. While this has its costs in that some fail (or fall behind, leading to income inequalities over time) because of a lack of managerial talents, this system does provide for self-discovery of such talents. The widespread opportunities for the development of entrepreneurial talents can have important positive consequences for the future economic development of the country as a whole.

A group farming system can provide a similar positive impulse for the development of a nation's managers and enterpreneurs. But much will depend on the nature of participation of individuals in the decision-making process. Where there is true and widespread participation in such a process, many individuals will gain experience and be able to test and develop their managerial capacities. In fact such experience may be of a broader and more significant nature than that developed in a family farm system in that it provides opportunities for dealing both with resource use and allocation questions as well as with human and social organizational issues. Second is the capacity of group farms to absorb individuals with a wide range of talents and interests. Accommodation can take place through specialization and division of labor. A person of less capacity, or simply less interest, is more likely to find a niche in a system of cooperative farming.

We cannot make a convincing case for the argument that larger units speed the adoption of new technology. They may do so, especially if decision-making is largely centralized. But, unless the system allows an appropriate degree of caution, sufficient flexibility, and enough local autonomy of judgment, it may be very difficult to correct (or even to discern) a mistaken technological judgment.

Certainly it will be easier for a government to provide support services of all kinds--credit, inputs, extension, marketing, etc.--to a small number of large units with centralized management, especially once the government has all the appropriate machinery in place and working. Moreover, large units are frequently capable, either alone or in combination with other units, of mobilizing both produce and the required labor force for establishing rural agro-industries which is next to impossible for an uncoordinated group of small farmers. These are pluses for a government.

However, in the area of credit and marketing, governments have had problems with debt repayment and marketing outside of official channels. When government agencies attempt to insure loan repayment by requiring that production be channeled through agencies which deduct loan repayments as first claim on receipts, obvious incentives are created to market through other channels; one procedure is to divide up the production among members and market it as the output of individual member plots. Similar problems of diversion out of official market channels arise if governments, faced with inflation, hope to have greater control over the output of group farms and to obtain that output at a lower price than the output of private farms, over which they have less control. However, the same kinds of difficulties are even greater when governments advance credit or try to control the marketing of the output of small peasant farms; what may be different is that individual peasant farms can usually survive even with government neglect, while group farms would in most cases need a favorable government attitude. A more basic point is that, outside of the socialist countries, the relations between governments and group farms are apt to be delicate and fragile. Group farms will need considerable government support, yet they are not likely to be obedient tools of government policies and priorities.

In contrast to the socialist countries, Third World governments should probably not expect that group farms will be levers for achieving state control over the agricultural sector. As indicated above, they are basically an attempt to include a much larger proportion of the rural people, on more equalitarian terms, in the development process. They are a form of development based on weaker members of rural society, and to achieve economic development goals (in contrast to social and political goals) they will have to be favored and supported by state action. At the same time, as organized groups, they will have a natural basis for political power which could be used to formulate goals differing from state objectives.

The favoring and supporting action will require efforts to provide markets, inputs, credit, and technical assistance. Guaranteed markets to group farms, with state agencies assuming the risks of changing market conditions, may be important. In general, favoring group farms probably means shifting at least some of the marketing and financing risks from group farms to state agencies.

Even with supportive action of state agencies, primary problems of group farming will be those of effective internal organization and of member commitment and morale. It is a delusion to expect that group farms have such obvious benefits to members or such decisive economic advantages to make it possible to overcome easily the organizational problems. These organizational problems are largely due to ambiguities in roles of both managers and members of group farms. Members are supposed to be both workers and participants in policy-making; managers are supposed to supervise the workers and at the same time to be responsible to them.

A common outcome of this dilemma is ineffective management on the one hand and poor work discipline and absence of effective participation in policy-making by the members on the other. Cunningham and Cliffe call this a "we-they" system in which there is mutual suspicion between members and managers, and the members have no feeling of identification with or control over the organization.[1]

Legal rights ("workers are owners") and legal rules ("workers elect policy-makers") do not by themselves create worker management and control. In this sense cooperatives are similar to publicly held corporations in the erosion of power from legal owners to management. There is apt to be much more participation by members in the period of organizing a cooperative farm than in a later period in which the organization is running more smoothly and routinely. In the earlier period the struggle to organize, the shared experience of setbacks or hostility from outside involve leaders in constant contact with members to build allegiance and loyalty for survival and cohesion. In the later, more smoothly running period, the interests of members and managers begin to diverge. The intense involvement in group activity decreases. Members become busy with their own affairs and managers with problems facing them--problems which tend to be too technical and complex to be of much interest to most of the membership. Meetings become boring, attendance drops, members may feel alienated from cooperative officials, and membership loyalty has to be fostered by deliberate efforts resembling public relations. Members have little time and energy to keep abreast of issues facing the management of their own coops.

Management, on the other hand, becomes more concerned and more skillful in dealing with specialized issues--changes in the relevant technology, organization of production, and interaction with outside agencies providing inputs, credit, and markets, or those supervising the group farms. These are not the tasks for amateurs. The knowledge necessary for good day-by-day decisions and for long-run strategies becomes concentrated in the management. In terms of their preoccupations, they may well come to feel that they need to concentrate thought and energy on dealing with outsiders, that they can best function with passive support of membership but with minimum distraction from members' demands for attention (demands for explanation, grievances, opposition to policies, etc.). It would be very natural for management to come to feel that they know best, yet their authority and legitimacy come from members and periodically or when faced with major decisions they need ratification by the membership. We can almost imagine management cringing at the need to present complex issues for membership ratification, worrying that issues are too complex or will seem too costly and worrying whether they will be able to explain them adequately or promote them sufficiently to get approval.

Successful avoidance of the "we-they" system in group farming is not common. It seems to have been avoided most successfully in the North American Hutterite Brethren communities, in Israeli kibbutzim, and possibly in the Chinese communes. (See Chapters 3, 2, and 5 in this volume.) This kind of success seems to require an ideological or religious commitment ("consciousness building"?) and an effective organizational form. North American utopian colonies

(New Harmony and others) of the nineteenth century are good examples of the insufficiency of ideology alone.

In a majority of cases cooperative farms in the Third World will differ from the Hutterite and Kibbutz commune model in several important respects: (1) the identity of interests between members and leaders is likely to be less because managers are more likely to be outsiders and are more likely to differ from membership in training and background; (2) while in some cases the members may have constituted a social community prior to the formation of the group farm, they would not have evolved a structure of community behavior patterns and community leadership for dealing with the kinds of issues which must be decided in group farms; the necessity of acting as a group on issues of cooperative farming is likely to be imposed or at least has to be learned; and (3) for all of the above reasons a considerable gulf is likely to exist between managers and members, and member participation in group decisions will be difficult to achieve.

One possible way to approach these problems may be to recognize that a partial (though hopefully reduced) "we-they" dichotomy is inevitable and that the real challenge in setting up successful group farms may be recognition of the validity of these separate interests of members, managers, and the state, and institutionalization of ways to negotiate among them.

As a tentative start in such a process, it may be useful to recognize the divergence of interests of the state (increased production, greater investment which may require restraining consumption, widespread access to income-earning and other opportunities for the rural population, discouragement of the emergence of economically successful individual peasants as a superior social class, increasing the base of political support in the rural society for those holding national power) from the interests of members of group farming cooperatives (rewarding income and services, autonomy of personal life, scope for individual and local initiative, meaningful opportunities to influence decisions affecting individuals and local groups, and, for the most ambitious, opportunities to advance within the institutionalized structure of the group farm and ancillary organizations as well as possibilities for engaging in at least some private family farming). One way to institutionalize this separation of interests is to divide up functions and decisions between state agencies and group farms with the relations between them controlled by contracts.

Recognition of the separate interests of the state and the members of group farms still leaves the problem of the ambiguous role of managers. In fact, the division of functions more clearly accentuates the manager-director's difficulties in dealing effectively with outside agencies while at the same time identifying with the concerns and objectives of members within their sphere of responsibility. Nevertheless, specific arrangements which allow managers to negotiate effectively with outside interests and so minimize arbitrary pressure on the farm on the one hand, and to support positive, constant efforts to increase leader-member interaction on more equal terms (in both labor and decision-making functions) on the other, could minimize the impact of this apparent conflict. Such arrangements will not be identical in each situation, but in all successful group farming systems they have been achieved in some form.

The purpose of this paper was to explore some of the critical issues in the development of agriculture, and to examine the manner in which such issues can be addressed under a system of group farming (with some comparison to other systems). Even a tentative conclusion with respect to the advantages or disadvantages of various forms of economic organization in agriculture would have to be carefully hedged and related to the specific circumstances of a given country. A major hypothesis that does emerge is the following: the economic rationale for a system of group farming is likely to be based on the possibilities of effectively mobilizing labor, combining agro-industrial development with farming, making it easier for governments to service and have a somewhat greater degree of control over the agricultural sector, promoting a more egalitarian distribution of the benefits of economic growth, providing a more rational system of land use and conservation, etc., rather than on the prospects for more efficient farm production, and that the primary problems of group farming will be those of effective external support and internal organization, member commitment and morale.

In the chapters that follow, many of the group farming issues discussed in this paper are analyzed for countries which differ greatly in economic, social, political, and historical circumstances. We have attempted to highlight some of the key questions in the hope that they might serve as a guide in reading the country-specific experiences that make up the greater part of this volume.

NOTES

1. Lionel Cliffe and G. L. Cunningham, "Ideology, Organisation, and Settlement Experience in Tanzania," in Cliffe and John S. Saul, eds., *Socialism in Tanzania*, 2 vols. (Dar es Salaam: East African Publishing House, 1973), 2: 133-36.

Part I: Characteristics and Classification

18/24 Bodoni

Editor's Introduction

In this first chapter, Galeski discusses some theoretical issues and certain characteristics associated with group farming. These characteristics differ depending on the type of organization established. His typology of different models of collective farming is based on the nature and origin of group farms and the underlying purposes of those who establish them. Galeski's discussion serves as a useful framework for understanding the specific country case studies that follow. The grouping of cases in the several parts of this book correspond, with some modification, to Galeski's four types.

In Part II of this volume, two communal forms of group farming are analyzed--the Israeli *kibbutz* and the Hutterite Colonies of North America. These correspond to Galeski's type one, "collective farms created by believers in an ideology which places a higher value on noneconomic than on economic goals." Part III consists of an analysis of the national collective systems of the Soviet Union and the Peoples' Republic of China, corresponding to Galeski's type three, "collective farms organized by governments to achieve nationwide economic and social goals." Part IV consists of case studies of a number of countries with mixed collective/individual farming systems--Yugoslavia, Chile, Peru, Tanzania, Dahomey, and Sri Lanka. This grouping corresponds in a general way to Galeski's type two, "collective farms created by landless families who were able to acquire land but, for various reasons, were unable to establish and operate individual family farms." However, the correspondence here is not so close and direct as in the others. Part V of this volume analyzes cases of group farming in the industrial market economies of Japan and France. These correspond closely to Galeski's type four, "collective farms organized by farmers in order to achieve the advantages of a large operation in the more effective use of land, manpower, and capital and the resulting higher economic returns."

Two qualifications to this typology are suggested by Galeski. First, these types are not mutually exclusive. In some countries virtually every type of collective distinguished above can be found. And second, these are ideal types. In reality there are often several underlying reasons for organizing collective farms. For example, the Israeli kibbutzim could be regarded as farms founded by groups of believers with a dominant ideological goal. Nevertheless, collective farming in Israel was also created in order to settle people on the land, frequently on land requiring extensive investments in irrigation where individual farming would have been difficult to establish.

Galeski uses several criteria for comparing different forms of collective farming, the two most important being: (1) joint ownership or use of land and of other means of agricultural production; and (2) socialization of work, or, more generally, joint farming since it includes both productive labor and management and decision-making. On the basis of these criteria, he establishes a continuum from the least socialized (communal ownership of land but with individual farming and mutual labor exchanges) to the most socialized (communes and the Kibbutz) forms of collective farming. Other features or criteria for comparing group farming systems are also discussed: management and decision-making (real access of members to decision-making processes, the free election or appointment of managers), the organization of and degree of socialization of consumption, and the distribution of output. In dealing with the continuum, however, no automatic transition from lower to higher levels of cooperativization is to be expected. Galeski's continua merely depict different levels of integration of existing group farming systems, and not automatic stages of an evolutionary process.

Finally, Galeski relates his typology of collective farms to the socio-economic systems within which they can operate. He distinguishes two broad types of socio-economic systems--an interactive (market) system and a directive (planned) system. His conclusion is that collective farms of his types one and four could emerge and survive only in an interactive system, while his type three will be found only in a directive system. Only his type two could emerge in both directive and interactive economic systems. However, he believes such farms will evolve to resemble state farms in a directive system, while they are more likely to be transformed into individual farms with some shared property or commonly organized services in an interactive system.

1. The Models of Collective Farming

Boguslaw Galeski

TYPES OF COLLECTIVE FARMS: A CLASSIFICATION

I want to distinguish four types of collective farms. They are:

- Type 1) collective farms created by believers in an ideology which puts a higher value on noneconomic than on economic goals;
- Type 2) collective farms created by landless families who were able to acquire the land but not to start individual family farms;
- Type 3) collective farms organized by governments in order to reach national economic and social goals;
- Type 4) collective farms organized by farmers in order to get the advantages of a larger operation--lower costs of production, more effective use of land, manpower, and capital, etc.--and consequently higher economic returns.

The first type includes religious communes, collective farms created by believers in radical social ideas, communes or collective farms created by people who reject contemporary industrial civilization with its system of values, and farms created by groups of believers in a particular mission of their nation, class, or race. All these collective farms have one thing in common--they are created to reach ideological goals, although they may incidentally achieve spectacular economic results as well. Some peasants or people of peasant origin might create or join such ideological groups, and in fact religious communes in the sixteenth century were composed predominantly of peasants. But in general, in ideological communes of recent origin, participation of peasants is negligible. Today, such communes are usually created by urban people, radical intelligentsia, industrial workers, or people who oppose the existing socio-economic system. They have their own vision of what society should be and they set out to create an alternative to an unjust world or simply an escape from such a world.

The second type includes the collective farms created by new settlers on land which requires substantial investments, such as irrigation, before it can be effectively used. The investment required may exceed the resources of a single family, or it may be much easier to overcome certain difficulties through group action. Laborers who receive land from a land reform program often find themselves in such a situation. They have no agricultural equipment, or the equipment remaining on the farm is functionally related to the pre-reform large-sized farms and therefore difficult to divide, e.g., tractors, combines, buildings, etc. In some cases an existing irrigation system is geared to large-scale operations. Frequently such laborers are not prepared to start individual farming since they have had little managerial experience and are usually unskilled or skilled in only one farming operation. The simplest solution for them may be to operate the farm as a unit, much as it was operated before, and to share the profits. What is common to all these groups is that they are created by landless families without extended experience of individual farming, and that they are in a new and difficult situation. They may or may not believe in a special ideology, but they are in a situation which can be much more easily overcome by group farming.

The third type includes collective farms created not by peasants themselves but by governments. Peasants may or may not oppose the collectivization, but it is surely not their idea. To date collective farms have not been created by peasants voluntarily and spontaneously. The reasons for collectivization are usually to achieve goals of national development. Governments of countries which do not have the means for industrialization such as rich oil or mineral resources must extract a surplus for industrialization from the productive abilities of the people. And since the lowest economic stratum is the largest, governments extract the means from it. Collectivization provides some measure of control over production and consumption in rural areas and over prices for food; it facilitates the use of large masses of cheap manpower for the construction of industry and infrastructure; at the same time, collectivization provides the means for administrative and political control over the masses of former petty producers.

In some East European countries, however, like East Germany or Czechoslovakia, reasons for collectivization were different. These countries already had a highly developed industrial sector, and additional programs of extensive industrialization created labor shortages in the agricultural sector. Lack of manpower was a major motivation for the organization of large farming units. Large private, profit-oriented holdings could of course not be accepted in a planned economy, both because they could not be so well controlled by government and because of socialist ideological principles. Therefore, only two solutions are acceptable--state farms or cooperative (collective) farms. State farms are usually more expensive; the government must assume all the risks. The introduction of state farming could also meet very strong peasant opposition. Hence, collective farms may be chosen as the form of large holdings which best serves to achieve national development goals.

The fourth type of collective farm is quite different. It exists where there are competitive markets; in this setting, enterprises must strive to achieve lower costs of production in order to increase profits. For certain types of high cost, indivisible technologies, it is much more economical to adopt these on a large farm than on a small farm. Therefore, farmers join their resources of capital, land, and manpower in order to produce more cheaply and to gain higher profits on the market. The French GAEC (Groupement Agricole d'Exploitation en Commun) is the best example of this kind of collective farm.

Cooperative enterprises have a long history of helping small farmers achieve increased returns in competitive market situations. But they usually did not enter the field of production. Their activity was generally confined to marketing, processing, and sometimes to the provision of production inputs and services. This no longer seems sufficient, however, and new forms are being created which have many intermediate stages, e.g., *maschinenringe* in Germany with joint ownership of mechanical agricultural equipment, agricultural circles in Poland with much broader goals, and cooperatives in Yugoslavia which operate on the principle of co-production with individual peasant farms. These forms of cooperatives could be regarded as steps toward collective farming of the fourth type.

Before elaborating further on this typology, I wish to introduce two important qualifications. First, in some countries virtually all the types of collective farms distinguished above can be found. In Poland, for example, a few ideologically motivated collective farms were organized immediately after World War II. During the period of land reform (1945-48) not collectivization but distribution of land was the main goal of government. Collective farming was not even allowed to exist at that time. Following the reform, however, some collective farms were spontaneously created by former agricultural laborers, and this type of collective farming is still dominant in the West-Central region of Poland. At that time some temporary collective farms were created by settlers on regained territories. Later, Poland also saw an attempt by the government to collectivize peasant agriculture, but this was not successful at that time and the third type of collective farm is hence nonexistent today but could be created in the future. There is little opportunity for the fourth type to emerge, but there are in Poland a few such farms oriented toward the market, created by farmers themselves in order to increase earnings and to gain other advantages.

The second qualification is that these are ideal types. In reality, there are often several underlying reasons for organizing collective farms. For example, the *kibbutzim* could be regarded as farms founded by groups of believers with their dominant goal being ideological. At the same time, however, collective farming was created in Israel in order to settle people on new land, sometimes in the desert where extensive investments were required and individual farming was not possible. Italian collective farms created by Christian Democratic, Socialist, or Communist parties could also be regarded as groups of believers and classified as Type 1. But they could also be classified as Type 2 because the reason for their creation was land reform and members belonged to the social stratum of former agricultural laborers (*braccianti*). The typology proposed, therefore, does not presume that ideology be the only goal of Type 1 farms, but that such a goal be *dominant* and that it determine the character of the collective farm.

The usefulness of the typology derives from the fact that it is nearly impossible to analyze organizational differences or economic and social consequences of collective farms without taking into account the goal of the particular organization. Certain criteria of an economic analysis of a collective could be quite meaningless for its members if they created it in order to reach some religious, moral, or social goal. The organizational principles of *kolkhozes* would be considered absurd in a nonplanned economy, but they are not so absurd if it is recognized that the kolkhoz is a form which allows for surplus extraction from agriculture for use in extensive industrialization of the country. Economic analysis is not able to explain fully the functioning of collective farming systems because they vary in accordance with the goals of their organizers. In each of the types we have, therefore, different organizational patterns, different types of conflicts, different problems to overcome, different criteria of success, different patterns of interpersonal relations, and different prospects for the future. To clarify these points, I will now analyze antecedent forms and the development of the kolkhoz as an example of the third type.

The Kolkhoz and Soviet Agriculture

Collectivization started in the Soviet Union in 1927. Immediately after the revolution, particularly between 1917 and 1919, the dominant form of collective farm was the *commune*. Communes were ideological groups created by former agricultural laborers. In communes, ownership (land was nationalized during the revolution), production, and consumption were collective. After 1921, the number of communes significantly declined, but on the eve of collectivization in 1927, 1,800 remained.

The other form of collective farm, which slowly increased in number, was the *artel*. In 1927 there were about 8,000 artels. Originally this form started in the field of artisan production and in services, but it was also adopted by the peasants. In the agricultural artel, the means of production were in the main collectively owned. Each member, however, kept his own house, a small plot of land, and a few animals. Each member was obliged to work for a given time on the collective land, but had time for working on his own plot as well. Labor payment for work on the collective enterprises in the artels was distributed according to the quantity and quality of the work performed. Usually produce, not money, was distributed. Only part of the production was sold in the market, usually going directly to marketing cooperatives or to the state. Artels were created mainly by landless and poor peasant families, partly on their own land (30 percent of the land came from members) and partly on land received from the state.

The third form of collective farm was the Association for Collective Work on the Land (*TOZ*). The means of production were owned individually; each family kept its land. Some work was done collectively, particularly work with state-owned tractors after the Machine Tractor Stations were organized, and part was done by family members working their fields individually. Animal production was fully private. All the production remained the property of the individual families.

The reasons for the sudden decision to collectivize Soviet agriculture are the object of discussion among economists, historians, and sociologists, even in the Soviet Union. There is no doubt that there was more than one reason, and that political factors influenced the decision. But some analysts consider the economic situation of the Soviet Union at that time as the most important reason. Underdeveloped to begin with and destroyed by war and revolution, the industrial sector was not able to meet the demands placed upon it. This in turn dampened the incentives for farmers to increase marketable output. An ambitious program of industrialization, mass migration from rural areas to the industrial centers, growth of the urban population, and the need to expand exports for the purchase of modern technical equipment all required the strong control of prices (particularly of food). Measures were needed to increase the supply of agricultural products to supply the industrial centers. Since agricultural producers were not inclined to increase their marketable produce, and thus it was not possible to extract the production via market exchange, it was necessary to use other measures to control the producers, their investments, their resources, and their consumption in order to extract the surplus from agriculture. Collectivization is seen by these analysts as the tool to control small agricultural producers and to subordinate them to the program of industrialization which was needed not only for economic development of the country but also for the creation of a military force sufficient to protect the country and to keep political power in the hands of its new holders.

Collectivization was imposed (a so-called "revolution from above") via economic, administrative, and political means. Continual aggressive political campaigns, high taxation of individual farmers, confiscation of crops, searches for hidden crops, etc.--all these means were used simultaneously and thousands of party members (industrial workers, officers of the army, etc.) were sent to rural areas with the task of collectivizing the peasants by any means. The peasants opposed collectivization, some in an active way via rebellions which were easily suppressed by the army, but the majority in a passive way by slaughtering livestock, hiding seeds, refusing to go out to work, or working slowly. This resistance created serious economic difficulties and collectivization was stopped for a short period. Methods became more rational and less harsh, and in 1931 Stalin announced the victory of the establishment of socialist forms of production in agriculture.

The artel was chosen as a model for obligatory collective farming and other forms (communes, TOZ) were transformed into artels, very often contrary to the wishes of the members. The new form took the general name of kolkhoz, an abbreviation of *kollektivnoj chaziajstwo*, or collective economy. In general, the kolkhoz united small, individually held family plots, usually intensively cultivated, with collective holdings. The means of production were collectively owned, with the exceptions of the nationalized land, and of tractors and other heavy machinery, which until 1958 were in the possession of the state machinery stations (MTS). The kolkhoz was obliged to deliver a stipulated amount of production to the state; these products were sold at much lower prices than they could have commanded on the private market. Other portions of the production went to the MTS as a remuneration for the use of farm machinery. Money income was primarily for several categories of obligatory investments. Portions went to special social funds, with a percentage of net production being set for each fund. What remained was distributed among members according to quantity and quality of work, measured in day/units--*trudoden'*, a day of work. Managers usually received 2 such units for eight hours of work, tractor operators about 1.6 for plowing a set area, etc. The economic plans were prepared by management but all important elements --obligatory delivery, payment for MTS, etc.--were determined by state agencies. All plans and all major decisions, in fact, required the approval of the state authorities, and the government bank had full control over financial affairs of the kolkhoz.

The economic and organizational model of the kolkhoz has been in a process of continual change. The changes became very frequent after the mid-1950s. Kolkhozes received the right to possess heavy machinery, and the MTS were transformed into service stations. The economic incentives--prices, a greater supply of attractive consumer goods, etc.--have been increased. In the new (1969) model constitution for kolkhozes, a system of wages which had existed in practice since 1966 was formally introduced. This changed the whole organizational system. Wages are now part of the costs of production, hence a minimal level of income is guaranteed. Before this use of wages, income depended entirely on what remained after all obligatory subtractions. In some experiments, kolkhozes were united into bigger agricultural-industrial units; in others, they were allowed more economic independence. In a few experiments, attempts were made to reduce the role of individual plots and to bring kolkhozes closer to the state farms (*sovkhozes*). In others, the role of self-management was stressed, and the family plot was declared a very important element of collective farming. In a few experiments, fields and mechanical equipment were given to groups (*zveno*) for longer periods and such groups were remunerated by receiving a share of the production from such enterprises, giving them a direct interest in the levels of production achieved. As such groups are, in fact, often organized on the basis of the family (as is common in Hungary), this experiment means the reintroduction of the principle of family farming into collective farming.

There is now a wide variety of organizational forms of kolkhozes in the Soviet Union. They vary depending upon region, specialization of production, and internal and external pressures, which differ sharply in different parts of the country. There are kolkhozes similar to state farms, kolkhozes with some persisting tendencies to revert to communes, and kolkhozes which can actually be described as aggregates of individual family farms. In this last case, intensive production is concentrated on family plots. Cash income is usually not very large because the members prefer to take their salaries from the collective operation in kind. The members of such a kolkhoz could be regarded as part-time farmers: they possess individual farms, small but very intensive and important as a source of cash income, and they also work outside on a collective farm. As part-time farming is now widely accepted by the government, industrial workers could also acquire small plots of land and produce food for their families or for the market.

The growing industrialization of the Soviet Union, and the necessary reorientation of industry toward consumer and agricultural needs, gives new opportunities and increased significance to economic incentives. At the same time, such incentives are limited by the overall economic system of the Soviet Union. Without major changes in it, particularly in the marketing system, these incentives will not give satisfactory results. The rise of the cultural level of Soviet peasants, together with the rise of modern technology, will gradually increase agricultural production. With the rationalization of planning and decision-making, the system of management will surely be improved. But most important for resolving the permanent crisis in Soviet agriculture will be the emergence of peasants as an active force putting into practice the principles of self-management and demanding further changes in the overall socioeconomic system.

SOCIALIZATION OF LAND AND LABOR

A number of features could be taken as criteria for comparing different forms of collective farms. I have chosen two as most important: (1) joint ownership or use of land and of other means of agricultural production, and (2) socialization of work or, more generally, joint farming because it includes both production work and decision-making.

Joint Ownership of Land

Access to land can be secured in a number of ways. One way is outright ownership by an individual or group (extended family, kinship group, tribe, community or local group, etc.). Another way is by lease from a private owner, group, organization, or the state, with or without payment. Collective farms may use either or both of these ways to acquire land: they get their land from their members or from the state. In the Soviet Union land was nationalized during the revolution and was later given to the kolkhozes for "eternal use." In East European countries the large holdings were confiscated after World War II and land was given to the farm laborers and poor peasants. Part of this land, however, was kept in state ownership. On this part, state farms were organized. Smaller size units of state-held land, not sufficient for state farming, were given with or without payment to the collective farms.

In Poland, there are diverse forms of landownership. Many villages still have communal property (mainly pasture and forest lands), and if a collective farm is organized, it assumes ownership of these lands. Most of the land, however, is held in private ownership. If the farmer wishes to become a member of a collective farm, he contributes his land. This land must be given back to him if he is not satisfied with the collective farm and wishes to start private farming again. Therefore there is constant fluctuation of membership on collective farms. But this is not true for most other East European countries where practically, but not legally, all farmers lose their rights of ownership when joining the collective farm. All land to which a collective farm has access can be used in a collective way, while part (even that part which is rented or received from the state) can be given to members for individual use.

These various forms of access to land create a large number of potential combinations. In a particular country these features have different significance depending on regulations limiting the control of land and on specific cultural values attached to the land. For example, in one country land might be privately owned but without the power to alienate and therefore legal ownership has no meaning; in another country land might be legally nationalized though it had been used for centuries by peasant families. Although such families joined the collective farm, they still regarded the land as their own.

What has been said about the socialization of land applies equally well to other means of agricultural production--buildings, machines and tools, livestock, irrigation structures, etc. After collective farms have existed for a long time, it is common for all the basic means of production to be owned collectively if legal restrictions do not exclude some means (like tractors and combines in the USSR until 1958) from communal ownership.

All forms of ownership of or access to land and other means of agricultural production are strongly influenced by the preexisting land tenure system and by the prevailing socio-economic system. In turn, the forms of ownership strongly influence other organizational features of the collective farm--distribution of income, interpersonal relations among the membership, external ties with the state, governmental agencies, other institutions, etc. We can rank all these forms on a continuum according to increasing socialization of ownership. Of course such a continuum is a simplification since we are dealing with both the legal ownership and the use of land. At one end of the continuum we would have the individual farmer along with some forms of collective ownership or use of land or other means of production; in the center of this spectrum the major part of the means of production would be jointly possessed or used, but part of the means of production would remain privately held; at the other end we would have the situation where all means of production are jointly possessed or used. Here, however, a problem arises concerning nationalized ownership. This can be regarded as the final or highest form of socialization, but where it is administered by the state or a state agency, it is usually not regarded as "socialized" by the group of direct producers. Whether or not it can truly be called socialized depends on the character of the state and on the extent of real democratic participation in decision-making by producers.

Socialization of Work

The second criterion for comparing different forms of collective farms is the degree of socialization of work. Two important questions arise. First, what percentage of the work potential of the members is used on the collective farm, as contrasted to that which is used for individual production or work outside agriculture? Situations exist where members work individually on their farms and where only some particular tasks such as plowing are done jointly; other situations exist where members work mainly on the collectively farmed land but also spend some time on their individual plots (in some cases, e.g., Soviet kolkhozes, members are obliged to work 200 or more days on the collective farm in order to keep their membership rights). Some may combine work in industry or in a village handicraft cooperative with their work on the collective. In some cases, of course, members work full time on the collective farm.

It is perhaps more accurate to speak about member-families rather than about individual members, because, in fact, in many countries the family (which is regarded as the basic productive unit in peasant societies) is the real but unofficial member of the collective farm. In such a case the question could be formulated as follows: what percentage of the work potential of the member-families is used on the collective farm? It is appropriate to speak of part-time farming families when, for instance, the wife is occupied on the collective farm and the husband is working in industry. And it might be called private-collective farming when the husband works on the collective farm and his wife performs most of the work on the private plot. If we take the work potential of the members (or member-families) performed on the collective farm as opposed to the work performed on the private plot or outside of agriculture, we may construct a simple continuum of the socialization of work.

The second question regarding the socialization of work concerns the percentage of work on the collective farm performed by members versus that done by hired workers. Usually in East European countries the use of hired workers on the collective farm is strictly forbidden. But in many cases, particularly in Poland and in Yugoslavia, collective farms do use hired workers. How should one classify the use of the labor of wives or children of the members? Legally they are not members of the collective farm if they work only seasonally, but hiring them is permitted if manpower is needed. And how would one designate the labor of specialists? Officially they are members of the collective farm, but in fact they regard their job as work for remuneration only.

Another more common phenomenon--very often found in East European countries, as well as in the Soviet Union, in Cuba, and in China--is the special brigades of urban dwellers, students, or soldiers formed to "help" collective farmers during harvest. The extent to which this is really voluntary work is not important in the context of this paper; the important thing is that it is work coming from the outside which cannot be regarded as hired labor because it is not, legally speaking, paid for by the collective farm, but neither is it the work of members (or their families). The work of MTS was paid for by the kolkhozes and hence was really hired work, but the work of outsiders during the harvest is not. This external work may be absolutely necessary for the continued functioning of collective farms, particularly in countries where there is a lack of manpower in agriculture or (Czechoslovakia, East Germany) where the agricultural age structure is very unfavorable. Although managers of collective farms often complain that such external workers cost more than hired workers (work accidents, crop damage, etc.), the fact is that in some countries collective farms could not exist without such external help.

Again we can propose a simplified continuum of individual and joint work in different forms of collective farming. At one end of this continuum, we will have different forms of mutual assistance, then a rising proportion of joint work in the whole production process, culminating in a situation where members and their families devote full time to the work of the collective farm. Hired and volunteer work is, however, not included in this continuum. We must construct another continuum for it and we could speak about a higher level of socialization of work when the percentage of hired work is kept low, or we could differentiate the work from outside as socialized (specialists, volunteers, work performed by state agencies) and not socialized (hired workers).

Work on the collective farm can be seen from another point of view. The question is to what extent the work on the collective farm is organized as the work of families. The family is the basic production unit on a peasant farm. The family is also very often the basic producing unit on large holdings. For example, in feudal estates families worked together on the fields of landlords. Even at the end of the nineteenth century in Germany, and Central Europe in general, agricultural workers were expected to be accompanied by wives, sons, and daughters when needed for performing certain kinds of work. On many collective farms (e.g., Hungary), the family is allotted a piece of land within the collective fields and works this piece with the collective's equipment. The family keeps part of the production from their assigned land and gives part to the collective farm. In contrast to such organization of work is the factory where, even if the husband, wife, and children are all employed, the fact that they belong to the same family is of no consequence in determining their production roles, which depend on their physical and intellectual abilities and skills. In the factory they are treated as individuals, not as a working team. A separate continuum can be constructed moving from work organized on the principle of family farming toward work organized according to the principles of division of labor and specialization found in the factory.

The two continua introduced--that of socialization of access to or control over the means of production and that of socialization of work--are interrelated. A higher level in the socialization of use, possession, or control over the means of production implies a higher level in the socialization of work in its two dimensions, i.e., work on the collective farm versus work on the private plots, and work organized on the basis of families versus work organized on the basis of individuals and specialized tasks. Neither one of these two variables is more important than the other since the relationship between them is reciprocal. It is not only that the socialization of the means of production implies or even requires the socialization of work in the way we have specified this term, but the socialization of work implies or more often requires socialization of the means of agricultural production. There are two distinguishable continua, but eventually they converge in the "higher forms" of socialization of both the means of production and of work.

These observations apply as well to all fields of economic activity, not only production but marketing, processing, and transportation. All these economic activities, performed in the past by individual farmers, are today penetrated by the process of socialization in two ways: (1) they are absorbed by the general economic system, so that in many countries marketing and processing are now fully separated from the economic activity of farmers, who are increasingly specialized in the production of farm products; and (2) what was formerly the task of individual farmers becomes organized as a group or cooperative activity via consumer and farmer cooperatives--for credit, purchasing, processing, storage, and so on.

The socialization of these other economic activities does not, however, necessarily depend upon collective farming. It depends on the extent to which these economic activities were separated from farming before collectivization. In other countries where marketing or processing cooperatives existed previous to the establishment of collective farms, these farms are usually in close contact with them and concentrate their own efforts on production. But in many countries such a division of labor was not known or was not well advanced. Marketing and processing were among the activities performed by the peasants themselves. In such a situation, collective farms must perform all these functions, and joint economic activities or jointly organized services could be ranked in the continuum which is related to the socialization of work.

In speaking about these continua, two qualifications are necessary. First, neither of these continua should be treated as a process of spontaneous social transformation. The mutual assistance groups, marketing cooperatives, and some forms of communal ownership are in fact very often regarded as the first step in the transformation of traditional peasant agriculture into collective farming. In general, the lower or simpler forms of agricultural cooperation are assumed spontaneously to transform themselves into collective farms, i.e., into higher forms of agricultural cooperation. It is true that each form has its own dynamic mechanism or potential for transformation. For example, we might expect that the participation of hired personnel in consumer cooperatives or in collective farms will increase, and that locally oriented cooperatives will transform themselves into larger, nationwide organizations, controlling or even monopolizing some fields of economic activity. But spontaneous transformation of "simple forms" of cooperation into collective farms has never been observed. Quite the contrary, these forms of mutual assistance and of communal ownership of land are usually relics of traditional, precapitalistic farming based on the cohesion of local village communities and, with the disintegration of such communities, these relics gradually disappear. On the other hand, the bureaucratization of cooperatives moves rather toward large corporate or state organizations and not toward the collective. It is true that in some countries simple forms were transformed into higher forms, but this was not a spontaneous process. TOZ in the Soviet Union, mutual assistance groups in China or East European countries, communal property in developing countries, and irrigation, machinery, and other types of cooperatives were all used in the initial phases of collectivization. They facilitated collectivization and were useful in helping farmers adjust to the idea of collective farming, but they were never transformed spontaneously into collective farms. Our continuum is not a description of a process, but a typology of existing forms ranked according to the increase of a certain variable or set of variables.

The second qualification is that both continua--socialization of ownership and socialization of work--are measured by using not a single criterion but rather a set of criteria. Socialization of ownership, for example, should be measured by the proportion of collectively owned versus privately owned and/or state-owned means of production. If we introduce the distinction between legal ownership, control, and use of the means of production, the choice of an adequate scale corresponding to the particular conditions of a given country becomes even more difficult. The same complexities apply in attempting to measure the level of socialization of work. Therefore, the continua must be treated as simplifications, and quantitative measurements must be developed to fit the conditions of individual countries.

OTHER ORGANIZATIONAL FEATURES

Until now, we have been speaking about two organizational features--the relation of people to the means of production, and the relations among individuals in the process of work. These are the basic features included in the definition of collective farming. But the collective farm as an organization can be characterized from other points of view as well. The size of the group, the scale of production, and the type of and specialization in production influence the system of organization. There are at least three other organizational features which are of special importance, keeping in mind the basic typology formulated earlier: (1) the system of management and decision-making, (2) the system of consumption organization, and (3) the system of output distribution.

The discussion concerning the system of management is not limited to the participation of members in planning and decision-making. A common contrast is that between decisions made by the entire membership (general assembly) and those made by hired specialists or even by elected management, i.e., management which has become specialized in its organizational function and which retains its dominant position in decision-making. There are different ways to secure democratic management, e.g., obligatory rotation of all positions of management. But a legal system itself is not enough to secure real democracy; special preparation of members is necessary before they can participate competently in management, if at all. There are instances of legally constituted democratic management, as in Yugoslavia, where the producers are not able to take part in decision-making. They sit silently in the general assembly, listening to what specialists say, sometimes not even understanding. Thus, not only the legal and formal provision for membership participation is important in evaluating the system of management. We must also know if members are actually able to participate. What is their level of education? What information do they have about subjects on which decisions must be made?

The second issue concerns the type of decisions which management is permitted to make. If obligatory deliveries are required, and if prices, the system of marketing, financial control, etc., are all decided by state authorities, the management of a collective farm has no more decisions to make than the management of a state-owned enterprise. In such cases, in addition to some internal production decisions, the main tasks of management are to assign people to working teams, delineate the overall farming operation into single production processes, control the performance of producers, and resolve problems leading to social tensions among members or between management and other members.

We conclude, then, that systems of management cannot really be placed on a single continuum since at least three dimensions are involved. First, real access of members to the decison-making process must be determined. Not only is the level of information on issues which the members are expected to resolve important (and through information members could be manipulated by managers), but of equal importance is the possibility to formulate the questions to be discussed and to express freely their opinion and to vote freely pro or con on all issues proposed by management or by members. The second dimension concerns the degree to which managers are freely elected. This could be evaluated in terms of managers being either elected or appointed, or a more refined classification could be used: freely elected, elected by members and confirmed by authorities, appointed by authorities without opinion of members, or other combinations. The third dimension concerns the significance of decisions management (or members) are allowed to make. If all issues of basic importance are decided by national agencies, local managers have very restricted alternatives for decision-making.

Using only a two-point scale for each dimension, a number of types could be identified, of which the extremes would be: (1) restricted access of membership to decision-making, appointed management, and restricted alternatives in decision-making; (2) full access of membership to decision-making, management freely elected, and nonrestricted alternatives in the decision-making process. Systems of management do, of course, reflect the existing conflicts and difficulties in collective farming. If they are analyzed in conjunction with the basic typology, together with other features of collective farms, and interpreted within global socio-economic and political contexts, systems of management are among the most important organizational features of collective farming.

Another important organizational feature is the organization of consumption. The concept used here refers primarily to material consumption, but this concept could be extended to cultural consumption as well--organization of recreation, for example. Variations in the organization of consumption fall somewere between strictly individual consumption in family households, and totally organized social consumption--groups of families living in the same building, preparing food and having meals together, organizing all recreational time, etc. This extreme form is very rare, but in some collective farms consumption as well as production is socialized. Such collective farms are usually called "communes."

Of course, socialization of consumption essentially defines the system of distribution. To the extent that all consumption is shared, special distributive measures are not needed. In the majority of communes, for instance, consumption is basically socialized, meaning that food is prepared in the common kitchen and everyone comes for meals at the same place and time, although in some cases meals are carried home. In kibbutzim, for example, houses are owned by the kibbutz but individually used by the families;

clothes, or coupons for clothes, are distributed equally or according to particular needs.

Together with consumption, the fulfillment of some other individual needs could be organized in a social way--medical care, education, and general care of children and aged persons. In extreme cases, the commune acts like a family and all functions otherwise performed by the family are organized and performed by the group. Needless to say, those social functions important for all members of the group or connected with the ideology motivating such a group --religious ceremonies, political meetings, military training, etc. --are organized by the group, as are other events such as weddings, holidays, etc.

When consumption is not fully socialized, different systems can be identified in which a part of the output is distributed among members. A word is in order here concerning the ideals of equality. The ideal of equality is present in all collective farms, but its meaning is different depending on the type of collective farm and the nature of the system of values shared by its members. We can distinguish at least four meanings of equality. First is the most literal: each member of the group (or each family) receives the same volume of the benefits independently of his/her work (or contribution). This idea (in Russia called *uranilovka*) is common among some religious groups and unskilled workers. Second is the idea of "equal reward for equal work." It is particularly common among former hired workers whose contribution is exclusively in the form of labor. As it is very difficult to formulate criteria of "equal" work or performance, the realization of this ideal usually creates tensions among members performing different production tasks.

The third idea of equality enlarges upon the meaning of contribution and can be expressed as "equal reward or remuneration for equal contribution," where the latter is measured in terms of land and capital contribution as well as labor performance. This idea of equality corresponds to the social situation of small producers --farmers or peasants--and can be accepted as just if the membership of a collective farm is composed of such a group. Of course, these measures accept preexisting inequalities and continue them on the collective farm.

Finally, the extreme and most sophisticated meaning of equality is "to each according to his needs" and "from each according to his abilities." But it is very difficult to put this idea into practice. The realization of this ideal may have to be limited to the benefits which are in abundance. For other benefits, equal distribution in some cases could literally mean that everyone gets the same share (modified or not by special needs) or it could mean "equal reward for equal work." There are also some social benefits like power or prestige which simply cannot be readily distributed even if the group would try to do so, because if they are distributed equally they cease to exist. In effect, then, inequalities among members are created and compromise solutions lead to the acceptance of more realistic meanings of equality. Usually the principle of "equal reward for equal contribution" is regarded as least socialistic and the principle "to each according to his needs" as most socialistic or even as the feature of a truly communist society.

I have outlined here only the most important organizational features. There are many more organizational features or dimensions of collective farms, but the present analysis was concerned mainly with the socio-economic characteristics linked directly to the basic typology. To highlight the organizational variety within collective farming, the following section discusses some examples of different forms, mainly from East Europe, moving from so-called "simple" forms toward more complex or "higher" forms of collective farming.

Simple and Complex Forms of Collective Farming

The simpler forms are of two kinds. First, there are relics of traditional village life--mutual assistance, communal or village ownership, and in some countries ownership by the extended family. The Russian *obszczina* would be classified among these forms. It consisted of communal or village ownership of land which was periodically (about every seven years) divided equally among families or according to the size of the family. In Yugoslavia, there was the *zadruga*, which was a form of ownership by a kinship group. In East Europe, these forms have disappeared completely except in some remote, especially mountainous, regions, where communal ownership of grazing and forest lands still exists. Zadruga and obszczina were in fact very complex forms of collective life and seem simple only when restricting the analysis to the level of socialization of ownership or work. As these forms no longer exist, they need not be ranked on the continuum shown at the end of this section. As representative of these forms, however, we shall list communal ownership and mutual assistance at the lower end of this continuum.

A second simple form of cooperation is that associated with technological development. Frequently individual families cannot afford to purchase tractors or other pieces of heavy equipment, so neighboring farmers buy and use them either individually and in sequence or jointly with some form of labor exchange arrangements. Usually, these are informal agreements. There may be no formal written rules, but members know exactly what these rules are and how they are changed with the accumulation of experience. In a sense, these could be viewed as a revival of earlier forms of group ownership and group work stimulated by socio-economic and particularly technical developments.

As the next form on the continuum, we may place the marketing, credit, and consumer cooperatives, which are usually organized to provide these services for farmers. On-farm production per se is usually excluded from their functions. Yet it is interesting to note that quite recently some of these types of cooperatives (e.g., those in Yugoslavia) have begun to perform important functions in vertical integration (contracts for production, processing, marketing) and are penetrating more and more into the area of production itself. This may be called "co-production"--including cooperative use of private farmers' land who are then paid partly in produce and partly in cash wages for their labor. All kinds of production services--plant protection and spraying, sometimes plowing, cultivating, and harvesting--are performed by such cooperatives. As the most advanced form of this kind we may include the Israeli *moshav ovdim.*

Closely related to this type of organization are the temporary cooperatives--land settlement cooperatives in England, cooperative farms organized after the land reform in Italy, cooperative settlement farms in Poland organized on regained territories after World War II, etc. This kind of cooperative farming is intended only for a determined period to help new settlers, or people who receive land as a result of a land reform, to overcome the first difficulties and to prepare them for family farming. In such cases, even after conversion to family farming, some forms of group ownership and services may continue to be organized by the cooperative, but the family works its land individually. In fact, even during periods of collective work families are accumulating capital for their own farms.

Moving up another stage on the continuum are what can be called partly collective farms. The TOZ in the Soviet Union belonged to this category, as do the so-called collective farms of Type I and Type Ib in Poland (where the cooperative assists private families in their farming operations). These are the instances in which there is collective farming in one field of production combined with individual farming in other fields. In all these cases of partly collective farming, the principle of family farming is dominant.

Next on the continuum are integral cooperative farms, the simplest of which are the French GAEC and former Bulgarian cooperative farms where profit is distributed according to contribution of land, capital, and work. In this form collective ownership and work are dominant, but such farms can be treated as agreements between unequal partners who may contribute only work or land or capital. Some organizational features of family farming can be maintained at this stage. Particularly in Hungarian cooperative farming, the principle of family farming is still maintained both in the form of an individual family plot and in the organization of work on the collective land.

The next entrant on the continuum is the kolkhoz and kolkhoz-like collective farms in East Europe. On these farms the level of socialization of ownership and work is much higher--in the Soviet Union, for example, land is nationalized, equipment is collectively owned, and theoretically at least, collective work is dominant. Some social services are organized by the kolkhoz, but consumption remains individualized and small individual farming in the form of individual plots remains. As mentioned previously, competition between collective and individual farming within the kolkhoz is expressed in terms of intensified cultivation of the individual plot.

In the kolkhoz, participation of a larger organization--the state--is very extensive, especially in management. Members of the kolkhoz can be called upon to perform some national or regional tasks--construction of roads, assistance in building industrial plants, etc. This is even more prominent in Rumanian and Bulgarian cooperative farms. The latter are organized in larger units within the overall administrative structure; they are more subordinate to central administration; and the sense of a collective is much weaker. The Czechoslovakian and East German creation of agro-industrial complexes subordinated the collective farm to these state-organized units and, in general, collective farms became similar to state farms. This penetration of collective farming by a larger organization, particularly by the central administration of the state, is very important for future prospects of collective farming.

The final grouping on the continuum includes communes where both production and consumption are socialized. Other aspects of social life are also often organized as group activities. In these extreme forms the transformation of single families into one large group family is the logical consequence of the ideal of collective life. This extreme form, however, is very rare because the full socialization of life in communes requires many sacrifices and highly motivated individuals. The commune as a form of socialization of consumption as well as production could be located beyond our continuum, as is the case also for the communal ownership-mutual assistance forms on the other extreme. But since ownership and work are fully socialized in communes, we shall include them at the upper end of the continuum.

The continuum proposed here is similar to one proposed by Ulrich Planck.[1] Planck ranked cooperatives according to the functions they perform. His continuum moves from credit coops to moshav ovdim, collective *ejidos*, kolkhozes, kibbutzim, and ends with communes. The similarities between Planck's continuum and the one proposed here suggests that perhaps a universal ranking of cooperatives and collective farms could be found.

The analysis of different organizational models is often undertaken for purposes of determining which form might be most advantageous. But, as mentioned earlier, the best model for reaching certain ideological goals may be quite different from that needed to maximize profits or to prepare people for individual farming. Economic success may even be a factor causing erosion of ideology. The goals of collective farms must, therefore, be taken into account. Collective farms function within given socio-economic, cultural, and political systems, and in different natural conditions. The appropriate organizational model for one set of these conditions may be completely absurd in another: maximizing profits may be a necessary and fitting goal for collective farms within a system of competitive markets, but in a planned economy the idea of maximizing profits would not make sense. To evaluate the prospects of collective farming, therefore, one must first classify these systems according to their stated goals, the various continua or

FIGURE 1

Continuum from Least Socialized to Most Socialized Forms of Collective Farming

Communes
Kibbutz

Agro-industrial
complexes

Kolkhoz

Integral collective
farms: GAEC

Partly collective
farms: TOZ, Type I
and Ib in Poland

Temporary collective farms

Contracting, processing,
and production-services
cooperatives: moshav ovdim

Marketing, credit, and
consumer cooperatives

Equipment or
irrigation cooperatives

Communal ownership-
mutual assistance

organizational models, and at least simplified typologies of social, economic, cultural, and political conditions. These external factors, however, require a separate analysis. In the following section, the interdependence of overall economic systems and forms of group farming is discussed.

COLLECTIVE FARMS AND SOCIO-ECONOMIC SYSTEMS

It will be useful to distinguish two socio-economic systems, using as the criterion the relationship between consumption and production. We shall call the first system an interactive system: the behavior of consumers directly rewards or punishes the producers. When the producer delivers products which are needed or desired by consumers, at prices which are acceptable to them, he will sell his products and be rewarded by profits. If his production does not meet the needs of a clientele and if the price is not acceptable in comparison to other prices or to the hierarchy of needs, he is punished by not being able to sell his products and get his profit. The behavior of consumers can, of course, be manipulated to some extent by producers. Advertising, monopolistic pricing, creation of new needs, fashions, etc.--all serve as means for manipulation of consumers. Even more important is the general system of values--consumption today is often seen as a symbol of social status or as the source of a universally accepted meaning of happiness. The system of values is an expression of an economic system which can function only by developing new needs. Manipulation of consumers can be more or less efficient, but the general principle of such an economic system is based on the interaction between consumers and producers. This interaction operates via the market and the consumer is the central figure in this functional relationship.

The second type of socio-economic system we shall call a directive system. It has no direct interaction between producers and consumers. The distribution of goods can be via the market, or goods can be distributed directly to consumers as happened during the war in many European countries. The basic distinction is that producers are rewarded or punished by the central steering body, planning committee, or other state agency for their contribution to fulfillment of the national plan, not for the acceptance of their production by consumers. Basic goods may be sold to consumers for money, but some especially prized goods could be distributed not for money but as a special privilege to people who are of strategic importance to the steering body--specialists, top managers and administrators, or individuals performing particularly important functions in the country such as military officers or police agents. The degree to which producers are isolated from consumers depends, of course, on how the plan is constructed. But even if everything is planned--kind of production, production practices, salaries and number of workers, standard quality, etc.--the system offers many possibilities to receive rewards for producing goods which were not necessarily demanded by the consumers.

The two systems described above are, of course, simplifications. First, each has been characterized by only one kind of relationship, and usually there are other important features as well. For example, in directive systems the production of such means of production as machinery, new industrial construction, etc., could and usually does have priority over the production of goods for consumption, reinforcing the relative independence from the behavior of consumers. Basic consumer goods must, of course, be produced, while on the other hand the consumer cannot easily refuse to buy something which he needs even if its quality is considered unsatisfactory. But the main regulator of the directive system is the national plan, and in this plan the needs of the system and not of individuals within it have a decisive priority. Contrary to the interactive system, which is oriented toward an increase in consumption (or to put it more strongly, whose basic condition for continued functioning is a permanent increase in consumption), the directive system usually keeps, or tries to keep, consumer demand at a stable level.

Another characteristic is that in directive systems employment, like production, is planned and unemployment is unacceptable. The economy must utilize all its potential manpower. As salaries are kept low in order to maintain a balance with the available supply of goods, there are often no real incentives for technical innovation and increased efficiency. This is not necessarily a consequence of a directive system, but in reality these features usually accompany the basic traits of such a system.

Both socio-economic systems as described above are ideal types. In reality, countries with directive socio-economic systems are increasingly introducing elements of interaction between production and consumption, while countries with interactive systems are introducing more elements of direction, especially via intervention of the government in economic life. We have called these systems directive and interactive, not socialist and capitalist. It is evident that a directive system could be more easily and fully adopted by a country where the means of production are nationalized. But, as in the case of Yugoslavia, planning can mean stimulation via the market rather than by directive orders to producers. The directive system is not particularly socialistic in itself. It is characteristic of most of the less developed countries which are trying to create the basis for heavy industry by vigorous governmental action and control. At the same time, the concept of a capitalist system based on private ownership of all the means of production belongs partly to the past since ownership by itself no longer assures the degree of power that it once did. State control continues to grow and economic life today is controlled more or less efficiently by governments. Private ownership can coexist within the directive system, and capitalist societies are in fact so organized during wartime. Socialized ownership can also coexist within interactive systems. But in general, the directive system is more common to a socialized economy and the interactive system to an economy based on private enterprise.

Collective farms located in either of these two socio-economic systems will have major differences in their organizational patterns, systems of operation, criteria of success and failure, source of social conflicts, etc. Even organizational forms which appear identical will in fact be different.

The collective farm in a directive system is part of a planned economy. The amount of production to be delivered is decided by the central steering body, and most investments and purchases of agricultural inputs are decided as well. The financial plans of collective farms must be consistent with the overall, national plan. This also holds for decisions concerning the amount of investment goods to be produced as well as their distribution. The plan of a collective farm could include, for example, the purchase of new tractors for which it has the money. The purchase could even be approved by national planners. But if the output of tractors is not sufficient to cover all planned purchases, or if the tractors were allocated to other regions, the plan of the collective farm could not be reached. If the country has an international trade agreement to exchange its export products for combines, it could happen that the exchange will be based on an overestimate of needs so that the country will receive more combines than were requested by the collective farms. Instead of the tractors which it needs, the collective farm might receive combines which it does not need or for which it has little use. It is, of course, theoretically possible to avoid such mistakes, but the examples are taken from current practice. If too many details must be decided by the central steering body, such mistakes are inevitable.

What is important is that in the planned economy a great many actions are taken by agencies other than the collective farm. The principle of self-management may be granted to all types of collective farms. In a directed economy, however, most decisions are made centrally in accordance with the global plan and there is simply no place for self-management. The economic activity of collective farms depends on the fulfillment of plans of other enterprises. Even if collective farms could produce more and hence get more income, they may have little incentive to do so if the production of other goods and therefore the possible purchases they can make is limited. Calculation of inputs and outputs takes on a new meaning in such a situation, and it is purely a question of accounting procedures whether or not the collective farm has a deficit.

In an interactive system all is different. The collective farm must be competitive--lower its costs of production, increase production, find buyers, etc. The field is much more open for economic decisions, but the market is competitive and the risk is much higher. It is much more difficult to decide what to do with manpower which is liberated from production by use of cheaper and more efficient technology. The economic conditions require quick, effective decisions, and these requirements could create difficulties for democratic decision-making processes. Subordination to the market requirements may create obstacles to truly democratic management.

Nevertheless, the collective farm in such a system is the private business of its members, and it will continue to exist so long as its members find this organizational pattern advantageous. On the contrary, in a directive system decisions of collective farm members are much more dependent on the national economy and in consequence they may feel less responsibility for the results of production. A common attitude is: if the government requires good results, the government must give all that is needed to achieve them. The role of the manager is such a situation, particularly when he is elected by the members and not appointed by the authorities, is very difficult. Of course, much depends on him. If he is very influential and can protect his collective farm from plans that are too demanding, if he can get more from the government and give less, and particularly if he can protect the collective farm from close control by authorities, he will be considered a good manager.

The two socio-economic systems just described are external factors which determine the functioning of collective farms. In some countries, however, particularly in developing countries, peasant agriculture cannot be described in terms of directive or interactive economic systems. A large part of the agricultural activity is devoted to subsistence production and is not oriented toward the market. In such a setting either of these economic systems is something new and imposed from above. It may well be that for peasants in a given country the interactive system is less familiar than the directive system if, for example, they were formerly obliged to give some part of their production to the landlord. The interactive system could seem too risky for them. They may be quite willing to produce for the government if they feel they are under the government's care.

In looking at these two systems, we should therefore determine to what extent members of the collective farm consist of farmers who were previously oriented to producing for the market, or of peasants producing mainly for their own needs. Where the latter might adapt quite easily to a directive system, the former could find such adaptation very difficult. All their former habits and values seem inappropriate in a system where there is no free market competition. The virtues which were formerly successful as norms of behavior--calculation of input and output, accumulation of money, saving, extending business, flexibility in response to market stimuli, etc.--are in the new situation meaningless, less important, or less subject to their control.

An ideal situation for new collective farms in a directive system occurs when the collective's members are former landless laborers. For them, the directive system is easy to understand. They leave all economic decisions to the management and are mainly interested in norms of work and in wages. But such a background of experience and attitudes is not the most useful under the conditions of an interactive system where the members of a collective farm take all responsibility for decisions. Of course, they could hire a good manager and place their trust in him, but their situation would then be much the same as before, only their wages might be higher. Such a situation is, in fact, the case on some collective farms in Latin America where after land reform large estates were transformed into collective farms. In such a situation, there exists the potential for democratic management, but it will likely be realized only by the next generation which will receive more education and more training for the business.

In an interactive system the best situation is when the collective farm is organized by farmers previously oriented to market conditions and economic management. But this is usually possible only for small groups. The experience of the French GAEC shows that in such a case both successful collective farming and the full realization of democratic management can exist.

The most common situation in developing countries with interactive systems is when former peasants, used to subsistence agriculture, are pressed to transform themselves into collective entrepreneurs. Commercial farming and the principles of collective life are both new to them. The only experiences they can bring to bear are the patterns of mutual assistance in the village and perhaps that of communal property. Without the help of specialists, they cannot take full advantage of collective farming. And it is rather difficult for them to accept the help of specialists, since frequently they regard them as outsiders. When they do accept managers, they lose their own initiative, feel frustrated, insecure, alienated, and powerless. Under such circumstances, collective farms are usually not very successful and it is difficult to turn them into really functioning businesses.

In describing collective farms in these two socio-economic systems and organized by individuals with different backgrounds and experiences, we did not mention our basic typology. There is, however, a link between this typology and the kind of socio-economic system in which collective farms are organized. Type 1 collective farms could not be expected to exist for long in a directive system. In directive systems the role of collective farms is defined by the needs of the national economy and not by the particular goals or beliefs of its members. Only in a situation where the beliefs of members are identical with national goals as formulated by the government in a directive system would it be possible to expect the continued existence of the first type. If the members believe in the national goals or if they believe in a kind of ideology which is accepted as official doctrine, they could be quite successful within the directive system. Such a situation apparently exists in Chinese communes. Usually, however, a government is very pragmatic and does not necessarily follow the behavioral doctrines it sets forth in official pronouncements. Thus, the principle of equality may be sacrificed for that of efficiency. The enthusiasm generated by an ideology during a revolution cannot readily be maintained after the revolution reaches its initial goals. Even the most radical government will appreciate obedient citizens more than citizens who strongly believe in ideological dogma and may on this basis oppose certain governmental decisions. Collective farms could be created by groups of believers, but they could not function for long in a directive system and might even be regarded as a threat and a source of deviation.

The fourth type of collective likewise could not exist in a directive system because the idea of getting small farms together in order to achieve higher profits is inconsistent within a planned economy. There is no competition between small and large farms;[2] production is planned; prices and supplies of seeds, fertilizers, construction materials, etc., are determined by the government. Large farms might, in fact, have lower costs of production and better economic results in general, but for the members of a collective farm this success is of secondary importance. Members generally feel more alienated and more suppressed by management in the larger collective farms. Even if some incentives to enlarge collective farms could be devised, they would not likely be considered sufficient reasons by peasants, who only agree to join collective farms, or to unify existing farms because of external pressure.

As the fourth type of collective is not suited to directive systems, so the third type does not fit interactive systems in which decisions of producers are influenced primarily by the market. Producers organize collective farms in order to achieve the advantages of the market and not for the general interest of the national economy. The government in an interactive system could encourage collective farming or could grant special privileges for collective farms, but it could not create collective farming on a mass scale without changing the entire economic system of the country.

Only the second type of collective could emerge in both directive and interactive economic systems. In the directive system collective farms will receive full economic support, will be likely to survive their internal crises, and usually will have a tendency to bring their organizational patterns closer to state farming. In the interactive system, the centrifugal tendencies will, after the first stage of collectivization, strongly manifest themselves, and most probably the collective farms will dissolve or transform themselves into a kind of corporation or cooperative composed of individual farms retaining some shared property or commonly organized services. The most promising situation seems to be where the organizational patterns of the resulting individual farms are similar, as in some of the Israeli moshavs.

NOTES

1. See "Individuelle oder kollektive Entscheidung. Versuch einer Systematisirung personaler Kooperation," *Zietschrift fur Agrargeschichte und Agrarsoziologie* Jg. 15, H.2 (1967).

2. If this type of collective farm, defined as a response to the competitive market, makes no sense in a directive economic system where the market is strongly controlled and competition is excluded, there is still a place for competition between small family farms and collective farms in such a system. This competition is, however, not in the market; rather it is expressed in the income of members of collective farms. If members' wages or income from collective farms are higher than the income possibilities from private plots, it is probable that the members of collective farms will reject their individual plots or transform them into small kitchen-gardens. If, for example, a member of a collective farm receives a sufficient amount of milk from the collective farm, he would not bother to keep a cow for his family. If his salary is better than the income from vegetables produced on his private plot for the market, he would probably spend more time on the collective fields than on his private plot. If he could get enough meat from the collective farm or in the butchery, or if the market is well supplied and there is no private market for his products, he

would certainly prefer to earn a good salary than to produce meat on his own. This competition is regulated by the constitution of the collective farm. Usually one is not supposed to produce more than is required to meet one's own needs, but this depends on the general economic situation in the country. Up to now, we have only had examples of countries with directive systems where there are permanent shortages in market supplies of food and consumer goods in general. It could be that the directive system in a country with an abundance of consumer goods will create different conditions for collective farming. The advantages of collective farming will become clear and in consequence the competition between individual plots and collective farms will diminish. But as of the present, such a country is not known to us.

Part II: Communal Forms of Group Farming

Editor's Introduction

The Israeli *kibbutzim* and the Hutterian Brethren in North America are the two truly communal forms discussed in this volume. Both correspond to Galeski's Type One. They also have a relatively long history of survival, especially the Hutterites. Both are dedicated to the principles of communal property and communal life, and both operate large-scale agricultural enterprises. It is not surprising, then, to find that both have similar problems of management, and that both face the need to adjust constantly to the institutions and changes in the surrounding society.

The Israeli kibbutz has a larger population than the Hutterite colony. In 1967, the 234 kibbutzim averaged 426 people per kibbutz, with a total population of almost 100,000. The Hutterites number about 24,000 on 232 colonies, averaging slightly over 100 per colony. They live in Canada and the U.S. Northern Plains, spread over an area about 3,000 miles in diameter.

The Hutterite colony is kept small by deliberate design. When a colony reaches a population of about 120 people, a new farm is purchased and the colony divides. Lots are drawn and half of the colony moves to the new site. This has provided continual opportunities for upward mobility. If colony division and expansion should cease due to land shortages or other reasons, other means will have to be found to ensure this mobility since it does create a system of rewards for hard work and dedication.

The kibbutz does not face the same problem for a number of reasons. Its people are more in tune with the general culture of the larger society; indeed, they have greatly influenced the social norms of Israeli society and provided much of the leadership for that society. In contrast, the Hutterites are frequently in conflict with the larger society's rules on taxes, land titles, education of children, military draft, etc. There is more movement off the kibbutz than in the case of the Hutterites. Also, the kibbutz economy has evolved to include industrial developments which provide more diverse employment opportunities.

In his discussion of these cases, Professor Lewis[1] notes that the experience of both the Hutterites and the kibbutzim shows that it *is* possible for human beings to so order their lives in an agrarian system that for a period of generations, even centuries, they can live communal lives with a high degree of consensus, cooperation, apparent harmony, and satisfaction. In addition, they may achieve considerable material and economic success. On the other hand, their very rarity and their special features lead to serious questions regarding the possibility of propagating similar forms elsewhere on any substantial scale. These cases suggest that there must be a strong ideological commitment on the part of the people who must live by the commune's rules. *They* must believe in the experiment and want to change their lives and be willing to submit to the necessary exacting discipline. These communal structures are in no sense "traditional" and are quite foreign to the farmers of Africa, Asia, and Latin America.

The farmers and artisans who became the Hutterites did not merely create a collective farm; they created a total communal society, not for economic gain nor in order to develop a country, but to construct for themselves a whole new life--one based on religious premises in which economic well-being was subordinated to the general aim of de-emphasizing "the self and material goals and ambition." The kibbutz had its origin in a secular attempt by well-educated idealists to rebuild a people, a country, and themselves. But their personal material goals were similarly limited. Both the kibbutzim and the Hutterite colonies feature very special total ways of life, and their members must conform to those ways or the systems cannot continue. Those born into Hutterite and kibbutz society may continue to belong and believe in it because they have been socialized to these ways, but the original pioneers who created these systems made the choices for themselves and designed their own "new worlds."

Professor Weintraub,[2] a second discussant, also emphasizes the ideological motivation and commitment necessary to make a commune into a going concern. Yet this ideological commitment creates some very special problems. To maintain itself and provide opportunities for a growing population, a commune must become economically viable. This requires increasing investments, technological change, commercialization, and the knowledge necessary for these developments. But these changes create tensions within the communal organization. At the same time, such developments require increasing contact and transactions with the institutions in the outside world. Thus there are both internal and external pressures and contradictions that threaten the ideology which forms the basis of the communal society.

How does the commune, with its austere mode of life, deal with the contradictions that result from "economic success"? How can it maintain the basic nature of its ideology, which is geared to preserving a certain way of life, on the one hand, yet adjust to these tensions and contradictions on the other? Weintraub suggests four possible reactions:

1. To shield the group from outside contact and communication as much as possible--i.e., withdrawal and seclusion. Internal pressures remain to be dealt with, but at least those outside influences bearing on the ideology are minimized. This is a course of action at least partially adopted by the Hutterites. The kibbutz cannot shut itself off since its ideology provides it with a sense of mission and active participation in the affairs of the nation.
2. To resist both internal and external pressure through a variety of social sanctions and legal controls and thus to prevent members from changing the ideologically based structures. But this reaction tends to deepen the contradictions between generations.
3. To yield to the existing pressure and give up the integrity of the ideology. This means, in effect, the discontinuance of the common venture.
4. To introduce sufficient ideological and structural flexibility to resolve at least to some extent the contradictions. Both the Kibbutz and the Hutterite colony have thus far been able to strike some measure of balance between the maintenance of what is considered absolutely essential to the integrity of the ideology, and the absorption of change and innovation necessary to meet the growing pressures.

In his analysis of the Hutterite colonies, Bennett makes reference to the distinction between the production cooperative and the commune. The significance of this distinction will become more apparent after studying the case studies in succeeding parts of the present volume. We would point out at this time that there are at least four important reasons for making this distinction.

1. In any collective, noncommunal enterprise there is the inevitable and ever-present issue of accounting: not only of costs and returns, and reserves for investment which are common to any economic enterprise, but of labor time and input of individuals and/or families, classification of tasks, differential earning rates for different tasks, etc. All this is required to achieve a reasonably acceptable system of distribution. Private plots become a key issue. Under a communal system with equal distribution for the necessities of life and sharply constrained consumption, these issues do not arise or certainly not in their stark and visible form as they do in most production cooperative enterprises.

2. In a production cooperative, owned and controlled by members with management elected by the members (this does, of course, not hold for all types of collective enterprises), the role and position of management is ambiguous. There is a tendency for a growing-apart of management and members, with the latter coming to look upon themselves as "we the workers" and viewing management as "they the bosses." This tendency of a we-they dichotomy is characteristic of most production cooperative enterprises, especially after

they have been in existence for some years. In any commune which indeed functions as a commune, the entire group is involved, in one way or another, in decision-making. Communalism requires the consultation of a very large number of individuals in the decision-making process.

3. Where the larger, outside society is governed by behavioral rules that differ widely from those that govern the behavior of the commune members (which is always the case since there are no communal nation-states), it is necessary for the commune to emphasize a special and unique type of socialization of children.

4. Finally, the distinction between the commune and the production cooperative helps to highlight the stern discipline required to achieve continuity and assure survival of the commune.

NOTES

1. Professor Herbert S. Lewis, Department of Anthropology, University of Wisconsin, Madison.

2. Professor Dov Weintraub, Department of Sociology, Hebrew University, Jerusalem, Israel.

2. Dynamics of Development in the Israeli Kibbutz

Yehuda Don

Israel has developed two principal forms of collective agriculture--the *kibbutz* and the collective settlement (Collective *Moshav*). Of the two, the kibbutz is dominant quantitatively and also has a longer history. At the end of 1973, 238 kibbutzim were registered at the Cooperative Registrar as compared to 38 collective moshavim.[1] The first kibbutz, Degania, was founded in 1909 and the first collective moshav, Kefar Hittim, in 1936.[2] Furthermore, the kibbutz is undoubtedly the more original, as well as the more interesting social experiment of the two. Nevertheless, it seems that, as a result of the rather unique circumstances which led to the establishment and consolidation of the kibbutz, the collective moshav may be of greater pragmatic importance for possible group farming models.

This paper deals with the kibbutz, as the chief representative of group farming settlements in Israel.

Historically, the foundation of the kibbutz is one of the principal expressions of the modern Zionist Renaissance and its efforts to return to the soil. Return to the soil was, during the first decades of this century, much more than mere colonization. Agriculture and physical labor in farming was the most explicit symbol of social regeneration and professional transformation. The Zionist movement mobilized its active followers mainly from the young generation of the Jewish middle class in Europe. The Jewish population in those countries was "composed in such a way as to resemble a pyramid whose broad base consists of merchants, their employees and commercial middlemen. These are followed by industrialists, by professional men . . . and by artisans, and it is only when we reach the narrow apex of the structure that we find farmers and industrial laborers. The order of this pyramid must, in Palestine, be exactly reversed. . . ."[3] Thus wrote Dr. A Ruppin, the head of the "Palestine Office" in the early 1920s, which was considered the early take-off stage of Jewish settlement foundations in Palestine. Agriculture was therefore closely associated with the normalization of the occupational structure and it acquired at the same time symbolic values for national rejuvenation.

The establishment of workers' settlements was also a necessity. The employment situation in Palestine during the first quarter of the century was such that the Jewish pioneer immigrant who insisted on "a European standard of civilization"[4] found it almost impossible to compete with the native workers whose wage requirements were well below the acceptable minimum of the Jewish worker. Independent colonization was, at the beginning, the result of weakness rather than of strength. The requirement of engaging in physical labor, the "Conquest of Labor," was feasible only in newly established colonies by the immigrant laborers themselves.

The basic facets of cooperation were also more the result of external circumstances than of premeditated planning. It was assumed that untrained immigrants, unaccustomed to local weather and conditions, needed the supervision of agricultural experts, and the colonizing authorities assigned them such experts to manage the agricultural affairs of the colonies, and to train them to become skilled farmers. The workers, according to first hand evidence, "felt that they had in them [(*sic*), i.e., they had the capacity] to work on their own responsibility," and "resented being treated as hired workmen. . . ."[5] The immigrants were too strongly motivated for independent action and too anarchistically inclined to accept guided management; thus self-management, carried to the extreme of absolute direct democracy with heavy self-imposed discipline, resulted from the failure to apply economically rational measures to a public with higher than average altruistic motives.[6]

This very intensive, idealistic approach to immigration had also been nourished on socialist and revolutionary ideas of social justice and an economic structure free of exploitation. Thus, together with the stress laid on the value of physical labor, "self-labor" was strongly emphasized for both the creation of classless subsystems and the avoidance of exploitation through benefiting from the surplus value of hired labor. Therefore it was a combination of reactions to new and strange circumstances, together with reflections of past social ideals, that established three of the most important fundamentals of the present joint farming system, viz. self-labor, self-management, and direct democracy.

The fourth fundamental, collectivism, was also the outcome of pragmatic considerations rather than of a pre-determined plan or constitution. Most of the basic characteristics of the kibbutz were originally conceived as temporary measures to overcome certain difficulties, e.g., devices for saving labor and reducing costs.[7] Eventually, after World War I, the collective elements of the kibbutz gained strength and became permanent features. The adoption of collectivism as a permanent way of life was certainly encouraged by its success over more than a decade in the first kibbutzim.[8] However, the socialist ideas which prevailed among the young immigrants during the post-World War I years were a major factor in making the collective settlement the dominant form of colonization in Palestine for almost 30 years.

The ideological transformation during the early 1920s was rapid. In 1924, one of the outstanding leaders of the kibbutz movement, Y. Tabenkin, already spoke of the "Commune" which was a "communistic settlement" in which everybody contributed "to society according to his capabilities" and drew "from society according to his needs."[9] This has been the fundamental principle of production and distribution in the kibbutz ever since. Moreover, Tabenkin emphasized the communistic feature of the movement, associated it with Marx and Bakunin, and claimed that the commune was not confined to the members of one particular kibbutz, "but it was the authority and property of the labor class in general."[10]

Collectivist ideology thus became the focal point of the young altruistic immigrants, who, by combining their new country and new profession (agriculture) with their new socio-economic norms, brought to new revolutionary realization the drastic change they sought in their very act of immigration to Palestine. In due course, the kibbutz assumed a dual role: that of the principal promoter of the objectives of Zionism, mainly by frontier colonization; and the establishment of a new society,[11] based on equalitarian conceptions of social justice which are different from those of capitalist society. It became a rallying point for the elite groups of the young generation, immigrant and native-born alike, as the optimum life pattern for self-realization as well as for the service of society. The kibbutz has been attractive only to elite groups, who have many alternatives besides kibbutz life, and has throughout been a strictly voluntary organization. Consequently, it has experienced, since its early beginnings, a very considerable turnover of members. Many candidates and members join and leave the stable nucleus whose relative size is one of the best criteria of social stability.

The following table summarizes the development of the kibbutz population at three different stages in the history of Israel, in absolute terms and relative to general demographic changes. The selected years represent the beginning of independent statehood (1948), the end of the post-independence wave of colonization (1961), and the recent past (1971).

The table clearly indicates, among others, the following trends:

1) The number of kibbutzim increased considerably during the first period and almost stagnated during the second.

2) The kibbutz population increased during the first period, though its rate of growth was remarkably lower than that of the Israeli population in general. During the second period the rate of growth of the kibbutz population almost kept pace with the general trend.

3) The kibbutz population as a share of the total population of Israel has declined, though the rate of decline is diminishing.

TABLE 1. POPULATION CHANGES IN THE KIBBUTZ AND IN ISRAEL

	1948	1961	1971	$\frac{1961 \times 100}{1948}$	$\frac{1971 \times 100}{1961}$	$\frac{1971 \times 100}{1948}$
Israeli population[a]	872,700	2,234,200	3,095,100	256	139	355
Israeli rural population[b]		297,900	275,200	-	92	-
Kibbutz population	54,200[c]	77,000[b]	99,700[b]	142	129	184
Kibbutz pop. as percentage of total pop.	6.21	3.45	3.22	56	93	52
Kibbutz pop. as percentage of rural pop.	-	25.85	36.23	-	140	-
No. of kibbutzim (end of the year)	177[c]	228[d]	234[e]	129	103	132
Average pop. per kibbutz	306	338	426	110	126	142

SOURCES: a. *Statistical Abstract of Israel 1972* (Jerusalem, 1972), p. 21; b. Ibid., p. 30; c. D. Giladi, ed., *Agriculture in Israel and in Workers' Settlements Statistical Abstracts* (Tel-Aviv, 1970), p. 2; d. *Report of the Registrar of Cooperative Societies in Israel for 1962* (Jerusalem, 1963), p. 254; e. *Report of the Registrar of Cooperative Societies in Israel for 1971* (Jerusalem, 1972), p. 33.

4) The kibbutz population as a share of the rural population is sharply increasing as a result of the continuous decline in the rural population.

5) Average number of people per kibbutz has been continuously growing, with an increasing rate of growth (over 2 percent per annum during the second period as compared to less than 1 percent during the first).

The kibbutz movement can be classified by age, size, and political affiliation;[12] each classification is significant enough to explain remarkable differences between subgroups of kibbutzim. Nevertheless, there is a very considerable similarity among all 238 kibbutzim in Israel today. This similarity extends well beyond the physical, spatial resemblance, and it is the direct result of those common objectives and denominators which ideologically and constitutionally define the kibbutz. The objectives of the kibbutz are: "to cultivate the personalities and capacities, individual and collective alike, of the kibbutz members, in the life of the kibbutz, of the kibbutz society as well as in the spheres of economy, culture, science and arts."[13]

The constitutional foundations of the kibbutz, as recently defined, are:

> The kibbutz is a voluntary organization for settlement objectives. Its outlines are: a collective community organized on the basis of society's ownership over the property, self-labor, equality and collectivism in the fields of production, consumption and education.
>
> Every kibbutz member puts at the disposal of the kibbutz his total labor force and resigns to the kibbutz all his incomes and assets which reach him from any source whatsoever. The kibbutz determines the work of every member, whatever this implies, and provides for all his needs, including the needs of his dependents who lawfully reside in the kibbutz, according to its abilities.[14]

The basic implication of these foundations is the virtually complete detachment of material rewards from the physical or mental efforts exerted in the course of producing goods and services. The principles which direct distribution are not connected with those involving production considerations. Application of the behavioral propositions of standard economic theory, implying profit-utility maximizing motivations, will be useless for explaining the economic efficiency of kibbutz production. To obtain consistency with economic theory, we must assume extraordinary utility functions for kibbutz members, by which total welfare of individual (A) is heavily affected by the welfare of individuals (B), (C), . . . (N). Thus, the strong influence of the community on the individual and the deep involvement of the individual in community life increases the weight of mutual interdependence in the utility functions of the kibbutz members. The greater the weight of the community's welfare in the individual's well-being, the easier it is to comprehend how large-scale production units can operate efficiently without due reward for efforts.[15]

The managerial mechanism is structured accordingly, the purpose being to maintain and intensify the sense of involvement of each individual member. The sovereign decision-making body is, and has been, the General Assembly of all members. In this respect, effective, direct democracy has probably had its classical manifestation in the kibbutz, which continues to make great efforts to foster the effectiveness of the General Assembly.[16]

Actual day-to-day deliberation and decision-making is in the hands of the "Secretariat." It is composed of 3 to 5 members, with equal right and authority, who are annually elected by all kibbutz members. It is assisted by permanent as well as ad hoc committees, also elected annually, which have both consultative and executive functions. In these bodies general involvement is induced by rotation in executive functions and committee work. Another device toward the same end is the complete lack of material reward for such public service.

The economic advantages of the kibbutz have been mainly attributed to its size, and to the fact that production of goods and services is carried out under conditions of increasing returns to scale. This contention has been confirmed empirically by the measurements of H. Barkai, who found that in both agriculture and industry production was carried out with increasing returns.[17] Consequently, size correlates positively with productivity, and therefore with profitability. On the other hand, size, apparently, correlates negatively with the social values which made the kibbutz the unique social experiment that it is.[18] There should thus be a certain trade-off between profitability and the effectuation of democratic institutions in the kibbutz, which operates, other conditions equal, through size. It is suggested, therefore, that the objective of production in the kibbutz is not maximum profits, or some combination of profits and revenue. It is rather a complex benefit function which includes elements of psychic income, derived, in the course of production, from ingredients like self-management, realization of professional ambitions and even, prima facie, purely altruistic motives of promoting the well-being of society.[19]

In view of such considerations, optimum size depends on such variables as: (a) relative weight attributed to profit versus social experience; (b) background, predisposition, etc., of members; (c) the standard of technology applied in production.

In the 1920s and 1930s, the issue concerning a desirable trade-off between returns to scale and the intensity of social experience was a matter of conceptual difference between various kibbutz movements. The basic question was whether the kibbutz should become a relatively small elitist enclave with a highly intensive social life, or an open organization for every worker who was willing to adopt its constitution and live accordingly. At present most of these differences have disappeared and a more or less general consensus has been established permitting a rather wide range for optimum size. It is agreed that optimum is certainly above 100 and below 600 members. Distribution by size and movements is presented in Table 2.

TABLE 2. NUMBER OF MEMBERS IN KIBBUTZIM ACCORDING TO MOVEMENTS[a]

No. of Members	Kibbutz Artzi		Ichud		Hameuchad		Religious	
	No.	%	No.	%	No.	%	No.	%
100 or less	7	9.5	10	12.5	10	17.5	4	26.7
101-200	27	36.4	37	46.3	16	28.1	7	46.7
201-300	14	18.9	18	22.5	16	28.1	2	13.3
301-400	19	25.7	7	8.7	7	12.3	2	13.3
401 or more	7	9.5	8	10.0	8	14.0	-	
Total no. of kibbutz	74	100	80	100	57	100	15	100
No. of members	17,807		16,529		13,940		2,487	
Total population	32,061		30,236		25,609		5,304	
Members/kibbutz	240		207		245		166	
Population/ kibbutz	433		378		449		354	

[a]The figures represent the situation in 1967. For the socio-political meaning of the "movements," see note 12.
SOURCE: D. Giladi, ed., *Agriculture in Israel and in Workers' Settlements Statistical Abstracts* (Tel-Aviv, 1970), p. 63.

The differences in size distribution between the various movements, despite historical controversies, were in 1967 unimportant. A Chi square analysis showed no significant differences.

As indicated, the kibbutz is directed by its members through the democratic instruments of assembly-secretariat-committees. Membership is acquired through votes of the General Assembly.[20] Two temporary stages lead to membership: the "guest" stage, lasting up to six months, which ends in secret ballot; the stage of "candidacy," which lasts up to one year and also ends in secret ballot. Absolute majority of the voting members (or of all members) in favor of the candidate admits a person to full membership. Revocation of membership is subject to the same secret ballot mechanism, usually of 66 to 75 percent of all members. Thus, the General Assembly exercises its authority as the sovereign body of the kibbutz in all matters of management and membership.

Golomb, in his essay on "The Kibbutz in the Seventies, Facing the Eighties,"[21] suggests eight typifying characteristics for the kibbutz today, the first of which is "maximum collectivism and equality."[22] The term equality is subject to various interpretations. In standard economic theory, equality is usually considered as equal opportunities, and the resulting income distribution system is that of marginal productivity under conditions of perfect competition in the markets of both final products and inputs. Marginal productivity theory therefore implies reward according to a factor's contribution to production. This conception is in explicit contradiction to the kibbutz norms, although it is apparently supported by a small minority of members.

Divorcing rewards from efforts still leaves the interpretation of equality and distribution open to two diverging principles, the principle of "individual needs" versus that of "equal needs." This very issue--the conception of equality and just distribution--was recently researched through interviewing a large sample of members (1,264) from 50 different kibbutzim.[23] The study, conducted by N. Cohen, investigated the way in which equality is conceived in the kibbutz. Four answers were suggested. The first suggests that people are born unequal and thus rejects altogether the need for equal distribution; the second offers a conception equivalent to the marginal productivity theory; the third suggests the principle of equal needs ("community should provide for each equal quantities and qualities of goods"); and the fourth supports the principle of individual needs ("community should provide to each according to his specific needs"). The answers received are summarized in Table 3.

TABLE 3. CONCEPTION OF OBSERVED EQUALITY IN THE KIBBUTZ

	No.	%
No equality	56	4
Marginal productivity	31	2
Equal needs	792	63
Individual needs	383	30
Total	1,262	99

In practice, the distribution system is dual. Most basic needs, such as food, health, education, and housing, are provided according to individual needs. Other goods and services, like clothing, footwear, entertainment, semi-luxury items, etc., are rationed in equal amounts. In the equally rationed items recent developments widened the scope for individual tastes as freely disposable cash was, through popular demand, substituted for the distribution of goods.[24]

In the absence of wage differentials, incentives for devoted effort and capability in work are explained by: existence of nonpecuniary rewards; high rate of altruistic motivations. Nonpecuniary rewards are of two kinds: material; social.

Though there are no differentials in those goods, services, and money distributed to the general membership, functionaries, directors, and executives do get indirect material rewards such as telephones, easy access to motor cars for personal needs, opportunities to travel (often abroad), etc. The extent of these rewards, however, does not nearly reflect the contribution of these members to the kibbutz economy. Thus they apparently play a rather insignificant role in so far as the working members are concerned.

More important are the nonmaterial, social rewards. Power, influence, improved social status, appreciation, and probably even affection are of a higher premium in the kibbutz community than marginal material benefits. The extent to which influence, appreciation, and affection are indeed acquired by public or directorial functions in the kibbutz was also investigated by Cohen.[25] His study showed that just about every member (94 percent of the 1,302 interviewed) attributed great or decisive influence over the fortunes of the kibbutz to central functionaries.

Though prestige through appreciation is apparently attributed much less frequently to these same functionaries, and affection even less (about 14 percent of 1,109 interviewed),[26] power and influence are apparently sufficiently strong factors to explain a great deal of extra effort without material rewards. Nevertheless, altruism also accounts for a great deal of the explanation. It is partially based on the assumption that the utility functions in the kibbutz include elements of interdependencies that are weighted more heavily than in the nonkibbutz society. Yet, one should not overlook the rather high sense of duty toward public objectives which has been carefully cherished and emphasized by kibbutz education. Such self-indoctrination about being the pioneering stratum of society, which must claim responsibility for the maintenance of the existing social norms and their institutions,[27] has until recently been very successful and has prevented any strong tendency toward negative selection of membership, i.e., no apparent erosion has been observed through the desertion of the elite elements. This is a remarkable achievement since the fast-growing urban economy of the country has provided ample material temptation for capable kibbutz members. And indeed outmigration from the kibbutz has not occurred in great numbers until recently. The general quality of the kibbutz population remains above the national standard. The average level of formal education is probably the highest, compared to any relevant subgroup, since virtually every child is given 12 years of institutionalized full-time education.[28]

In recent years, there has been a very sharp rise in the participation of the kibbutz population in university education. The university student/population ratio in the kibbutzim is probably higher than among the population in general. In one of the large movements, Kibbutz Artzi, there were, in 1972, 1,000 university students[29] out of a population of 35,940.[30] The crude student/population ratio of 2.77 percent compares very favorably with the crude national ratio of 1.17 percent.[31]

Consequently, there is no basis for the proposition that the absence of a proportional reward/contribution system induced negative selectional processes. The contention that altruism and very special utility functions play a significant role is supported by the unique labor relations in the kibbutz. Strikes or slow-downs are unknown among kibbutz members, and management-labor disputes are usually transformed into disputes among members concerning communication methods and channels in the working place.

As a result, the labor relations constraint which often hinders, rather effectively, the introduction of new technology in industry does not exist in the kibbutz. Obviously, internal labor relations are not the only reason for technological improvements; reluctance to employ hired labor and the high intellectual and technical quality of the average kibbutz member are also relevant. In any case, keeping pace with up-to-date technology and production methods has been one of the more typical characteristics of the kibbutz economy.[32]

The kibbutz has left its marks on all aspects of life in Israel. Its impact in politics, in economics, in the military command, and recently also in universities and other research organizations has been far beyond its quantitative weight. Four members of the present Israeli Cabinet, including the Foreign Minister and the Minister of Education, are active kibbutz members. The Parliament, and especially the Labor Alignment--the backbone of present and past coalitions--has a rather heavy representation of kibbutz members. That is also the situation in the Histadrut, which is simultaneously the Israeli trade union movement and a giant economic conglomerate. Its former Secretary General is also a kibbutz member. In the military command, kibbutz members are of very great importance, particularly in the Air Force and among medium-rank officers, as well as among the select commando units. In the political parties kibbutz members are among the ideological avant garde. They are among the more prominent "hawks" and the most militant "doves." The central institutions of the three big movements established teachers' training colleges, research institutions, and a classical music orchestra. There are also several university professors who are active kibbutz members. This is undoubtedly an impressive performance. The motivation behind it is a very deep urge for involvement in the affairs of the state. This urge is probably as old as the kibbutz itself.[33] The kibbutz is, in the view of its members, the perfect manifestation of unselfish pioneerism. Thus, it is not only the right of the kibbutz to leave its imprint upon the Israeli political and social scene, it is its obligation to do so since the norms and values it stands for in the view of its members are of absolute validity.[34] Most institutionalized youth movements have educated their members toward "a life of fulfillment," i.e., kibbutz life, and the Army has consented to the establishment of special units, which spend over half their compulsory military service in kibbutzim.

In historical perspective there is no doubt that the social and political influence of the kibbutz is on the decline. The great bulk of immigrants were either unable or reluctant to adjust to collective life patterns. With the consolidation of the State, most self-imposed national functions were transferred from the kibbutz and other voluntary bodies to state-managed institutions.[35] The kibbutz thus suffered from a process of "overfulfillment." Fast economic growth, which induced materialistic thinking "was saturated with disintegrating elements for both solidarity and collectivism, for society in general, as well as for the kibbutz itself."[36]

The activities which have been favorably affected by these processes are production, efficiency, and profitability. The relative weight of the kibbutz in the economic sphere has increased, and it has probably become the most dynamic and innovative sector. In agriculture, the share of the collective settlements[37] in the total agricultural output from the Jewish sector increased from 32.8 percent in 1960/1, to 36.7 percent in 1964/5, and to 38.2 percent in 1971/2. At the same time total agricultural production in fixed prices increased from I£ 654.3 million in 1960/1, to I£ 922.3 million in 1964/5, and to I£ 1,386.4 million in 1971/2.[38] Thus, production in the kibbutzim has more than kept pace with the very rapid expansion of total agricultural production. The labor force in kibbutz agriculture has remained about the same for the last fifteen years, while output has doubled. This is the combined effect of a sharp increase in capital intensity, and of continuous improvements in the efficiency of input utilization. Barkai calculated capital/labor and capital/land ratios in kibbutz agriculture for the years 1954 to 1965. His results show a dramatic increase of 148 percent in the capital/labor ratio from I£ 9,406/man-year in 1954, to I£ 23,339/man-year in 1965, and a rise of 218 percent in the capital/land ratio from IL 80/*dunam* (1 dunam = 1,000 square meters) in 1954, to I£ 280/dunam in 1965.[39] Correspondingly, the total productivity index rose from 100 (1957) to 146 (1965) and increased to 190 in 1973. Similar results are shown in another study[40] which attempted to construct an "Economic Success Index" with variables such as profit ratio, capital account ratio (profit + depreciation + interest/production), net income per economic unit (family of two members), and capital structure (net worth/fixed assets). The findings indicate a rise of 32 percent in that index from 38.0 points in 1957 to 50.0 points in 1968.[41]

These various measures show that the relative decline of the kibbutz in the political, ideological, and social sphere did not carry over to its economic activities. Indeed, the relative economic power of the kibbutzim may well have increased. It is suggested that the second phenomenon may be the result of the first, i.e., the declining political and social influence diverted energies from national responsibilities to economic efforts on the kibbutz. Since the kibbutz operates under clear-cut conditions of increasing returns to scale,[42] with a very high quality labor force, economic success is to be expected.

Two major consequences are inherent in these developments: (1) a continuous rise in the marginal product of labor in agriculture, which is a rather reliable indicator for the standard of living under conditions of egalitarian distribution; (2) due to the opportunity to apply highly capital-intensive production methods, the kibbutzim gradually withdrew from the more labor-intensive enterprises and specialized in the production of those products where a high capital/labor ratio insured a high reward for labor.

Regarding potential returns to labor in kibbutz agriculture, we shall resort again to the data compiled by Barkai.[43] His figures indicate a real growth of 270 percent in average annual income per unit of labor from I£ 3,016 in 1954, to I£ 11,128 in 1965. These figures were 2.5 to 3 times higher than average annual wages in agriculture (including fringe benefits). Since 1967 this gap has widened as productivity and the capital/labor ratio on the kibbutz continued to rise, while unskilled labor in the rest of agriculture did not improve its relative share.

A gap of 200 percent or more between the potential or shadow wages for members and the price of unskilled agricultural labor offers a strong temptation to extend production by hired workers. This has not been done, however--and the very high marginal product of labor is an indication of it--because of ideological and internal social reasons. Instead, kibbutz agriculture has gradually reduced production of labor-intensive produce, such as vegetables and poultry, and increased the production of commodities such as cotton and dairy products which, especially in the case of cotton, lend themselves to greater mechanization.

A further, relatively recent development in the economy of the kibbutz is its gradually intensified industrialization. In 1973, there were 246 industrial plants belonging to the kibbutz movement, employing over 11,000 workers. Thus, the average size of the kibbutz plant--45 workers per plant--is considerably larger than the average plant in the country. Likewise, it is more dynamic in its growth pattern and export drive than industry in general.[44] And, indeed, the rate of industrialization has been most remarkable. The entire addition to the labor force in the kibbutzim, from 1958 on, has been diverted to industry. Consequently, the ratio of labor in industry to that in agriculture rose from 15 percent in 1958 to 65 percent in 1973, and the comparable ratio for capital grew from 7.5 percent to approximately 30 percent. Finally, the productivity index has grown even faster in kibbutz industry than in kibbutz agriculture, showing the following pattern:

1957 = 100
1965 = 157
1973 = 210 - 220.

As a result, shadow wages, returns to capital, and thereby profitability significantly surpassed agriculture's performance.

"The kibbutz entered the seventies under conditions of increasing community problems. A sense of fatigue is felt towards the kibbutz way of life among the first generation, while the second generation is not sufficiently determined to take upon itself the comprehensive responsibility for the kibbutz with all its problems."[45] Thus wrote N. Golomb, himself a first generation kibbutz member of the most doctrinaire movement, "Hakibbutz Haartzi," in a retrospective survey written for the ideology seminar of his movement.

Following Golomb, let us summarize the achievements of the kibbutz in the perspective of over two generations of operation. These achievements reflect the kibbutz' contribution to its members as well as to Israeli society as a whole. Internally, the kibbutz created for all its members a socially intensive way of life, as well as a way of life with near total security, reasonable equality, high moral standards, high educational and cultural levels, and a standard of living which ranks within the seventh decile of the Israeli population.

If we look at the kibbutz' contribution to the state and its objectives we may suggest the following achievements:

1) the kibbutz created a network of frontier colonies for the security of the country;
2) it developed one of the most outstanding agricultural technologies known today;
3) it produces over 35 percent of Israel's agricultural products and 5.5 percent of its industrial products;
4) the export coefficient of the kibbutz production is above the national average;
5) the share of the kibbutz in the combat and commanding staff of the army is above average;
6) the social norms provided by personal example to Israeli society in general is still of great importance.

Considering the failures of the kibbutz in its obligations toward society, two major shortcomings should be mentioned.

1) The kibbutz has failed to absorb immigrants from developing countries. Consequently, the ethnic structure of the kibbutz is totally unrepresentative of that of the country--over 52 percent of the Jewish population in Israel is of Afro-Asian origin.
2) The kibbutz population failed to keep pace with the national growth in population and thereby it runs, in the long run, the risk of becoming an enclave of collectivism with growing irrelevance to its surroundings.

In the perspective of sixty-five years of dynamic existence, and considering the fact that in the older kibbutzim control has been passed to the second and third generations, it seems clear that the kibbutz is a permanent fixture in the Israeli social milieu. Its economic focus may change, together with its role in the direction of the political and social life of the country. It will, however, remain as a major ideological force radiating its norms and way of life to the public as a whole.

NOTES

1. "Annual Report of the Registrar of Cooperative Societies for 1973," *Labour and National Insurance* (December 1974), p. 412 (Hebrew).

2. A Bein, *A History of the Zionist Settlements* (Tel-Aviv: Masadah, 1954), pp. 427, 430 (Hebrew).

3. A Ruppin, *The Agricultural Organization of the Zionist Organization in Palestine* (London: Martin Hopkinson and Co., 1926), p. 1.

4. Ibid., p. 3.

5. I. Baratz, *Degania* (Tel-Aviv: Palestine Pioneer Library, 1944), p. 9.

6. For this reason, the well-known experiment of the theoretician, social reformer Prof. F. Oppenheimer, failed. He conceived the "perfect" model for Jewish colonization in Palestine, but forgot that the human element at his disposal was of a unique type and disposition.

7. This contention is by no means accepted by all students of the kibbutz. However, in this respect the author accepts the views of Dr. Ruppin, who was considered the patron of the first kibbutz, Degania, and followed closely its development.

For a rather detailed description of the stages by which the first kibbutz became a collective, see Ruppin, *Agricultural Organization of the Zionist Organization in Palestine*, pp. 129-37. The same conclusion is reached by reading the personal accounts of I. Baratz, one of the founding members of Degania, about the trial and error of their commune with regard to the role of women, as well as consumption and education patterns. See Baratz, *Degania*, pp. 24-32.

8. Not everybody in Degania agreed to the perpetuation of collectivism. A group of dissenters left the kibbutz in 1921 to establish the first smallholders' cooperative settlement, Nahalal.

9. From Tabenkin's speech in 1924, quoted in *Hakibbutz Hameuchad, Collected Documents* (Tel-Aviv: Hakibbutz Hameuchad, 1961), p. 31 (Hebrew).

10. Ibid., p. 35.

11. H. Darin-Drabkin named his book on the Kibbutz *The Other Society* [(Sifriat Hapoalim: Merhavia, 1961), Hebrew] to distinguish it from the rest of the Israeli society.

12. There are four socio-political organizations which have affiliated kibbutz movements, stretching from the right-of-center religious organizations, through the central social-democratic "ichud" movement, and the left-of-center "Hameuchad" movement to the left-wing Marxist-Socialist "Kibbutz Artzi" group. The breakdown according to affiliation and period of establishment is as follows:

Period*	Artzi	Hameuchad	Ichud	Religious	Total
To 1920	2	-	7	-	9
1921-1930	6	7	7	-	20
1931-1940	17	18	18	2	55
1941-1948	23	17	27	5	72
1949-1974	27	18	27	9	81
	75	60	86	16	237

*Only the presently functioning kibbutzim are counted.

13. Quoted as the last and most comprehensive version of the long-debated "kibbutz objectives," by Y. Ben David in "Change and Continuity in the Kibbutz: Research and Theory," *The Kibbutz, Interdisciplinary Research Review* (July 1973), p. 16 (Hebrew).

14. This is the text submitted to the Parliament within the framework of the new Cooperative Bill. As of this writing, the Bill has not yet been passed.

15. This is totally different from the nonpecuniary rewards of the Stachanovici pattern. In the kibbutz we assume that the reward for effort is reached in the final consumption stage by augmenting social well-being.

16. Among the main adverse effects of the recently introduced television broadcasting was that "the kibbutz meeting, assembling Saturday night, is in danger" of members' attention being diverted to the T.V. Quoted by M. Gurevitch and Z. Loevy, "Television in the Kibbutz, A Study in the Diffusion of Innovations," *The Kibbutz, Interdisciplinary Research Review* (July 1973), p. 106.

17. In a Cobb Douglas production function for agriculture with three inputs, the aggregate value of production elasticities exceeded unity in all twelve years investigated. In manufacturing, the Cobb Douglas function with two inputs showed increasing returns to scale for nine out of the ten years investigated. See H. Barkai, "An Empirical Analysis of Productivity and Factor Allocation in Kibbutz Farming and Manufacturing," Mimeo., Discussion Paper 748 (Falk Institute, 1974).

18. In situations where productivity depends heavily on the attitude of the worker toward his workingplace, as it does in the kibbutz, profitability will depend on the intensity of social ties, and thus correlate negatively with size of plant and the kibbutz itself. This issue was developed for the general case of cooperative production in Y. Don, "Management Patterns and Economic Results in Agricultural Cooperatives," *Journal of Rural Cooperation* (October 1973).

19. This rather unique objective function is indicated in N. Golomb's paper, "The Social and Organizational Structure of the Industrial Plant in the Kibbutz," Mimeo. [Givat Haviva: Unit for Kibbutz Management, 1975 (Hebrew)], p. 2: "Profitability and satisfaction of the workers are two objectives of equal value in the management of the industrial plant in the kibbutz."

20. This rule does not apply to founding members, who automatically acquire membership.

21. Mimeo., issued by the Unit for Kibbutz Management, Givat Haviva, 1973.

22. Ibid., p. 3. The others are: (2) Voluntarism, (3) Democracy, (4) Detachment between members' efforts and rewards, (5) Public property, (6) Collective production, (7) Collective education, and (8) Collective consumption.

23. N. Cohen, "The Kibbutz in the Eyes of its Founders and Continuers," Mimeo. (Rehovot: Settlement Study Center, 1974), pp. 85-88 (Hebrew).

24. Ibid., p. 85.

25. Ibid., p. 106.

26. Ibid., p. 116.

27. For a careful analysis of this issue, see Y. Don, "Adaptation of Cooperatives to Economic Change," *Journal of Farm Economics* (1967), pp. 119-30.

28. In the Jewish Israeli population, the rate of participation in the 12th grade--compared to the 8 compulsory years of education--is 43 percent. The dropout rate is reflected by the diminishing rate of participation in high school. For the academic year 1971-2, with a reasonably stable age structure, the rate of participation was as follows: 8th grade - 100, 9th grade - 94, 10th grade - 72, 11th grade - 59, 12th grade - 43.

29. This figure was given in a report by M. Chizik about higher education: "Studies at the Kibbutz' Expense," *Hedim* (*Echos*) (October 1973), p. 194 (Hebrew).

30. "Annual Report of the Registrar of Cooperative Societies for 1973," p. 432.

31. Calculated from the *Statistical Abstract of Israel* (Jerusalem: Central Bureau of Statistics, 1972).

32. For a general survey of this issue see F. Kanovsky, *The Economy of the Israeli Kibbutz* (Cambridge, Mass.: Harvard University Press, 1966). See particularly Ch. 5, pp. 74-86.

33. Y. Tabenkin in an address as early as 1924 (see note 9, supra) spoke of the kibbutz as the army of the Histadrut. "The Histadrut needs a pioneering army at its disposal. That is the kibbutz." Quoted in *Hakibbutz Hameuchad*, p. 34.

34. M. Rosner, a prominent kibbutz sociologist, saw the necessity of active involvement because "the kibbutz is an elite society on the background of the material and quantitative criteria dominant among the rest of the society." Y. Paz spoke to the same issue of involvement, in terms of a "Kibbutz Paradigm" to be suggested as normative and organizational rallying points. Y. Don, ed., *The Kibbutz Under Changing Patterns of the Israeli Economy* [Tel-Aviv: CIRCOM, 1975 (Hebrew)], pp. 34, 37.

35. This process is analyzed in detail in Y. Don, "Adaptation of Cooperatives to Economic Change," in Don, ed., *The Kibbutz Under Changing Patterns of the Israeli Economy*, pp. 126-9.

36. M. Rosner, in Don, ed., *The Kibbutz Under Changing Patterns of the Israeli Economy*, p. 34.

37. Figures included about 25 collective moshavim.

38. Source: S. Pohoryles and A. Szeskin, *F.A.O. Research in Contemporary Changes in Agriculture* (Tel-Aviv: Ministry of Agriculture, 1973), p. 8; D. Giladi, ed., "Agriculture in Israel and in Workers' Settlements, Statistical Data," Mimeo. [Tel-Aviv: The Histadrut, 1970 (Hebrew)], p. 49.

39. Barkai, "An Empirical Analysis of Productivity and Factor Allocation in Kibbutz Farming and Manufacturing," p. 11. Ratios are calculated in fixed prices of 1958.

40. L. Shashua and Y. Goldschmidt, "Economic Development of the Kibbutzim in the Sixties," *Economic Quarterly* (April 1973).

41. Ibid., p. 155.

42. Average landholding per kibbutz is 5,330.5 dunams (1 dunam = 1,000 square meters) as compared to 50-60 dunams holding per unit in the moshav sector and 89.5 dunams in the private sector.

43. "An Empirical Analysis of Productivity and Factor Allocation in Kibbutz Farming and Manufacturing," p. 11.

44. In 1972, 5.6 percent of total industrial output originated from the kibbutz industry, whereas 7.5 percent of total industrial exports came from those sources.

45. N. Golomb, "The Kibbutz in the Seventies," p. 7.

3. The Hutterian Colony: A Traditional Voluntary Agrarian Commune with Large Economic Scale

John W. Bennett

INTRODUCTION

This paper is concerned with the Hutterian Brethren of the northern Great Plains of North America, a people who originated in the Austrian Tyrol in the sixteenth century, and who migrated through southeastern Europe, eventually to settle in the Ukraine. Their move to North America occurred in the 1870s, when the Czarist government attempted to renege on its earlier promise to exempt the Brethren from military service. While traditional in cultural background and social organization, these people have responded to the changing economic and technical demands of Great Plains agriculture to forge an efficient, mechanized, large-scale operation. They achieve this with the help of a strong moral order, reinforced by communal child socialization and by a carefully controlled, partially withdrawn posture *vis-à-vis* the majority society. It is the opinion of the writer that the only agrarian communities which bear close comparison to the Hutterian colonies are the *kibbutzim* of Israel, which, although with very different historical traditions and duration, are organized in virtually identical fashion.[1]

The Hutterian colony is in essence a village of approximately 100 persons, carrying out a variety of production tasks with a complex table of organization to be described. While each colony functions as a separate enterpreneurial unit, it is obviously closer to a corporate farming enterprise than it is to the typical family farm. However, in some districts of rural North America, neighboring farmers, related by kin or by friendship ties, may form linked systems of mutual aid and exchange which resemble a colony spread out, so to speak.

I have designated the colony as a "traditional voluntary agrarian commune," and a word of explanation of this label is desirable. The essential characteristics are as follows:

1. Collective socialization of children, with minimal participation by the nuclear families in this process.
2. Nuclear family units exist, but their primary function is childbearing and shelter; their activities and interests are subordinated to the collective needs and objectives.
3. All important property, especially capital and capital goods, are owned in common, and personal property is held to a minimum.
4. The surplus created by productive activities is retained by the collective as capital, with controlled redistribution.
5. Remuneration of labor and services is withheld, or is minimal and variable. Compensation for labor is mainly motivational: the collective good. Necessities and "luxuries" are distributed equally among all members, with marginal recognition of differential needs by age, and privileges for leaders.

These characteristics define the typical institutional pattern of the agrarian commune as it developed in Middle Eastern, European, and, more recently, North American civilization from early Christian times to the present. The pattern is probably to be considered distinctive for Judaeo-Christian civilization, and may have its beginning in pre-Christian Palestinian dissident Jewish groups like the Essenes, from whence it was transmitted to early Christian communities; thence to appear and reappear throughout the Christian era, usually as a heretical, anti-Church movement, and as opposition to worldly tendencies and the accumulation of wealth. Secular rationale for the template was introduced in the eighteenth century in the form of the Utopian Socialist movement, and by the nineteenth century was reinforced selectively by Marxist ideology. However, in both the secular-socialist and religious versions, strong moral sanctions, emphasizing brotherhood, altruistic sharing, and equality were present. The historical experience of the movement attests to the fact that idealistic components like altruism were never sufficient in themselves to ensure socio-economic continuity of the communities: the enduring forms always manifested complex social and legal sanctions to encourage conformity with the rules.

All of the ideals of the traditional voluntary commune are accessible to the majority society in the standard Christian and Jewish religious and philosophical literature. In a sense, the commune is simply that social form in Judaeo-Christian civilization which takes the stated ideals of brotherhood and altruism literally, and attempts to construct a viable community on their model. That is, the traditional voluntary commune is not an exotic plant, but simply the pure species. The persistent attempts by the majority society to persecute or harass the movement are based in large part on the guilt and anxiety generated by the co-existence of a true believer group in the midst of a world which has apparently fallen away from the teachings.

The enduring communes are often agrarian because they find it convenient to avoid the socially differentiating influences of urban industrialism, and also because, in time, they came to define agriculture, and agricultural labor, as desirable values. The traditional commune views agricultural production as the most moral economic posture, since it minimizes exploitation of other human beings. Some of these themes are, of course, to be found in other collective communities: the monastic orders of Europe and the Orient, and the agrarian collectives of China come to mind. However, these other forms have special origins and purposes which differ from the private, self-constituted, "intentional" societies of the voluntary commune.

The distribution of modern communities partaking of this tradition can be suggested by the accompanying chart. The left-hand column contains a list of group-farming communities, presented in an order from what I consider to be the most "communal" at the top (the Hutterian colony) to the most "cooperative"--or, perhaps, least communal--at the bottom (the Saskatchewan cooperative farm). However, all groups on the chart own their basic agricultural resources and property in common, and all groups were formed on a voluntary basis, with adherence to existing traditions and templates. Below, and off the chart, would exist other forms of cooperative farming lacking full pooled ownership, or limiting cooperation to special functions, like irrigation. These would escape the definition of cooperative-collective or group farming as I have conceived it.

There are, however, significant differences between the groups on the chart. In the Hutterian colony, no individual or family is considered to hold equity in the establishment; in the cooperative farm, individual or family equity may be calculated on the basis of amount of resources contributed to the pool at the time of joining (as well as some other devices we need not go into). If there were no other differences between these two extreme forms, this would be sufficient to make the typological distinction between communalism and cooperation, since the true communitarian society rejects individual shareholding. I am not suggesting that this particular difference is the most critical: actually it varies, depending on other factors, in particular groups. However, it is a generally important difference because it distinguishes both types of groups from individual enterprise, and creates continual problems for these groups in their dealings with majority-society institutions.

If we inspect the institutions which create the columns on the chart, we can observe various gradations of communal-cooperative patterns. By reading the descriptions for each group *across*, one obtains an institutional profile of that particular form; by reading each column *down*, one gains a sense of the diminishing force of the full communal frame. The burden of the exercise is to suggest that cooperative farming represents a compromise between communal and individualistic institutions. Thus the tradition of the voluntary agrarian commune has changed and adapted to the institutions of the Western world in the course of its centuries of existence.

The chart contains a heavy line which marks off the cleavage between what may be usefully considered communal and cooperative

CHART: FROM COMMUNALISM TO COOPERATION

(All institutions above the heavy line represent true communal patterns.)

	SOCIALIZATION OF CHILDREN	NUCLEAR FAMILY AUTONOMY	PROPERTY OWNERSHIP	DISTRIBUTION OF SURPLUS	REMUNERATION	CONSUMPTION
Hutterian Colony	Family to age 3; colony schools thereafter, but children always live at home.	Private residence but in multi-apt. building: compromise. Family must submit to colony interests.	All communal; land, equipment, capital, etc. Minimal personal property; no individual equity.	All communally retained: minimal redistribution.	No "wages" or cash remuneration: work is for good of colony. Necessities equally distributed.	Rigorous austerity in both personal and collective consumption.
Left Kibbutz	In Kibbutz schools, nurseries; children never live at home save for brief periods.	Same as in Hutt. colony; formerly in dormitory. Family more autonomous than in colony, however.	All communal, as in Hutt. colony, but more property, and some residual sense of equity.	All communally retained, like Hutt. colony.	No wages like Hutt. colony, but paid vacations, etc.	Less rigorous, especially in collective sphere and generally more property--higher living levels.
Moshav Shitufi	In Moshav schools, but children live at home, in separate domiciles, at all times.	Families have separate homes, much autonomy.	Communal land and much equipment, but families own more personal property and some tools of production.	Same as above.	Same as above.	Same as above, though possibly more family-personal property.
Moshav Ovdim	Same as *Shitufi*.	Same as *Shitufi*.	Tools of production and land communally owned, but land allotted to individual family on share.	Most produce communally pooled and proceeds shared. Surplus sold by individual families.	Wages for labor on time basis.	Family autonomy limited only by finances and some social convention.
Matador Cooperative Farm in Saskatchewan	In public school.	Same as *Shitufi* and *Ovdim*.	All land and tools of production owned communally--no individual share though amount of land contributed to pool governs individual shareholding in enterprise, or equity.	All produce communally sold and proceeds shared equally after wages.	Wages for labor on time basis.	Same as *Ovdim*.

*For accounts of the Israeli groups, see D. Weintraub, M. Kissak, and Y. Asmon, *Moshava, Kibbutz and Moshav: Patterns of Jewish Rural Settlement and Development in Palestine* (Ithaca: Cornell University Press, 1969). For a description of the Matador farm, see H. Cooperstock, "Cooperative Farming as a Variant Social Pattern," in B. Blishen, ed., *Canadian Society* (Glencoe, Ill.: Free Press, 1961).

institutions. The list of five characteristics of the traditional agrarian commune presented earlier generalizes the pattern defined. Note that in one respect the Hutterian colony is "less" communal than the Left Kibbutz: the colony has permitted its children to live in the nuclear family domicile, even though it socializes them in collective schools and restricts the amount of punishment the parents may provide. In essence, then, the Hutterites and the kibbutzim represent the most communal groups.

Cooperative institutions seem more varied. Since property and resource management are less rigorously administered by the community, a variety of management procedures can exist. This implies different sociological emphases as well.[2] The communitarians seem further removed from the majority society: their behavior differs due to the intensive socialization in affiliative processes. In this comparative sense, the *family* becomes the crucial element: the cooperative farming societies preserve the essential features of rural family structure in which nuclear families enjoy considerable autonomy and children grow up in a looser structure. The greater constraint in communitarian groups produces a personality pattern which demands affiliative experiences and often avoids spontaneous decisions, as well as a possibly larger quotient of positive affect.

HUTTERIAN HISTORY

The history of the Hutterian brotherhood has been covered adequately in other publications,[3] and so we need only provide an outline here. Modern Hutterites--sometimes technically labeled "Christian Brethren called Hutterian"--are a revitalized body of adherents who migrated from the Ukraine, along with many other German-speaking farmers, in the 1870s. This southern Russian group represented the terminal population of a movement which originated in the Austrian Tyrol in the mid-sixteenth century. They are considered affiliates of what is sometimes called the Radical Reformation[4] to distinguish it from the better-known Reformation originated by Martin Luther. Anabaptism is the proper name for this collective movement, consisting of various groups dissenting from Catholic, and later Protestant, orthodoxy and who were consistently persecuted by both. These groups differed on many issues, and sometimes fought each other harder than their orthodox opponents, but had three major beliefs in common: (1) a denial of the validity of the bureaucratic organization of the sect and the church; (2) a rejection of the practice of infant baptism (the name, Anabaptist ["re-baptism"], derived from the early period when the new sects required their new members to undergo a second adult baptismal ceremony, is indicative of the fact that these groups wished to adhere to the stricter interpretation of the Galilean sect led by Jesus which seems to have practiced baptism at the age of consent, not at infancy); and (3) the de-emphasis of self and material goals and ambition--a tenet which led both to communal brotherhood and a certain withdrawal from worldly affairs. A fourth key doctrine, but followed by only a few Anabaptist groups, was that of communal property, based on certain descriptions of the early Galilean sect in the Book of Acts, which appears to have required its adherents to give up all property for the benefit of the congregation. Among the Anabaptists, the groups later called Hutterian were the only thoroughgoing followers of the communal property rule.

Intermittent persecution drove the original band of Hutterian brothers--named from their most effective early leader, Jacob Hutter--out of the Tyrol, and then from Moravia; and finally, across southeastern Europe and into southern Russia, where they became part of the large German agrarian minority.[5] However, most of the basic organizational strategies of the Hutterites, including their agrarian regime of diversification and efficient management, were forged in several flourishing decades in the last half of the sixteenth century in Moravia, where some Bruderhofen became attached to noble estates until the Church required the aristocracy to expel them. It is important to note that while the Hutterites were devoted to spiritual matters, they also had extensive practical experience with large-scale agricultural and craft-industrial organization--the only Anabaptist group which did so. The present-day Anabaptist body includes the Mennonites, and their conservative wing, the Amish, all of whom practice individual farming with inter-unit cooperation substantially similar to their non-Anabaptist farmer neighbors in North America and Europe.[6] Mennonites are often considered to be members of the Protestant order, and they have also moved out into the larger society to accept the standard roles of urban and rural institutional life.

By the 1870s the Hutterian communities in the Ukraine had mostly fallen away from true communalism, and many had joined Mennonite communities. At this time, the Czarist government withdrew its guarantee of immunity from military service, and since the Anabaptists and other German-speaking groups refused military service, they elected to immigrate to America. The Hutterian families decided to reconstitute the original communal template in the New World, and were assisted in this by Mennonite relatives and neighbors after arriving in South Dakota. Attempts by the U.S. Government to draft these people in World War I led to another exodus, this time to Western Canada. A backwash of Hutterian settlement into the Northern Plains of the United States has since occurred, and something less than one-third of the entire Hutterian population of approximately 22,000 souls is now located in the states of Montana, North and South Dakota, with a few colonies in other northern states.

The adaptation of Hutterian life to the Great Plains since the 1870s has involved a combination of the basic sixteenth century template of communal agrarianism with the expanding technological and economic practices of commercial grain-livestock farming. The need to support a rapidly growing population (a rate of increase of approximately 2.5 percent) on the specialized and constrained resources of the region led the Hutterites to adopt modern technology despite initial resistance (the Amish still attempt to hold out[7]), and therefore to enter into the growth-oriented, high-cost, high-returns cycle of commercial agriculture. Hutterian colony farms are usually the largest enterprises in most of the districts they inhabit, and often the most mechanized, efficiently managed, and conservationist. They are permanent units, productively geared to the outside market, but not to the speculative market in farm and land sales. No Hutterian colony, to the writer's knowledge, has ever sold out or disappeared, even though a number have had severe social problems and had to be supervised by councils of Elders of the sect. This permanence lends a timeless quality to Hutterian social and economic life which manifests itself in a steady, cautious innovativeness and orderly growth patterns.

THE AGRICULTURAL REGIME

The basic pattern is diversification. The original format was self-sufficiency, and included commercial crops, food production, all necessary craft industries, and several professional fields, including medicine. Translated into the terms of modern Great Plains agriculture, this means mainly commercial agriculture on a grain-livestock basis. However, the Hutterites cling to the original custom of engaging in as many activities as possible, in order to promote survival in the event that the outside world experiences an economic collapse, or other disaster. In addition, the self-help crafts, like homemade clothing, foster a sense of tradition and continuity. There is, moreover, a basic distrust of the Outside, and a withdrawal from it, although this self-imposed exclusion has been changing in recent years, as Hutterites move toward greater participation in the larger community and in organized producer groups (in itself a kind of survival technique insofar as it mitigates prejudice and accusations of "not pulling their oar").

There exists no large-scale published survey of Hutterian production, colony by colony.[8] The writer has made a limited one of his own, including over half of the colonies in the triangle consisting of northern Montana, southwestern Saskatchewan, and southeastern Alberta--a group of about 40. In this group, 70 percent of the colonies received over 50 percent of their gross agricultural income from commercial cattle and grain crops, but the range varied from 25 percent to 90 percent. Other important crops, comprising substantial pluralities, and in some cases majorities, of received income, were: hogs; sheep (wool and lambs); milk and cream; chickens, eggs; turkeys; and ducks. All colonies received significant fractions of their income from chickens, dairy, and eggs (from 10 percent to 25 percent). Turkeys were highly variable: only a few colonies specialized in them. Ducks are produced mainly for domestic consumption by all colonies. Some vegetables are sold locally but are not important in income. Grain crops sold on the market are mainly wheat and barley; oats, rye, and other crops are used mainly for livestock feed. A few colonies have gone into the feed business, selling products ground and mixed in their own large mills to local farmer feeders and ranchers. A few have developed small, part-time machinery-repair businesses, but with a few exceptions these are carried on mainly as a "neighboring" service to local farmers. Custom planting and harvesting is also done on a local basis.

These data suggest that Hutterian enterprise remains a diversified business, although it changes or fluctuates with market conditions. A study of the regimes over a decade showed that colonies vary the extent of their investment in particular crops as profitability changes. The Hutterian capacity for rapid shifts in production is high, due to the possession of abundant machinery and the capital or credit necessary to effect the shift and to minimize opportunity costs. Currently, a possible secular trend in colony agriculture is toward increasing specialization in particular crops (but not abandonment of basic diversification) which means that colonies are beginning to differ in the proportions of income and investment devoted to particular activities more so than in the past. This has also resulted from a gradual learning process concerning the best uses of particular soils, topographies, moisture regimes, and other resources. Agricultural land is becoming increasingly scarce, which has meant that the traditional expansion to desired acreages is no longer possible for many colonies, which requires them to experiment with intensification of particular crops suited to smaller and more specialized areas.

All of the processes outlined above are visible as well among individual enterprisers in the northern Plains; the Hutterites differ only in the large scale of their operations and their impressive ability to manage capital costs. Despite this, it should not be concluded that Hutterites seek out boom-and-bust opportunities eagerly, and alter their production regime without due consideration and experimentation. They manifest an unusual combination of conservatism and innovativeness, and will not change their strategies unless they are sure they will pay off. Watching experiments of other colonies is a prime technique; we shall have more to say about this later, but it can be noted now that communalism among Hutterites is not only a matter of the organization of the individual colony, but is also expressed as a system of information exchange among all colonies.

THE ECONOMIC POSTURE

The Hutterian colony as an enterprise is a free, independent entrepreneurial unit, competing with other units (including other colonies) in the market agricultural economy of North America.[9] This competition is technically manifested at the level of sales of produce like grain and livestock on the commercial market, utilizing existing sales entities like the Wheat Pool in Canada, the local livestock auctions, private buyers, contract production, and the like. However, it is more easily visible in the form of local sales of farm produce like vegetables, chickens, and eggs, sold from trucks in the towns and villages, from farm to farm in the country, or to local grocery and meat outlets. That is, Hutterites participate in both the national wholesale agricultural market and the local retail sales market. Studies by the writer of some of the latter activities indicate that little or no profit is realized from them, but they are carried on vigorously by many colonies in order to provide outlets for garden and dairy production surplus, and as useful activities for retired persons, women, and boys. However, many colonies restrain their retail activities in deference to local non-Hutterite suppliers who could easily be frozen out; Hutterites are increasingly concerned over their superior competitive position in the community and seek to mitigate it. They also may avoid intensive competition between colonies for similar reasons. The independent entrepreneurial posture of the individual colony creates a structural situation similar to that of the individual private farmers and other producers.

This means that there exists no *formal* overall Hutterian economic organization. There is no central cooperative, no credit union, no formalized system of inter-colony loans; no central direction of the economy; no or at least minimal reliance on Government grants and incentives (Hutterians everywhere seek to reduce their ties to government and organized agriculture, although this is beginning to change). There are no centralized repair depots; no organized channels for the purchase of commodities, machinery, or building supplies. In this the Hutterites offer a marked contrast to the kibbutzim of Israel, which display extensive centralized organization for credit, production, and maintenance, and are closely tied to Israeli governmental production schedules and procedures. This lack of formal inter-colony organization is also duplicated in the social sphere: each colony has its own "German schools," and uses local teachers for standard elementary education. Each colony is its own religious entity; there is no central sect headquarters or "pope," although the exigencies of life in North America have brought forth a legal body called the United Hutterian Brethren which meets occasionally and deals mainly with large-scale legal issues affecting the sect. More important bodies are the three separate councils for each of the informal subdivisions of the sect, called *Leute*. The Leut council deals with social, doctrinal, and financial questions. The Leute are not comparable to the kibbutz federations, since they are mainly kinship and marital networks, with minor differences in religious custom and dress, and have only limited economic functions. The councils do, however, assume managerial responsibility in cases of financially and socially disordered colonies.

In spite of the absence of formal large-scale economic organization, the Hutterian world has considerable economic unity. Individual colonies loan money to each other when needed, usually at very low interest; they "neighbor" on all agricultural and construction tasks; they exchange ideas, information, and commodities; sell each other materials and machines. Hutterites are careful to point out that none of this is required by belief or rule; it is all a matter of friendship, and even more, is based on the fact that colonies are enterprises of similar scale, and that therefore such informal cooperation is "natural." But there is more to it: the writer's studies leave no doubt that many Hutterites feel beleaguered and "different"; they also cooperate out of a "consciousness of kind," and to protect the well-being of the sect.

In the most general sense, the Hutterian colony exemplifies the advantages of large economic scale (in itself a factor not associated with voluntary communal farming elsewhere in the Euro-American world, except for the case of the kibbutz). As noted previously, colonies as enterprises are usually the largest in their districts in the northern Plains, which gives them a commanding position in the land, produce, commercial-crop, and credit markets. They are known to banks as absolutely safe risks, and can enjoy prime rates or better. Their large purchases make it possible to enjoy the discounts available for bulk sales. In many districts colony production in certain commodities has become the largest in the locality; in southwestern Saskatchewan, for example, the colonies are the largest producers of feed grains and hogs, and are beginning to take leadership positions in provincial producer associations in these and other fields. Commercial companies producing feed, farm machinery, and housing and utility buildings have begun to develop special lines for the colonies, and many of these companies have special salesmen and literature designed to fit the large-scale operations and the Hutterian desire for buying necessities at wholesale rates.

Scale is not therefore merely a matter of the conferred advantages in the market economy, but also refers to the expanded world of Hutterian colonies as a large economic system. Once again we should note that while the Brethren lack formal centralized economic institutions, they have extensive informal networks of information and capital exchange. The moral sanctions are as important as the economic; no Hutterian colony is allowed to "fail"; the Leut ministers will bail it out, move its population elsewhere if discipline is needed, and replace it with a new group of families. That is, the Hutterian scale is composed of the expanded linked colony system and its managerial extensions into the Leut council, and this provides an almost inexhaustible (and largely incalculable) reservoir of capital to finance starts, expansions, and recoveries.

If information is a form of capital in modern large-scale agriculture, then the Hutterites have an equally large reservoir of this important ingredient. Since the sect exists in part in a state of self-imposed isolation from the Outside (as Hutterites sometimes refer to it), they have a problem with respect to information: they know full well that they cannot maintain their agricultural system without it; yet their own beliefs and customs create boundaries between them and the information sources. Hutterites have developed a series of devices to overcome this handicap. The close informal network ties between colonies mean that if one farm boss acquires important information from the agricultural extension service or other agency, this nugget will quickly be communicated to all other colony executives and production managers. Most colonies will possess at least one highly literate manager or boss who makes it a point to spend time in town talking to important government "information brokers," merchants, and political officials, and to an increasing extent colonies have authorized the purchase by its managers of technical manuals and agricultural periodicals. The large scale of Hutterian operations means that purveyors of sophisticated machinery and other equipment seek out the colonies as suitable sites for experimentation with new items, which makes it possible for colonies to become the spearheads of new practices and management techniques.

Moreover, the Hutterian culture contains profound respect for education and training--even though in the need to maintain boundaries, formal education is restricted to the 8th grade (with a controlled tendency toward correspondence high school courses, technical training, and even some teachers' college attendance). To substitute for more advanced education, the colony institutes intensive informal training for the young men who will later become managers and executives--"learning by doing." A Hutterian boy becomes a working adult at 15 years, after his 8th grade graduation from the public school situated on the colony premises and staffed by a teacher sent out by the local school board. At that time, the 15-year-old becomes apprenticed to one of the production managers. In the majority of cases, he must choose a field different from that of his father. While Hutterites do have problems resulting from the lack of formal technical training, they manage to overcome the handicap to a surprising extent by these and other methods.

Some figures on the finances of Hutterian agriculture will be useful. The following are condensed and rounded balance sheets for a typical colony in the Alberta-Saskatchewan-Montana region for two years about a decade apart:

TABLE 1. INCOME AND EXPENDITURES FOR ONE COLONY, 1965-73

	1965	1973
Colony gross income	$250,000	$500,000
Operating expenses	200,000	350,000
Net income	50,000	150,000
Debt retirement	15,000	--
Domestic expenditures	10,000	100,000
Cash savings	5,000	10,000

These figures show a doubling of the gross income received from agricultural production, but also a steep increase in the costs of operation. Steep increases also occurred in domestic consumption, but the large figure of $100,000 for 1973 reflects the construction of new buildings and installation of central heating and other conveniences, long delayed in the case of this colony. In the mid-1970s, colonies all over Western Canada and the United States were making large expenditures on such improvements, showing that consumption austerity as practiced by Hutterites does not mean that the Brethren do not appreciate and aspire to comfort and convenience on a collective basis.

The colony selected here was also only a decade old in 1965, and it was still in the process of paying off the large debts acquired at the time of establishment. By 1973 these debts had been amortized, and the colony was in a position to purchase what it needed on a cash or trade-in basis. Or, the colony had reached what the writer called in his book on Hutterian economy [10] the "plateau of affluence," and was in a position both to afford many improvements as well as to make large contributions to its cash reserve, saved for financing in part the new colony at fission, only three or four years away.

The Hutterian economic system possesses a number of advantages over individual farming traceable directly to the collective organization and the rules which govern consumption. Hutterites pool their income; the necessities of life are furnished to the families on a strict collective basis, budgeted as a lump sum for the whole commune. Decisions as to improvements, e.g., indoor toilets, are made in the assembly (as a rule, any expenditure for any purpose over $100 must be voted upon). Each family also receives a small *wergeld* or monthly cash stipend of from $2 to $20, depending on the general financial picture, which it can use to buy little treats and other things, although even these items are taken from an approved list. No family can buy its own transportation vehicle: each colony has one or at most two passenger-type station wagons which must serve all needs for movement, such as shopping or visits to the doctor. Recreation is collective, and confined largely to Christian holidays, weddings, and similar group occasions.

Table 2 illustrates the per capita expenditures for selected items for a colony and a relatively prosperous individual grain-livestock farm about 1964. At the time of the study, the colony had 74 persons; the farm, 4. By 1972, the per capita expenditures for all domestic consumption items in the colony had risen slightly, reflecting the general rise in all western colonies in the affluent late 1960s and early 1970s.

Other gains in the colony system are evident with respect to labor costs. While the large scale of colony activities usually means that a colony experiences a labor shortage, in fact there are much greater opportunities for covering all labor needs than in the case of individual enterprises. Since Hutterites pay no wages, and labor is considered a person's due contribution to the Christian Brotherhood, it is not easy to calculate the actual savings, but there is no doubt that the colonies can respond to emergencies, and bring masses of labor to bear on any task at a moment's notice--something the individual farm is often unable to do, or do only at

TABLE 2. PER CAPITA EXPENDITURES FOR SELECTED ITEMS: HUTTERIAN COLONY AND INDIVIDUAL FARM

	Colony	Farm
Automotive fuel for trucks	$77.02	$703.75
Automotive fuel for passenger vehicles	6.76	250.00
Heating costs	24.79	150.00
Clothing	47.30	175.00
Household maintenance	25.94	62.50
Recreation, entertainment	33.78	375.00
Municipal taxes	75.80	186.00

great cost. Moreover, the colony can depend on its neighboring colonies to furnish labor in situations of real emergency, while the farm must make do with very limited services from its individual neighbors. In order to allay resentment stemming from this superior position, colonies often donate labor to local farms in critical instances, like branding, harvesting before weather changes, and the like.

In a real sense, Hutterian agriculture is *both* capital-intensive and labor-intensive. That is, it represents a system with the full advantages of mechanization and other expensive technologization, but also with the opportunities conferred by the existence of a large, loyal, hard-working labor force. Hutterites state that they do not buy machinery in order to create more leisure time (for that matter, who really does, in modern North American agriculture?), but simply to maintain their rate of production in order to meet the rising costs of operations. Moreover, the per capita acreage or other population-resource ratios of the typical colony are less favorable than those of the individual farmer. While the colony described above had about 15,000 acres, it also had a population of 74; the farm had 3,000 acres for 4 persons (about 202 acres p.c. against 750). Similar ratios are present for most other parameters. Most colonies do have to produce at a high rate in order to sustain the relatively austere, but comfortable consumption level, and accumulate the cash needed for branch colonies. This is not generally understood by local people in the Great Plains, who simply observe the very large acreages, large numbers of machines, gardens, and so on.

Taxation has been a source of continual negotiation between colonies and government in both Canada and the United States, but in most places the colonies pay taxes as corporations or cooperatives. That is, their collective identity makes it possible to acquire the tax benefits associated with large-scale and corporate structure. Attempts have been made to tax colonies as a series of individual families, dividing up the income among these for the purpose of writing returns, but the system has been effectively challenged in the courts. Recently a colony in Alberta successfully challenged federal taxation and secured a court judgment requiring the Canadian government to return a total of $1.4 million in taxes paid between 1961 and 1966. This litigation was introduced by the Darius Leut, who claimed they did not sign an agreement between the federal government and the other two Leute (Lehrer and Schmiede). Members of the two Leute have criticized the Darius action, stating that it can only generate resentment against Hutterites.

Actually, Hutterites are engaged in continuous legal action on some issue or other with federal and state-provincial government in both Canada and the United States. At several key points: taxation, landownership and title, eligibility for various economic and welfare benefits, and schooling, the Hutterian communal system is at odds with the legal arrangements of an individualistic culture, and there is probably no final solution to most of these issues. Enterprising states attorneys, or combinations of resentful local people can always find it possible to attack the colonies on some point or other. However, the Hutterites have usually come off well in these struggles. Attempts to force them to limit their land purchases, an endemic challenge, have usually failed due to their unconstitutionality in both Canada and the United States. Hutterites have a rule against direct political participation, due in part to these perennial attacks and legal actions, but to an increasing extent in recent years Hutterites have exercised their right to vote and influence local politics, especially when some major issue affecting them is at stake.

THE INSTRUMENTAL STRUCTURE

This is diagrammed in Figure 1 (reproduced from the writer's original version in his book on the Hutterian socio-economic system). The diagram is subject to some modification from colony to colony, and there are slight differences in the pattern among the three Leute, but all these differences are minor, and the chart can serve as a suitable introduction to the general structure.

Considering the colony as a commune, or agricultural collective, it is, first, a hierarchal system. That is, the allocation of authority for the male members forms three distinct levels: the executives; the managers; and the laborers. Persons in these echelons must take direction from those in higher positions; the direction is clearly formulated and not vague: a manager manages; an executive or "boss," bosses. The Hutterites make no important compromises with the need for formal authority and direction in their instrumental system; they know that you cannot run a large-scale enterprise, or an efficient enterprise of any size, without clearly demarcated patterns of authority and responsibility. This is in contrast to the many small agrarian communal experiments in many countries where a strong belief in equality has imposed severe lim-

FIGURE 1

INSTRUMENTAL STRUCTURE OF THE HUTTERIAN COLONY

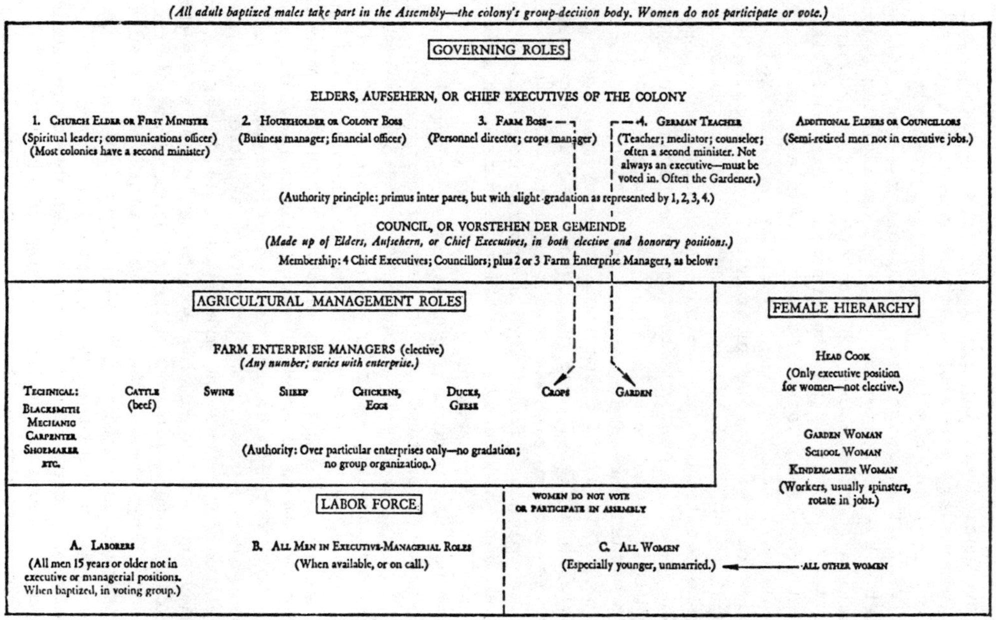

Reprinted with permission from *Hutterian Brethren: The Agricultural Economy and Social Organization of a Communal People*, by John W. Bennett (Stanford: Stanford University Press, 1967), Diagram 3, p. 144.

itations on life-sustaining and economic activities insofar as authority was vague or confused.

On the other hand, this system is not a tyranny. In the first place, Hutterites believe that all male members are equal in the face of God: they are brethren; all baptized; all with the same formal opportunities to assume responsibility and authority if they display the talent. The hierarchal system is, then, also a system of upward movement: every laborer in the colony knows that if he works and studies hard, he will eventually assume a position in the managerial echelon; and most of them can expect ultimately to become executives. The expectation of upward movement for nearly all men is assisted by the Hutterian practice of dividing colonies when their population reaches about 120 persons. At this point, land is purchased and a "new farm," as the Brethren often call it, is established. A year or so later lots are drawn and half of the colony population moves to the site. This has meant that the opportunities for upward mobility are substantial. In fact, a Hutterite *must* take such responsibility; he cannot avoid it, since he is elected to the position by the colony assembly. If colony division and expansion should cease, due to land shortages, other means will have to be found to ensure this mobility, since it does create a system of "rewards" for hard work and dedication. Hutterites know it is a technique to avoid dissension and envy, both of which are sins in Hutterian beliefs. One device discussed in colonies at the present time is earlier retirement for men, to make room for upward movement; some colonies, faced with a growing population and lack of room for expansion, are experimenting with regular rotational systems of executive and managerial assignment.

Considering this structure purely as a system of management for a large-scale enterprise, it contains few surprises. That is, there is nothing particularly "communal" or "cooperative" about it: it could describe the structure of General Motors, with the executive echelon corresponding to the Board of Directors and the Presidential group, and the management sector corresponding to the heads of the various automobile brand divisions; and the labor force simply to the labor force (not, however, unionized among Hutterites!). However, there is one difference between the Hutterian system and a modern industry: all Hutterians are on call as members of the labor force, if they are not performing some other role at the moment, or are not too old to function. In this flexible element, the Hutterian instrumental system is standard agrarian: all agricultural systems maintain this fluidity of roles since the tasks are extensive and the labor needs often unpredictable.

It is also apparent from the chart that Hutterian women constitute a special segment of the instrumental system. They are "below" the men in the sense that they do not vote and have no formal roles in agricultural management. However, they do have something of a hierarchy of managerial roles, as indicated by such terms as "Head Cook," "Garden Woman," and so on. These positions are subject to elective choice by the men, but only after careful consultation with the women. In general, women have considerable informal power in the colony system, and men almost universally say they make no important decisions about colony welfare or change without consulting the ladies first. The most significant female economic roles are of course the care of the garden and all food preparation and cooking tasks; in this sphere women can influence capital allocations and expenditures. For example, they may make it clear that the construction of a garden irrigation system must take precedence over a new hog barn. Since Hutterites seek almost complete self-sufficiency in vegetables, meat, dairy, and baked goods, gardening and other food production activities is a substantial component of the overall economy. In a typical colony in the 1960s, the value of home-produced food on the basis of wholesale prices equaled about 10 percent or more of the colony income; this figure has proportionately fallen in the 1970s due to increased values for commercial crops, but it still remains a significant fraction of the economic picture.

Now let us view this instrumental structure from the standpoint of the social system of the colony. In this context, we find that the roles and echelons are populated by kinship relatives. Each colony is a normal social community composed of nuclear families related by blood and marriage (although most colonies seek most of their mates from other colonies). The typical colony will have two or more kin groups represented, but in almost all cases a majority of members will belong to just one of these groups. That is, the typical colony consists of one large extended family with members of two or more extended families as a plurality. This means that the majority of executives, managers, and workers will be related by blood or marriage: the diagram of instrumental roles forms a simplified version of the colony genealogy. Moreover, the echelons have a rough correspondence to age levels: the executives are all older men, the "grandfathers," so to speak; the managers, the young and young-middle-aged, or the "fathers"; and the workers, some "fathers" but mainly the "sons" or "grandsons." The age differentials are similar for the female roles. In other words, the system of role upward mobility is also a progression in seniority.

This partial fusion of kinship and instrumental roles has both advantages and limitations. It makes management more efficient insofar as communication between relatives can often be empathic and considerate; it makes for difficulties since relatives--among Hutterites as among everyone else--often fight over issues extraneous to economic or instrumental considerations. To guard against such difficulties, Hutterites insist that in the context of instrumental role-playing, that is, management, the members must consider themselves as Brethren, not relatives. They must suppress their personal or family feuds for the sake of the collectivity. A number of formalized procedures are available to enforce this: if two agricultural managers get into a fight over family rivalries, accentuated, perhaps, by competition for resources for their respective enterprises (e.g., the cattle boss may want a new feed lot, but the chicken boss a new automatic feeding system, and only one of these can be afforded), they can be required to apologize publicly and enter into an amicable settlement. The apology is ritualized and carried out in a regular church meeting and assembly.

Much more could be said about this, but the point is clear: the commune, the communal principle, emerges precisely at the point where it is needed: *as a set of sanctions to enforce harmony and sensible settlement of the inevitable disputes that emerge in a complex enterprise staffed largely by relatives*. Such disputes are not only disruptive, but are also sinful for Hutterites. I have no doubt that the evident success and efficiency of the Hutterian agrarian commune is due to the use of communal brotherhood beliefs and practices to modify or eliminate dissension. Again, such dissension has been a marked feature of many experimental agrarian communes which have been powerless to avoid it. The Hutterites seem, on the whole, to have solved the problem by using their beliefs in a highly practical manner. Hutterites are quite articulate about this: they say that communalism is not just an ideal, but an intensely human system of survival--and one that requires all the ingenuity that humans can devise to keep it on an even keel. The mere *belief* in communal principles will not suffice.

The system has many other features of note. The chief executive role in the colony is the First Minister, who is the spiritual shepherd and counsellor of the community. That is, major authority of a broadly charismatic nature, is lodged not in the general manager or "householder," or the "farm boss," but in the pastor. This calls attention to the fact that the colony is not merely a big farm or economic enterprise--however important this may be. It is, first and foremost in Hutterian doctrine, a religious community, a band of brothers in Christ. Hutterites do not "give" on this; they know that the colony in trouble is the colony without an effective pastor; it is this element that tops out the system of hierarchal authority and social mobility--that unique combination of patriarchism and equality which the Hutterites specialize in. This institution of the spiritual leader is not necessarily a communal feature: it is a facet of late medieval community social organization, analogous to the head rabbi, perhaps, in a *shtetl*, or even to the abbot of a monastery. But it signifies that Hutterites have grasped another important principle of communal effectiveness: the *need for someone with the authority to establish the meaning and authenticity of the important symbols of social existence*. Hutterian charisma, however, is highly regularized, and is strictly elective--it is not a matter of self-appointment or revivalism. It is, like their economy, a managed, rational system.

COMMUNALISM AND COOPERATION

Although the Hutterian colony is classified as a commune, because of the combination of collectivization of property and the raising of children by the community, the conduct of daily affairs also has cooperative features. Certain spheres of Hutterian life present us with an opportunity to discuss the interplay of communalism and cooperation in the same social framework: in fact, communalism can be seen to have both facilitating and inhibiting effects on cooperation.

With respect to the facilitation, we have already described the cooperative networks between colonies which function to exchange goods, information, and personnel. These networks come into being as a result of the Hutterian decision to remain apart from the outside world because of the communal frame of life, and they are reinforced by the large economic scale of the colonies, which are required to exchange with each other because of the lack of any comparable productive units in their districts. In addition, there exists inside the colony a series of informal exchange dyads and networks which develop out of the need for personal resources and to get tasks done efficiently. Central communal direction goes a long way, but it does not encompass all problems and decisions, and the Brethren develop their own exchange relations of a cooperative nature. This is seen particularly between siblings: brothers and their families.

However, in the background there are inhibiting forces at work. Communalism does require centralized direction, and an argument can be made for the fact that there is less free cooperation between individuals and families in a colony than one might anticipate given the possibilities and needs--or in comparison with families in a cooperative farming community like an Israeli *moshav*. The curtailment of personal property, the non-payment of wages, and the miniscule monthly cash allotments tend to generate a feeling of closed possessiveness and jealous autonomy which tends to drive families (and individuals, although the family cleavage pattern is the more common) apart, into little "private" worlds--in the midst of communal association. Hence there is little free exchange of property, and probably less exchange of services and favors than one might expect given the value emphases on sharing. The Hutterian colony is a kind of cooperative, particularly at inter-colony and production levels; but it is perhaps less than a cooperative at informal social exchange levels.

DECISIONS, INCENTIVES, COMPETITION, AND ACHIEVEMENT

These matters are important because the colony, like any commune, seemingly lacks the conditions believed to be necessary for adequate performance in Euro-American instrumental contexts: personal ambition and open competition. The problem presented by competition and individual striving in a commune is of course a classic one; it is not only a problem for Hutterites, with their explicit ideology of harmony and brotherhood and suppression of acquisitive values, but a problem for all communal groups, since the individual is expected to suppress his desires and needs for the good of the collectivity. At the same time, the individual is expected to perform at a consistent level of productivity and efficiency in order to permit the group to survive. The ideological issue is clear enough: the idea that the individual must be paid or in some way rewarded for this consistent performance presents a challenge to the core beliefs of communalism.

That it is a challenge, and also a breaking point for many communes is equally clear. Only a handful have survived it since the sixteenth century, and many of those who have managed to survive for a few decades, or even a century, have done so at the expense of viability or peace of mind, or more commonly, at the expense of the ideology.

Some have relapsed into little tyrannies led by a charismatic figure; others have survived by handouts; many have simply died slowly, inch by inch, abandoning their self-sufficiency and their ideals in the process, failing to achieve the dedication and industriousness that collective life requires. The Hutterites, and the kibbutzim, have undoubtedly survived, and even flourished. Therefore they are important; they can tell us how the agrarian commune can survive in an individualistic civilization and how it can do so without major disruption or corruption of its major beliefs.

The crux of the issue is simply that *communal instrumentalism requires the consultation of a very large number of individuals in the decision-making process*. In a typical family-farm enterprise, a maximum of perhaps two persons must be consulted in the making of a particular resource or managerial decision. The figure is larger, of course, in corporate-commercial agriculture, but even in these cases a single manager may be made responsible for all decisions--although he also may be held responsible before a board of directors if the decision is wrong. In such cases, which are typical, the social field of the decision process is even narrower than it is in the family enterprise. In the commune--that is, the commune which functions as a commune--the social field is always much larger. The entire group is involved, in one way or another, in the decision. In the Hutterian colony, any expenditure over $100 (excepting a few emergencies) must be discussed and voted upon; no production manager can assume that he can build his enterprise without prolonged consultation with the entire group; or without their consent for any improvement or capital investment.

That this system functions well is testimony to the solidarity of the commune, the efficient methods of child socialization, and the existence of a basic consensus in the group that this is a legitimate and valid procedure. But it is far more than this: it also means that human tendencies in other directions are being effectively suppressed or diverted. For Hutterites are humans, not paragons; in fact, the central theme of Hutterian doctrine is that humans are difficult, "failed" creatures; they cannot expect Christian perfection; they can only strive to approximate it. This is the central meaning of Christ's *persona*: man cannot be Christ, he can only hope to come close. Hutterian belief is less optimistic than most standard Protestant versions, and certainly less egocentric.

The dilemma of the commune (and of social life generally), and the one which Hutterites have probably come closer to solving than anyone (in or out of communes), is that in order to meet the demands of life-sustaining activities, there has to be differentiation of function among humans. And since humans are not identical copies, but to a greater or lesser degree individuals with differing capacities and interests, *differentiation of function also means differentiation of status and prestige*. And this leads directly to envy and competition.

The agrarian structure of the colony produces differentiation in two senses: first, the technical and economic difference between the various production sectors, and the varying tasks and responsibilities that these activities require. Such differences in responsibility inevitably lead to differential importance or status on the part of the individual persons involved in them. Second, the system creates differentiation insofar as the activities inevitably begin to compete with one another with respect to resources. Their managers, and others involved in these activities, thus are placed in a de facto position of personal competition. In both sectors of differentiation, there are severe conflicts with Hutterian beliefs in brotherhood, equality before God, suppression of ambition and "idolatry," as Hutterites call unduly strong attachment to material objects or personal activities and goals. In Hutterian doctrine, this is, in fact, the fate of Man: a constant overcoming of temptation.

Among the many practical consequences of this situation is the question of recruitment for positions in the colony, and preparation by individuals for such positions, assuming they have an interest in acquiring them. In the first place, the individual must not, by Hutterian rules, openly display his interests; that is, ambition. Yet he must prepare himself, since these tasks and responsibilities are complex. If he is intelligent and ambitious, he must train himself in literacy and mathematics, skills needed for the managerial and executive positions, and he must also gain practical experience in a variety of tasks and enterprises. Since he cannot do all of these openly, he must do them quietly, often on his own time, so to speak; and he must, above all, do them in the interests of the collectivity, for the good of the group. He cannot announce that he is preparing for an upward track from field hand, to manager, to Farm Boss or Minister; he can only say that he is preparing himself to serve the colony. Nor can he seek votes, when it comes time for an election to office. He must trust in his preparation, and in the honesty with which his preparatory activities have been carried out.

The colony system is built meticulously to permit this type of personal training and movement. In the first place, in the developed colony a man officially works just 8 hours a day; this gives him time for study and practice. Church services are mandatory in attendance on Sundays, but on other days it is possible to skip them, if it is not done too often. The nuclear families have private apartments; they live at least as individual an existence as the kibbutz family, and see their children more often than do parents in the left-kibbutzim, since Hutterian children always live and sleep in the family apartment even though they spend the day in the childrens' collectivity.

When a man is elected, say, to a managerial position, he is given considerable autonomy and privacy. No one is supposed to interfere with his decisions or impose advice; if there is something wrong with his management it will generally be explained to him privately, by the executives, and if he simply cannot handle it, another job will be found for him and another colony member will be elected to the post. Hutterites covertly recognize that a man derives personal satisfaction from doing a job well and making his own decisions--commune or no commune--whether it is the operation of a tractor by a field hand, or the keeping of the income tax records by the Householder. As a matter of fact, Hutterites have told the writer in so many words that human beings require rewards --a remark seemingly in direct contradiction to the explicit dogma, but not really so, since the recognition of these human needs for satisfaction, recognition, and so on are covert, played down. And by keeping them covert, the Hutterites manage to balance the two sides of the equation: the ideals of brotherhood and selflessness on the one hand, and the human needs for recognition and aspiration, on the other. In other words Hutterites, more than anyone else, show us that the two key components of the Protestant ethic--achievement motivation and individualism--are in fact two different things. Achievement, if it is directed toward the good of the group, from early childhood and then repeatedly through life in the context of institutions, can be socialized. That is, it is actually possible to induce these individualistic humans to achieve largely not merely for the self, but for the self and others, or perhaps, for the self *as* others.

THE COLONY AS A MODEL FOR GROUP FARMING

In this final section I wish to comment on the suitability of the colony as a model for agrarian development.

At first sight the question seems meaningless since it is difficult to visualize transfer of institutions from a traditional religious sect to any other society: the motivations, the beliefs, and above all, the painstaking socialization of children and adults are impossible to replicate. However, it is possible to view the Hutterian colony as a collective agrarian institution which has succeeded in working out adaptive responses to a number of problems which most group farming experiments experience. Regardless of the colony's unique history and culture, the solutions have a certain commonality which is worth careful study.

Some of these have already been discussed: the problem of meaning and motivation in a collective where considerable selflessness and acknowledgement of the role of other people is required; the careful balance between the autonomy of management in the separate enterprises, and the centralized decision-making; the use of extensive cooperative networks between colonies as a means of focusing capital allocation and distributing information.

Some other things also deserve comment. The first of these is the control over consumption. This has its implications in an agrarian world of increasing constraint and rising costs. Clearly material abundance is a will o' the wisp which cannot be caught by the majority of the world's population; they will have to do with much less, despite the "revolution of rising expectations." Controlled consumption or planned austerity is something we shall all be required to cope with in the future; the Hutterites represent the world's most successful effort. Parts of their system cannot be duplicated, but it contains devices which may be transferrable: for example, the substitution of a relatively comfortable, but low-cost *collective* consumption, for individual property accumulation. That is, one need not emphasize the communal property feature quite as rigorously as do the Hutterites, but one can copy their emphasis on collective abundance and personal austerity.

A second feature of note is the conservationism inherent in communal management of physical resources, especially in a marginal environment with friable and easily abused resources. Hutterites are the most capable adaptors to the semi-arid northern Great Plains: their economy of scale, and the impressive control over consumption, prevents them from pushing against their resources. They are able to take the time required to develop poor and light soil, whereas the smaller farmer must abuse it in order to make a living. They can supply their own soil-building materials, like manure, from their livestock operations, which also means they can be excellent recyclers. These and many other conservationist measures are made possible also by the social *continuity* of the colony: it is a permanent institution which has no problems of succession. Its development is steady and consistent, and is not disturbed by land sales and the alternation of operators with different, and often destructive, resource practices.

The third and final topic concerns the problem of change and adaptability.[11] The Hutterian collective system simultaneously changes and remains the same. This is accomplished by defining any necessary change in procedure required by adaptation to the world as legitimate, so long as it does not seriously alter the traditional fabric. Since the Hutterites accept a full range of modern technology as a means, not as an end in itself, they find it possible to maintain a prosperous developmental posture without serious internal injury. This is so because Hutterites do not distinguish between *society* and religion, or *beliefs*. Society, the way one lives and exists, is for Hutterites equivalent to the belief system: you manifest your faith by doing what is appropriate and necessary in the Hutterian agrarian commune. Again, one must admit that this is a neat trick if one can do it, and not all of us can, but certainly the Hutterites provide a standard. Culture--ways of life and belief systems--are not easy to invent or to change, but they are not completely resistive, and some changes in local systems along Hutterian lines, in order to foster a group farming system, are certainly worth trying. The writer recommends that the attempt be made, or rather, that agrarian planners make a serious and careful study of the Hutterian world in a search for technique.

NOTES

1. David Barkin and John W. Bennett, "Kibbutz and Colony: Collective Economies and the Outside World," *Comparative Studies in Society and History* 14 (1972): 456-83.

2. For sociological discussions of historical communes, see my "Cultural Integrity and Personal Identity: The Communitarian Response," in W. Dillon, ed., *The Cultural Drama* (Washington, D.C.: The Smithsonian Institution, 1974) (revised version published in *Theory and Society* 2 [1975]: 63-94); and Rosabeth Kanter, *Commitment and Community: Communes and Utopias in Sociological Perspective* (Cambridge, Mass.: Harvard University Press, 1972).

3. Robert Friedmann, *Hutterite Studies* (Goshen, Ind.: Mennonite Historical Society, 1961); John Horsch, *The Hutterian Brethren: 1528-1931* (Goshen, Ind.: Mennonite Historical Society, 1931); Victor J. Peters, *All Things Common: The Hutterites in Manitoba* (Minneapolis: University of Minnesota Press, 1965); John A. Hostetler, *Hutterite Society* (Baltimore: Johns Hopkins University Press, 1974); Paul S. Gross, *The Hutterite Way* (Saskatoon: Freeman Publishing Co., 1965).

4. George H. Williams, *The Radical Reformation* (Philadelphia: Westminster Press, 1962). The basic Hutterian religious testament, defining the communal rationale, is Peter Rideman, *Account of Our Religion, Doctrine, and Faith*, trans. K. Hasenberg (London: Plough Publishing House, 1950).

5. Georg J. Eisenach, *Das Religiose Leben unter den russland Deutschen: Im Russland und Amerika* (Marburg, Germany: Hermann Rathmann Buchdruckerei, 1950).

6. John A. Hostetler, *Amish Society* (Baltimore: Johns Hopkins University Press, 1963).

7. Victor Stoltzfus, "Amish Agriculture: Adaptive Strategies for Economic Survival of Community Life," *Rural Sociology* 38 (1973): 196-206.

8. For some economic data, see my *Hutterian Brethren: The Agricultural Economy and Social Organization of a Communal People* (Stanford: Stanford University Press, 1967); Hans D. Radke, *The Hutterites in Montana: An Economic Description*, Bulletin 641, Montana Agricultural Experiment Station, Montana State University (1971); Marvin P. Riley, et al., *South Dakota's Hutterite Colonies: 1964-1969*, Agricultural Experiment Station, South Dakota State University, Brookings, South Dakota (1969). For historical notes on Anabaptist economic attitudes, see Donald Sommer, "Peter Rideman and Menno Simons on Economics," *Mennonite Quarterly Review* 28 (1954): 205-233.

9. For details, see Barkin and Bennett, "Kibbutz and Colony," and Bennett, *Hutterian Brethren*.

10. Ibid., Chapter 7.

11. For some recent changes in Hutterian life, see my "Change and Transition in Hutterian Society," in A. W. Rasporich, ed., *Western Canada: Past and Present* (Calgary: McClelland and Stewart, 1975).

Part III: National Collective Systems: The Soviet Union and the People's Republic of China

Editor's Introduction

The Soviet system of collectives, with emphasis on the *kolkhoz*, and the Chinese system of communes are the two major national collective farming systems discussed in this volume. Given the importance of and interest in the Chinese experience, the present section contains two papers on the People's Republic of China. Professor Etienne's paper includes a commentary on some key points in Dr. Wong's analysis, but the major part of his paper is focused on agricultural practice and technique and the state of agricultural research in China. Wong's major emphasis, by contrast, is on China's organizational transformation and agricultural strategy.

National collective farming systems have been established in most communist countries and are of Galeski's Type Three--collective farms organized by governments in order to achieve nationwide economic and social goals. However, there are many operational and organizational variations among (and even within) these systems. This is especially pronounced in Eastern Europe, and of course Yugoslavia and Poland retain a system where most of the farm land continues to be farmed by small, independent farmers but with complex cooperative service structures.

Placing the Soviet and Chinese cases in the same section of the present volume should not lead to the conclusion that their systems are basically the same. There are wide differences both in the nature of the collective processes and in their present forms. Nor should it be concluded that the Chinese commune is currently a communal organization with characteristics shown for the kibbutz and the Hutterite colony. The early commune of the late 1950s did indeed have many of these characteristics. But the experimentation and evolution since that time has resulted in a unique structure with different levels of planning and authority.

A three-tier organizational structure has evolved in the Chinese communes--the team, the brigade, and the commune. Direct control over day-to-day farming operations (including accounting and distribution) rests with the team and to a lesser extent with the brigade, while at the commune level basic local governmental functions are carried out, including administrative responsibilities for planning and coordinating agricultural production. An effective distribution mechanism has evolved and is reinforced by moral incentives. This provides a balance between ideological requirements for greater egalitarianism and the economic requirement for work incentives. As commune life becomes more complex, further decentralization may be necessary to maintain the viability of this structure.

In both systems the individual family farming plot remains important. These had been eliminated in China when the communes were first established. Economic returns from the individual plots may represent 20-35 percent of the family's income. As Professor Etienne points out, individual interests are still so strong that the brigade is too large to be taken as the basic unit of production and accounting since it would lead to an egalitarianism which the richer production teams are not prepared to accept. Thus in both countries inequalities continue to exist; the teams (in the case of China) and the kolkhoz (in the case of the Soviet Union) with the most favored resource situation have higher incomes than those less favorably located.

Since the Chinese and the Soviet cases are the major national experiences with a history of 25-50 years, the authors were asked to give some thought to the lessons to be learned and possible transferability of certain aspects of these experiences to the less developed countries. In the Soviet case, one of the points stressed by Professor Wädekin, as well as by Professor Stuart[1] who discussed his paper, is the change that has occurred in the kolkhoz over the past 25-30 years. The average size increased from 500 hectares in 1940 to 3,000 hectares in 1970; quality of managerial and technical personnel has been improved; the introduction of better cost-accounting methods and substantially increased money incentives have helped to generate increases in labor productivity; there has been a major injection of state aid.

With respect to China both Professors Wong and Etienne agree that it would be most difficult for a developing country of today to emulate the Chinese organizational strategy without having first achieved a thorough social and political revolution. Furthermore, one cannot ignore the tremendous influence of China's long history and civilization on current developments: traditional intensive agricultural techniques, ability to mobilize the masses for rural works, sophisticated bureaucracies, etc. But the Chinese experience can serve to highlight for many Third World countries the problematic areas in their own agricultural development policies. The main thrust of the bold commune experiment in China has demonstrated that group-organized farming, even in the context of a peasant agriculture, can be a viable alternative to the time-honored system of individualistic, small peasant farming provided that group activities are properly organized with sufficient flexibility and pragmatism. The Chinese model, however, has also served to demonstrate that organizational strategy alone is not sufficient to achieve development. It must be supported by appropriate non-urban-biased government policies and by a leadership not alienated from the peasantry. Organizational strategy has to be sustained by real progress in technical transformation. The communes can furnish the structural framework for the mobilization of human and material resources, but they are not intended to be substitutes for modern agricultural production techniques.

Perhaps the idea of transferability is best summarized by Dr. Elliott R. Morss.[2] He relates the following story:

> An Asian educated in the West recently had dinner with Mao Tse-tung after having spent a week studying the agricultural system of China. Mao asked the visitor what was the primary lesson he had drawn from his examination. The visitor said he had been tremendously impressed by all aspects of Chinese farming and urged Mao to make some of his farm experts available to advise other countries that were not doing as well. At this point Mao looked very pained and said he was very disturbed if this was the major impression the guest had retained. Needless to say, this troubled the guest, and he asked Mao what his primary impressions should have been. Mao replied that there were no easy solutions, that the complex system the Chinese use had evolved over many years and had entailed many mistakes. Mao said it would be wrong to attempt to apply the Chinese approach elsewhere, and that the only thing that was replicable from the Chinese experience was the slow, painful, dialectical process of experimentation that the Chinese had pursued over a number of years.

NOTES

1. Professor Robert C. Stuart, Department of Economics, Douglass College, Rutgers University, New Brunswick, New Jersey.
2. Director of Research, Development Alternatives, Inc., Washington, D.C.

4. The Soviet Kolkhoz: Vehicle of Cooperative Farming or of Control and Transfer of Resources?

Karl-Eugen Wädekin

If cooperative farming of the Soviet type is to be viewed as a possible model or partial model for reorganizing agriculture in less developed countries, one must have a clear understanding of the purposes it was to serve in the country of origin within both agriculture and the economy and society at large. The history of Soviet collectivization may be presumed to be known in general terms,[1] but its essential traits relevant to present-day economic development strategies are rarely dealt with.[2] For a better understanding of the subject, the historical development of Soviet collective farming has to be placed within this special perspective.

THE HISTORICAL SETTING

At the time of mass collectivization of agriculture, the Soviet Union was in many respects what today would be considered a developing or less developed country. But it differed from most such countries because of its immense size. Vast regions within the Soviet Union varied widely in terms of natural conditions (climate), in ethnic and cultural background, as well as in the stage of economic development. It had some urbanized and industrialized areas and it was rich in mineral resources which, even before the October Revolution, had made it one of the world's largest producers of iron and some other raw materials (though not on a per capita basis). Nevertheless, four-fifths of its population was dependent on agriculture, whereas the industrial working class was numerically important only in the few industrialized areas.

Another special feature was that the country was run by the Communist (Bolshevik) party, which held direct political power in the narrow sense of that word. Additionally, through its members it indirectly controlled almost every important function in the economic, social, and cultural life of the country. A social basis of its own, a consistent and elaborate ideology, strict discipline, and a record of hard and victorious fighting enabled this party not only to exert such power over public life, but at the same time to act as an organization separate from the state--less involved in day-to-day affairs, collecting information from its rank and file, and transmitting the leadership's will to all levels of administration and society.

Marx and Engels had put forward some general ideas on the desirability, even inevitability, of cooperative and large-scale agriculture in place of small-peasant farming after the victory of the proletariat. But they gave no blueprint for the organization of socialist farming, nor for the process of change toward this new agrarian structure. Lenin, during the years immediately following the October Revolution, seemed to be convinced that the general idea of socialist (cooperative or collective) farming would materialize on the peasants' own volition, once the powers of the old regime were destroyed and benevolent help--in the form of model charters, among other things--was supplied by the party and the state. This expectation proved to be wrong. Under the impact of severe economic setbacks, the will of the overwhelming majority of Russian peasants to develop private and individual farming was respected. This formed part of the New Economic Policy from 1921 onwards. In his last major article, "On Cooperation" (January 1923), Lenin developed the idea of collectivization without compulsion through an increasingly comprehensive system of peasants' cooperatives of all kinds, which would eventually embrace all peasants and their marketing, processing, buying, and--finally--also production activities. For this process of change to materialize he was prepared to wait for "a whole historical epoch," albeit a rather short one, which might last only one or two decades if circumstances were most propitious.

History shortened even this mini-epoch to five years following the publication of his article. Lenin's successor made socialization of agriculture a forced and extremely rapid process, which put an almost unbearable burden on the peasants, in most cases infringed upon the principle of voluntariness, and violated Soviet law. How and why did this come about?

Socialization clearly was and is the ultimate goal for agriculture according to Marxist-Leninist ideology. But it is equally clear that the way it was achieved in the Soviet Union and the form of cooperative farming that emerged--the Soviet *kolkhoz*--were not predetermined by ideology, nor was the timing. The intensive economic and political debates on this subject among the Soviet leadership of the mid-1920s showed that even for Bolshevik Marxists various courses toward this goal were conceivable. What actually resulted after bitter fighting over this and the broader issues of economic and revolutionary policy was a collectivization drive which was more closely related to the objectives of rapid industrialization, maintenance of Bolshevik power, and control over Russia's peasant masses than to ideology.

At the time Russia was essentially an agrarian state. Its peasant population lived in tightly knit and functional village communities. These had (in Russia proper) a tradition of communal landownership, although juridically this had largely been abolished under the czars. It is difficult to assess--historians are not unanimous on this point--the extent to which communal landownership was still alive. In any case, the family had been and was the production and working unit, communal work being the rule in many villages only for haying and wood cutting. The family itself was in an advanced state of change from the extended to the nuclear family. As no private estates were left after the revolution, and other large-scale farms of the time (state or early collectives) owned less than 5 percent of the total area, Russia had almost an exclusively peasant agriculture. There were some 26 million peasant families, of which some 4 million were said to be landless. Holdings averaged 15-20 acres each. Most of them were on poor podzolic soil under a harsh climate. After 1917, all land was nationalized, but the peasant had a permanent use right, which in practice was equivalent to ownership, except that the land could not be sold or mortgaged.

Although living very much in a traditional way, the Russian (and other Slavic) peasants had proved to be open to innovation, if this was economically feasible and promised advantages to the individual or the family.[3] Nor was the peasantry wholly illiterate. By 1917, more than two-fifths of peasant children of 8-11 years of age were in elementary schools, and more than one-fifth of peasants of all ages in European Russia were literate. In the Asian parts of the country this proportion was lower.[4] Although the majority did not know about or could not afford modern production techniques, there was a minority who on their own had started on the road to modernization. Only the poorest and most backward still tilled their land with the wooden *sokha* instead of a plough. Significantly, peasant demand for machinery and fertilizer surpassed the combined supply from Soviet industry and imports. On the whole, agricultural productivity still was below pre-1914 levels, and far below what was attainable in a peasant economy of the time. Vast production potentials through a more intensive use of land, including fertilizer application, remained untapped.

At the same time, this peasant population, after the huge losses suffered in World War I and the Civil War, was again growing rapidly. Much of the increasing food production was thus consumed by the growing population; furthermore, the man-land ratio deteriorated. Industry had scarcely recovered from the damages of the war and the revolution; structural unemployment and underemployment were widespread and threatened to become worse. There was unanimity among Soviet leaders that a rapid expansion of industry was needed, but there was controversy over whether greater emphasis should be placed on consumer and agricultural input industries, or on the industrial investment industry. The first alternative promised more jobs in the near future and incentives for peasants to increase their production and marketings in exchange for more industrial products; the second favored the building of heavy industry as a basis for rapid and thorough industrialization. The second policy would create fewer jobs per unit of capital invested, and would further endanger the precarious exchange balance between agriculture and the rest of the economy. But it had the advantage of rapidly increasing the military and political power of the Bolshevik state, of relieving the pressing import needs for investment goods, and of creating for Soviet Russia its own basis for continued industrial growth. For the initial period, however, imported investment goods were needed to implement either alternative, and agricultural exports were the means of paying for them.

Within agriculture, there was also a dilemma: there was general agreement over the necessity to increase agricultural productivity, especially output per unit of land, as well as to expand the cultivated area and thereby provide at least part of the agricultural labor surplus with employment opportunities. The question was whether this should be done within the framework of individual peasant agriculture, or by reorganizing it into socialized, large-scale farms which were conceived of as technically modernized, capital-intensive enterprises. Such farms would have the disadvantage of saving on labor, which was abundant, and of requiring capital, which was in short supply and urgently needed by industry. In terms of capital cost, the most expedient way to increase agricultural production would have been to provide the existing farms with fertilizer, with high-quality seed and breeding animals, and possibly (this was done for a time but was discontinued in 1930) with cooperative machinery ownership and use. This, too, would have strained the capital resources, but it held promise of more immediate returns and it would have facilitated the use of available labor resources by more intensive land use. In addition, of course, it would have been necessary to offer the peasants incentives through improved terms of trade and more industrial goods as well as to provide the means for the more efficient peasants to expand their size of business.

But the output of industrial goods desired by the peasants was insufficient in quantity and quality and was produced at high cost. Therefore, the government was bent on fixing agricultural producers' prices below equilibrium levels in order to restrict peasant purchasing power and to reduce the cost of urban living (and thereby the cost of industrial labor). Likewise, the purchase price of agricultural products for export would be lower, and the government would benefit from selling at the higher international market price.

Whether or not the peasants, under the pressure of these policies, withheld their output (grain, above all) and/or purposely reduced the area sown to grain, whether or not the government's price policy (including suppression of the private grain trade) merely induced them to use more of their grain for feeding livestock, or whether they just did not have as much grain in stock as the government estimated need not be addressed in this paper.[5] What matters in the present context is that even before collectivization official policy was full of disincentives for peasant production and marketing, and that in part it thereby predetermined the choice of alternatives outlined above.

Moreover, the government was eager to prevent, for socio-political reasons, the development of an economically strong and socially assertive upper layer of the peasantry (the so-called *kulaks*), who had not been and were not likely to become Bolshevik. There were Bolsheviks, like Bukharin, who thought that this risk could be incurred and that balanced growth, favoring the peasants for a period of time, would in the long run yield better results in terms of economic development and political stabilization. But it was the radical solution through forced collectivization, endorsed by Stalin after he won supreme power, that was implemented after 1928-29, and on the whole completed by the mid-1930s.

WHAT THE KOLKHOZ WAS AND WHAT IT WAS NOT

One need not explore and discuss what Stalin really may have had in mind when he started the forced collectivization drive, i.e., whether he misjudged the overall economic situation or not. It is sufficient here to point out what Stalin's collectivization did and did not effect, since he later stuck to his course and thereby endorsed the result.

The result, as far as agricultural output is concerned, was a short-term decline and a medium-term stagnation.[6] In organizational and social terms the result was an agriculture in which the kolkhoz was the most typical and by far predominant form of organization. State farms played a special but on the whole minor role, and remaining private peasants in outlying areas became fewer and fewer in number. Politically and psychologically the result was an atmosphere where any potential resistance against Bolshevik agrarian policy was stifled as a consequence of the collectivization and the accompanying fierce campaign of confiscation, incarceration, and deportation (with a high mortality rate) against not only kulaks but all those resisting collectivization. There is no sign that traditions of communal landownership or the fact that juridically all land had been state property since 1917 made collectivization more acceptable to the peasants. They gave in to sheer force, and the intimidation effect was enhanced by the famine of 1932-34, which was at least in part a consequence of the collectivization campaign and abortive peasant resistance.

Statements similar to those above have been made by all competent Western observers and were confirmed, although in very cautious terms, by a number of Soviet authors of the de-Stalinization period.[7] To document them in detail here would mean repeating what has been demonstrated by renowned authors like Naum Jasny, Otto Schiller, Lazar Volin, and others.[8] While their descriptions deal mainly with the macroeconomic aspects of collectivization, the present analysis will center on the collective farm, i.e., on microeconomics. It does not apply, however, to the present state of affairs, which is different in some respects.

Allegedly, the kolkhoz was a voluntary union of peasants for common ownership and use of land and productive capital, including part of the livestock. In reality, it did not develop as a voluntary union, either during collectivization or afterwards. Significantly, by a decree of 1933, the kolkhoz management was authorized to refuse permission for leaving the kolkhoz. The kolkhoz Model Charter of 1935 contained provisions for entering the kolkhoz, but not for leaving it willingly (only through expulsion as a punishment). The mass of *kolkhozniks* (collective farm members), as distinct from the rest of the population, were issued identity cards (internal passports) only in exceptional cases, and therefore were not free to move outside their own rural district. Anyone who left without written permission from the kolkhoz board and the village soviet was not allowed to settle elsewhere, nor to be hired by a nonagricultural enterprise, even within the same rural district. (There were, however, legal and illegal ways to circumvent this.)[9] As a consequence, the kolkhoznik hardly resembled the cooperative owner of a farm; he had only a pro forma vote on the major issues of farm life and management elections. (There was no secret ballot; any voting had to be by a show of hands in the assembly.) He was an agricultural worker obeying the management's orders.

As an inhabitant of his village, the kolkhoznik was also very much under the direction of the collective farm management, which took over many functions of the village commune, and the village soviet, especially after 1950 when amalgamations of kolkhozes had made most of them larger units than the villages. The village soviet was economically weak while the kolkhoz management allocated the former communal usage rights (meadows and pasture, forest) and most goods and services the population needed: wood and other building material, transport facilities, social benefits, etc. During the early years, the village soviet was frequently important as a political lever of party and state in fighting against the kulaks and for collectivization. With the consolidation of the kolkhoz system it waned into insignificance, except for some formal functions as a civic registrar's office and with regard to school and other public services (though there were few of the latter in the kolkhoz village of Stalin's time). There was little distinction among the economic, the social, and the political aspects of rural communities. The village assembly and soviet could undertake few things without the kolkhoz management's consent or help.

Personal nonproductive property, including the family house, remained private, as did the permanent use of a garden plot (which legally was kolkhoz land owned by the state), limited numbers of livestock per household, and small farm implements as well as shelters for both privately owned tools and livestock. The right to a private-plot-subsidiary-economy was dependent on membership in the kolkhoz and more restricted for other segments of the population.

Collective work was paid for at the end of the year in proportion to individuals' labor contribution, evaluated in "labor-days" according to type and amount (norm) of work performed. Each such "labor-day" (*trudoden'*) was worth a share in the total amount of produce and cash left for distribution after all obligations in kind and money to state agencies and for current production needs were met, i.e., it was considered a share in the cooperative farm's profit, earned by labor. In absolute terms, such remuneration was much less than the wages paid on state farms, which, moreover, were paid at regular intervals during the year. As a rule, payment in kind was the main form of kolkhoz labor remuneration up to the mid-1950s, and served both for human consumption and for feeding the privately owned livestock.[10]

Work was performed in sections or brigades and sub-groups (*links*) under the direction of *brigadirs* or section chiefs and group foremen appointed by the farm's management.[11] Most rank-and-file members were employed only during the plant growing season. Although they were under a labor obligation, and after 1939 had to fulfill a minimum number of norm-days per year, the kolkhoz in its turn was obliged to provide employment only to the extent the production process required.

The kolkhoz management was elected and dismissed by the members' general assembly, which in practice could not deviate from the "recommendations" invariably given by district party and state authorities. Thus, managers were dependent on the confidence of these authorities rather than on that of the members. The paramount duty of the kolkhoz--that is, of its management--was to fulfill production and delivery plans imposed by the state. In this situation kolkhoz management by its very functions was made nearly omnipotent over the rank-and-file members, and had next to no power against the district authorities' will. The latter were supposed to represent the interests of the economy and society at large, and these had priority over possibly conflicting interests of the micro-units--collective farms and social groups.

Agricultural producers' prices were abysmally low. State prices paid for kolkhoz products were (as a country-wide average in 1940, which includes weighted premia payments) 11.6 rubles per metric ton of wheat, 3.8 for sugar beets, 4.6 for potatoes, 55.6 rubles for beef and veal (per ton, live weight), and 17.0 for milk (all prices converted into present-day rubles).[12] The true market situation was reflected, although in a distorted mirror, in the fact that food prices on the legal free kolkhoz markets were several times higher than those paid by the state; in a number of cases (e.g., milk and fruit) they were more than ten times as much. Since the kolkhoz delivered most of its produce at the low state prices, it had almost no cash profit to distribute among members for their "labor-days."[13] Hence, the kolkhoz members, who still constituted the majority of the Soviet population, lived mainly on the income from their private plots. Only the urban work force and their dependents (and a few other minor groups within the Soviet population) were supplied with basic food at relatively moderate state retail prices (though these prices were still high, in comparison to wages and salaries, because of the state's exorbitant profit margin).

During collectivization the peasants had slaughtered much of their livestock rather than let it be collectivized; more animals perished or were slaughtered out of desperation during the great famine of 1932-34. As a consequence, the remaining numbers totaled roughly one-half those of 1928, and did not greatly exceed the numbers permitted for private herds of kolkhoz households (18.7 million households in 1940). The collectivized animal industry in 1935 accounted for only 3 of 19 million cows, 7.3 of 20 million other cattle, 3.5 of 17.1 million pigs, and 12 of 40.8 million sheep and goats; as late as January 1951, the kolkhoz share was still less than half of the country's total livestock herds, including those of state and institutional farms.[14] Thus, most livestock farming remained in private hands, although it was in part dependent on the socialized sector for feed and fodder, including pasture. This fact alone, although there were others, would explain why a free food market, in addition to the state and consumers' cooperative retail trade, had to be readmitted as early as 1932.

With crop farming it was different. The production of grain and some other crops which were relatively easy to mechanize and/or needed extensive processing (e.g., sugar beets, sunflower seed, flax, cotton) became an essentially collective (or state farm) business. For this part of farm production, machinery (tractors, for the most part) was supplied. Even so, the quantities were insufficient, as Soviet agricultural machinery output lagged behind demand by a wide margin. With total horse and other draft animal numbers reduced by one-half as a consequence of the collectivization campaign, the supply of tractors did not suffice to bring effective traction power supply per hundred acres of sown area to pre-collectivization levels.[15]

Crops which were labor-intensive, perishable, or difficult to store and handle (e.g., potatoes, vegetables, fruit) remained largely within the domain of private production on plots of kolkhoz members and other population segments.

This division of labor between the private and the collective sector within the kolkhoz resulted in a close connection and interdependence between these two, which I have dealt with in detail elsewhere.[16] It encompassed the subsistence of kolkhoz peasants as well as part of the food supply for the rest of the population, the productive use of labor and land resources, the feed and fodder base for privately owned livestock, as well as the supply of young animals from the kolkhoz to private owners (and vice versa) and of animal products and fattened animals to the kolkhoz--often in hidden forms, but also legally. Without this interdependence the Soviet kolkhoz during the first decades of its existence would not have been viable.

The collective farms were large indeed, with an average cultivated area (including nonarable) of more than 3,000 acres in 1937, and more than 10,000 in 1953, but production processes were not large scale in the sense of modern techniques and specialization. This was only logical in view of labor surplus and capital shortage conditions, the latter having been aggravated by the huge capital losses during collectivization. As machinery was almost absent in the beginning and in very short supply later on, supervised manual work predominated. Collective animal farming was in most cases dispersed among several premises per kolkhoz, which had been hastily built or adapted, and were rarely large enough for more than 20-40 cows or a few dozen pigs.

As was common on the large East European estates of the nineteenth century, which the kolkhoz production process resembled in many ways, agricultural labor was organized in large units (brigades or sections) and subgroups instead of by millions of peasant families. In this way, social control over the peasantry and over the distribution of basic crop output was facilitated, though this was less true for other crops and for animal products. It was possible to extract at nominal prices the basic food needed for the growing urban and industrial population and for exports, while the kolkhoz peasantry made a subsistence living mainly on their private plot "mini-farms" without appreciable capital inputs. Only at a later stage--delayed by the destructions of World War II and by German occupation policy--did the kolkhoz become a vehicle of technical modernization and truly large-scale, specialized production.

THE IMPACT ON VILLAGE SOCIETY

Turning to the social aspects of collectivization, it is hard to draw a general picture, not only because of lack of information, but also because of the impact of other factors which left a deep imprint on village life without necessarily effecting long-term social change. These include Stalin's regime of terror over the whole of Soviet society, the war with its misery, and the hardships of reconstruction. As to rural-urban migration, these factors accentuated greatly what would have occurred in any case. Nevertheless, a few traits specific to collectivization proper shall be pointed out. They are abstracted from readings of very scattered eye-witness reports, from Soviet publications (especially belles lettres of the de-Stalinization period), and from insights gained retrospectively out of the much clearer picture of the Khrushchev period.

There is, first and easier to characterize, a negative effect: what little there had been of a peasant upper stratum in Russian villages had disappeared;[17] the peasantry as a whole was more than ever the mass of underdogs of Russian society. The goals and values of the smaller but dominant part of this society and its leaders were those of an urban-industrial society of the Soviet-Marxist type, with disregard not only for traditional peasant values but also, in the name of industrial progress, for the interests of the collectivized agricultural population. The old rural authorities--the village elders, the priest, the miller, the retail trader (if there had been one in a given village)--had largely disappeared along with the kulaks. Soon a new differentiation emerged: the rank-and-file kolkhozniks on the one hand, with the collective farm managers, the lower level cadres, and those with nonagricultural employment (in the procurement system, the consumers' cooperatives, retail outlets, etc.), on the other. Whereas formerly the "authorities" of state and party and the rural intelligentsia had largely resided in the central town or village of a district, they were now brought closer to the locus of agricultural production. But these social strata were very thin compared to the mass of kolkhoz peasantry, and remained so for the first two decades. One profession, not new but of smaller numbers in the old villages, gained in numerical importance, if not in social prestige: the rural school teachers, soon to be joined by the medical workers. This was a consequence of considerable efforts made by the government to raise the educational level and the supply of medical services for the rural population.

The peasant household, although stripped of many of its productive functions, in many other respects retained its former role. The patriarchal standing of the head of the household, however, was weakened by the change from an economy of peasants to one of agricultural workers with subsidiary plots. At times it seemed that the small subgroups (links) of the collective work force might in practice turn out to be a family or kinship group working together for the kolkhoz. That such an integration of the family into the formal labor and production organization of the collective farm, including the system of remuneration, is not inconceivable for Communists was demonstrated by the famous controversy over grain-growing links in the late 1940s, and is being demonstrated anew by recent Rumanian practice.[18] But in the Soviet Union it has always been rejected.

Although it was not given a role in the organization of collective work, the kolkhoz household was recognized as a primary social group and private economic unit of specifically peasant character. This role, which was inherited from traditional Russian law and retained in the Soviet Land Law of 1922, was not abrogated by the kolkhoz Model Charter of 1935.[19] In many respects the household, not its individual members, was the subject and object of law, whereas in Soviet civil law governing the rest of the population it was either the family or, more often, the individual. The kolkhoz household in this specific role was said, by a Soviet jurist, "to consolidate the kolkhoz system."[20] It was defined as a "family and labor community," for which joint living in the household, but not necessarily kinship of all members, was an essential prerequisite. Side by side with personal property of individual members there exists common property of the household (plot usage rights, livestock, living house, etc.), which may be inherited only by the remaining group (as a whole) of those who have qualified for membership through active participation in household and plot work. Whenever a household member dies or leaves, there ensues a hypothetical redistribution among the remaining members.

"In the actual life of the kolkhoz village" cash incomes of individual household members often were considered part of the household income and merged with the common income from the private plot.[21] Work on the private plot was performed mainly by women--as well as children, the aged, and invalids--and this freed men for the main tasks in the kolkhoz and for temporary occupations outside the farm (*otkhodnidchestvo*). Some of the able-bodied women likewise worked full-time in the kolkhoz livestock sector. This enhanced the economic role of women since up to the mid-1960s the income from the private plots made up the most important single item of the average kolkhoznik. So, paradoxically, and in contrast to the collective labor sector, female labor gained in importance for the family as a consequence of collectivization, while that of men declined. For decades, the collective work of unskilled men was near to unpaid socage and its primary family function was to assure plot usage rights.

To be sure, the general Marxist propagation of emancipation of women, the improved educational facilities for them, and the breaking up of the patriarchal peasant economy through collectivization had a great impact on the improved social standing of rural women. But the change of roles in generating family income must also have exerted considerable influence, although the weights are impossible to assess exactly, if only for lack of sociological data.

A significant symptom of the influence of the subsidiary private sector on family structure and life is the fact, sometimes mentioned in Soviet publications, that it has contributed to breaking up the extended family.[22] This is because only a household, irrespective of its size, has the right to a private plot; whenever a new household is set up, e.g., by dividing an extended family or by partitioning out a newly married member, the right to a separate and additional plot is created and land is provided from kolkhoz land specifically set aside for this purpose.

On the whole it may be said that the change in family patterns, which had begun before collectivization, was accelerated and acquired new traits with the changing roles of family and household in the collective as contrasted with the peasant farm. But the process remained a slow one and was delayed by the fact that technical modernization lagged behind organizational restructuring of the production process. Manual, unskilled work remained predominant for 2-3 decades after collectivization. During that time, the new occupations and skills, including those of managing, supervising, and accounting, were represented by only a small group in kolkhoz society, which was in fact rather above than within it. Many of those who performed these functions did not originate in the village.

This is not to say that no fundamental change occurred. But it was not an immediate consequence of collectivization. Only some time after the war did it accelerate appreciably, when the kolkhoz economy itself underwent the economic, technological, and organizational change of genuine modernization. Similarly, the worldwide phenomenon of the feminization of agricultural work--as far as collective work is concerned--made itself felt rather as a consequence of the war than of collectivization, and acquired a momentum of its own only when technological change and regional labor shortages began to be felt in the 1950s.

The one notable exception was the new eminently male profession of the "mechanizer," which for the first decades was essentially that of a tractor driver. Its central role was created as a result of the substitution of mechanical for animal draft power, which was as much a consequence of the unwanted loss of horses as of collectivization per se. In contrast to modernizing peasant or farmer societies, tractor driving became not a new skill of the farmer but a separate profession. And as a consequence, it was the "mechanizers" who were most disposed to and capable of leaving agriculture, thereby indirectly obstructing its modernization.

ASPECTS OF LABOR AND CAPITAL TRANSFER

It has often been said that it was collectivization that made possible, or at least facilitated, the transfer of labor from agriculture to industry. This persisting interpretation is misleading, to say the least.[23] Worldwide experience shows that in any sociopolitical system where there is an agricultural labor surplus, and where there are nonagricultural jobs available with rewards equal to or above the average income level of underemployed agricultural labor, the transfer proceeds spontaneously. The more so if urban employment and residence offers additional advantages, mainly of infrastructure and of education and career prospects, over the village, as was and still is the case in the Soviet Union. (Pre-collectivization industrial wages in the Soviet Union may not have been very conducive to rural-urban migration, but this was a special feature of Soviet industry, not of agriculture; and surely nobody today would argue in favor of cooperative farming as a means of depressing rural incomes because urban wages are low.) To make the forms of spontaneous rural-urban migration socially less undesirable, and in a vast country to facilitate regional manpower redistribution over the wider distances, government action may be needed. But such action does not necessarily affect the organization of agriculture, nor presuppose its socialization.

What Soviet collectivization did with regard to labor transfer was something different. First, it must be kept in mind that, preceding collectivization, Soviet industry experienced no shortage of labor, except for skilled labor, which usually does not come directly from the villages. Rather, there was latent urban unemployment, and there were millions of landless peasants as a potential source for labor recruitment. Second, by depressing rural living standards to misery levels (at least up to and beyond World War II) collectivization created a powerful motivation for rural-urban migration in almost any kolkhoz village (with the possible exception of a few prosperous kolkhozes in the South, and of the special situation in Soviet Central Asia and parts of Transcaucasia). The inhibiting psychological influence of individual peasant property was weakened (not wholly eliminated, because of the private house and plot). Third, and opposed to this push factor, the new kolkhoz order prevented the outflow from spontaneously sweeping the country against the planners' will by instituting administrative restrictions on leaving. But whenever and wherever it was decided to recruit labor for nonagricultural work, the government could get the necessary vouchers issued for leaving the kolkhoz, and at the same time could select among applicants, of whom there were usually more than sufficient. The most common form of doing so was the *organizovannyi nabor* (organized recruitment).

Although this system was never fully effective, it undoubtedly served to restrict and channel the labor outflow rather than to initiate it. The agricultural labor surplus which was not transferred to industry was more or less absorbed by the primitive private plot sector--at no cost to the economy at large. But it can be said with confidence that without the kolkhoz as a means of enforcement it would not have been possible to depress rural living standards to the extent actually experienced, to extract marketable produce in the required quantities, and at the same time to restrict and channel (though imperfectly) the potential flight--not just outflow--from the land.

It need not be discussed here whether the immense labor requirements of the new, often (in extractive branches) primitive and labor-consuming industries were fully met by the rural outflow. The point is rather that, under other socio-political conditions, peasant agriculture would very likely have supplied as much or even more labor--at higher wage cost, however--than did collectivization with its concomitant millions of rural dead as a result of the 1932-34 famine. Significantly, Soviet procurement policy in favor of the urban population prevented starvation, though not hunger, only in the towns during that famine.

Transfer of capital can with more justification be ascribed as one of the functions of collectivization in Soviet development strategy. In view of the strained relations with the Western industrialized countries and of Soviet hesitance to incur debts or to grant licenses which might bring about dependence on capitalists, agriculture was *the* major source, apart from the already existing industrial sector itself, from which to raise capital for the desired rapid industrialization of the country. As the peasants were poor and it was not proposed first to allow them to become well-to-do (kulaks) and then to extract the capital (by taxation, for instance), the only means of extraction was to get their produce cheaply. This system of "price-scissors" was in fact deliberate government policy even before collectivization, and it was pursued to extremes later on. Grain was the crucial product. The repeated grain procurement campaigns of 1926-1928 had demonstrated that it was very difficult to extract the required quantities at administratively depressed prices from millions of individual small producers. Here the kolkhoz organization, supplemented by the state procurement agencies, formed the instrument for low-cost procurements, and thereby for forced capital transfer.

These instruments, however, soon revealed a most serious disadvantage. The upheavals of collectivization disrupted the agrarian economy to such a degree that there remained less to be extracted. In subsequent years, labor morale was so low and capital inputs were so inadequate[24] that agricultural production up to 1940 did not regain the 1928 level, except for 1937. Even in that year, total *net* farm income (in 1928 prices) has been estimated by a competent Western observer as 26 percent below 1928 (10 percent below on a per capita basis since the farm population had decreased in the intervening time).[25] Thus, capital transfer did take place, but in absolute terms far less than expected, and at an opportunity cost that before 1929 nobody would have dared to predict, nor to consider a justified corollary of collectivized agriculture.

On the other hand, it has been argued that the new organization of agricultural production, including state farms and state-owned machinery stations for servicing (and supervising) the kolkhozes, the construction of facilities for collectivized herds, etc., necessitated also a flow of capital into agriculture, so that on balance there may not have been an appreciable capital transfer.[26] Even so, such a balance would have been a consequence of the greatly reduced productivity of Soviet agriculture, and resulting reduced capital outflow, not an intrinsic part of the collectivization strategy as such.

In a recent article, J. R. Millar has demonstrated that on balance there was either no transfer or, if calculated on a different price basis, not appreciably more than before collectivization.[27] But he also showed that there was capital transfer out of the collective farm sector, while state farms, especially where they were organized on formerly idle lands, and machinery stations were the recipients of substantial governmental investments thereby in part strengthening the balance in favor of agriculture as a whole. Machinery stations serving the kolkhozes were largely an offset to the losses in draft power. Another offsetting factor was the high free prices on the kolkhoz markets. Through these the kolkhoz population and in part the kolkhozes themselves (where they were able to enter this market to a sizable degree with above-plan produce) recouped part of what was extracted through the low procurement prices. So there was also capital transfer in the opposite direction, via the high prices on the residual free market, as a consequence of the reduction of output following collectivization. But it was the urban consumer, not the state investment budget, which paid this part of the bill.

In other words, given the huge losses caused by collectivization, the new organization still enabled the government to transfer more capital out of the collective sector of agriculture than would have been possible (in relative terms) had such extraction been directed at the pre-collectivization millions of individual peasant farms.

THE CONTROL FUNCTION OF THE KOLKHOZ

The point could be made that, with the help of a voluminous and powerful state apparatus of supervision and procurement, roughly the same amount of product extraction could have been achieved without collectivization, simply by forcing every single peasant to produce a certain amount of food and to deliver a fixed part of it at nominal prices (a percentage tax in kind combined with command production). Such an apparatus had not existed in Russia before and was difficult to organize, but since it proved possible to do so in connection with collectivization one might argue that it should have been possible without collectivization as well. In addition, the losses of collectivization would have been avoided. Such reasoning overlooks a fundamental aspect peculiar to collectivized agriculture: the penetration of control into the production process itself, combined with the role of the Communist party.

Except for the private plots and private livestock farming, kolkhoz production was directed by leading cadres, most of whom were placed into positions of farm management as well as rural district administration from the outside. Their well-being and careers depended much more on successfully executing orders from above and fulfilling procurement plans than on the prospering of the farm and its people. Subordinate to them were lower echelons of managers and supervisors (brigade, section, link head, village soviet chairmen, etc.) who often were local people. These may or may not have sympathized with the rank and file; in any case they were mere executives without any real decision-making functions. In any event, the parameters of action of both the higher and the lower functionaries were influenced by the prevailing atmosphere of terror under the Stalinist regime, which for the countryside had begun in 1929 as part of the collectivization campaign, and for the whole Soviet society continued under the "great purge" of 1935-37. And most importantly, they did not have to supervise the productive activities of millions of peasant farms, but an army of agricultural workers, hierarchically organized in brigades (sections) and links: socio-political control was part of the kolkhoz production process itself.

Even so, one could imagine that local cadres, over time, might have begun to identify themselves with, or at least adapt to, local economic and social exigencies, e.g., by withholding from the higher state and planning authorities information essential for decision-making. In the initial years, this was made very difficult by the prevailing terror and, possibly even more so, by the lack of a common social and regional background of the rural political and managerial elite. The war years created a special situation. But subsequently, as more peaceful times emerged, the main factor preventing such a development was the parallelism of state and party organs. The kolkhozes became larger units by amalgamation (on a mass scale during 1950-52), and the number of party members in the countryside increased, albeit slowly. Practically all local cadres of any importance, including a sizable number among the farm specialists, were Communist party members and therefore subject to party discipline in addition to that specifically related to their duties as administrative or management personnel.

The implications of this situation cannot be dwelt upon here; they have been extensively dealt with in the relevant political science publications. Suffice it to say that the party organization has on the whole proved successful as a control as well as a mobilization and information feedback system. (Some observers may be inclined to consider the secret police another such factor of control, at least at times, but it is doubtful whether it ever has had an extensive network down to the village and kolkhoz level.) It is my opinion that the relative success of control cannot be explained without taking into account the role of the party organization and its local cells. Success is to be understood here only in its own terms, i.e., in terms of forcing the leadership's will for capital transfer through forced savings on a recalcitrant majority of the population, not in terms of a rapid and sustained rise in agricultural productivity.

SOME CONCLUDING REMARKS

If collectivization did not really favor modernization and productivity, then what was its economic effect on Soviet agriculture? It remains a question, still to be explored in detail, whether in the long run the kolkhoz will remain a brake on labor productivity or will favor it, not to mention capital productivity. But a few general statements may be made:

- In its Stalinist version the kolkhoz was a brake on labor productivity. Not accidentally, one of the first steps of Stalin's successors was to change the kolkhoz system in a number of respects.
- Even after Stalin, and up to the present, labor productivity in the kolkhoz has not increased faster than in Western peasant or farmer agriculture. Since the outflow of labor from agriculture was faster in the Western industrialized countries, their farm labor productivity in fact increased at a more rapid rate.
- This faster increase in the West may be attributed in part to the greater capital inputs which are available in highly industrialized countries. But when Soviet industrialization reached the level where capital inputs were being made available to collective agriculture in greater amounts, the cost of agricultural growth also increased by leaps and bounds. This manifested itself, by the 1970s, in subsidies to the agricultural sector larger than those in any Western country, without the benefit of lower producers' prices.
- It may be argued that the expansion of agriculture into low-potential dry-farming areas or into high-cost new irrigation areas would have been too risky for individual peasant farm-

ers. Yet in the Soviet Union during the last two decades, and in part during the 1930s, it was not the kolkhoz which served as the main vehicle for such an expanison, but rather the state farm, where the risk was borne by the state budget.

Two statements are of a more conjectural nature:

- The continuing low overall labor productivity of Soviet agriculture may be explained in part as resulting from the combination of mechanized large-scale and primitive private plot farming as it developed over the first decades after collectivization. As long as there was--and in some regions still is--surplus of agricultural labor, the private sector counterbalances underemployment in the collective sector, and moreover makes good some of the productive shortfalls of the latter. Yet it may be that the viability of this combination presupposes a high consumer (scarcity) value for the specific products of private plots as a return for what would otherwise be nonrewarding inputs of great quantities of labor, i.e., a society which lives neither at famine nor at affluence levels. Moreover, a private agricultural sector is a socio-psychological factor working against collective-mindedness, and as such has often led to conflicts of goals in the ideologically oriented Soviet system.
- The second conjectural point is in favor of collectivization as such: it created large farming units, which are not subject to the structural problems besetting most Western peasants and farmers under the impact of technological progress and the concomitant need for large-scale specialized production. The future may throw the advantages of the Soviet large units into bolder relief. But up to the present, i.e., for almost half a century, this advantage of the Soviet collective farm has been offset by many brakes on its economic and technological progress, apparently inherent in the way in which the Soviet system of agriculture developed. The late Jerzy Karcz subsumed these persisting--although now somewhat mitigated--negative peculiarities under the term "command farming" and went on to say:

> The institutions and attitudes generated by the latter have proved to be quite difficult to eradicate, not only in the USSR but also in other socialist countries of Eastern Europe. These factors, including the distortion of incentives, of patterns of decision-making, and hence also of specialization and production, as well as trends in the supply of labor, were developed in various efforts to increase the short-run contribution of agriculture to economic development, and they must therefore be viewed as part and parcel of what is often called 'the Soviet development strategy.' . . . In this sense, the emergence and consequences of command farming due to its impact on efficiency must be viewed as long-run capital costs of Stalin's

> basic decision on the rate and nature of industrialization and of these on agricultural organization deemed necessary to implement the former.[28]

In the present paper I have tried to put the kolkhoz system as a form of cooperative farming during its first quarter century of existence into proper historical perspective. If the circumstances of the 1920s and 1930s are compared with those of present-day less developed countries, with regard to the question of applicability of the Soviet type of cooperative farming, the following similarities emerge:

1. The Soviet Union was an agrarian state with the large majority of its population employed in small-scale peasant agriculture.
2. There were traditions of communal landownership, although not in all parts of the country; yet it is hard to assess to what degree they were still alive.
3. There was a high rate of population growth and a surplus of agricultural labor.
4. The economic and social infrastructure was underdeveloped, except for some parts of the country.
5. Political democracy, although not wholly absent (before Stalin became the sole authority), was less than fully developed.
6. The ruling elite was preponderantly of urban origin.
7. The need for rapid industrialization was urgently felt not only by this elite but also by part of the nonagricultural population.
8. Capital was in short supply; as much of it as possible had to be transferred from agriculture to industry.

In contrast, a number of dissimilarities, i.e., of circumstances peculiar to the Soviet Union between the wars, also emerge:

1. In parts of the country there already existed a number of industrial growth centers with a developed economic infrastructure and social base (industrial working class).
2. The peasantry, at least in parts of the country, had already shown its readiness and capability for innovation; it was far from being wholly illiterate.
3. Production of basic foods, although not abundant in view of the population increase, was sufficient to allow substantial exports.
4. Vast unused stretches of land were still available for the expansion of agricultural production, although most of them were rather dry steppe regions, or needed irrigation or drainage to become cultivable.
5. The country possessed deposits of almost all raw materials (either already being exploited, or to be opened up) necessary for an industrialized economy, especially an autarchic one.
6. An administrative structure, though far from perfect, existed for the whole country. It was supplemented by the

unique Bolshevik party organization with its uniform and strictly enforced ideology. Both were directed by one of the most able and ruthless dictators history has known.

7. Capital aid and loans from Western industrialized nations were not available and/or not asked for.

These dissimilarities reveal that, except for the availability of foreign aid, Soviet Russia was generally in a much better position than most developing countries today. The government had the power to enforce a fundamental reorganization of agriculture. And, as history has shown, the country could stand a substantial decrease in agricultural output, caused by forceful rapid collectivization, for a number of consecutive years, although this meant starvation for part of the rural population. Given the average food/population ratios in less developed countries of today, it is more than doubtful that many of these could stand a similar output decrease without catastrophe.[29]

It seems obvious to this writer that these dissimilarities preclude an imitation of Soviet collectivization as it developed historically, except perhaps for the combination of cooperative large-scale with individual mini-scale (private plot) activities, as pointed out above. This combination was not part of the Soviet development strategy, but rather came into existence as a consequence of the shortfall of collective production. Nor do the Soviet leaders of today overtly recommend collectivization in its original form for the less developed countries, although some ambiguity in this respect can be sensed,[30] especially with regard to countries with a socialist orientation of government.[31] On the other hand, the Soviet kolkhoz of today, with the changes that have occurred, functions in a very different setting, no longer that of a less developed country. It is essential to note that in the present-day Soviet Union, capital can be transferred from highly industrialized to very distant, less developed regions with a non-European historical and cultural background, within the national territory and through a common state budget.

It is the actual functions, as distinct from the often alleged ones, of the Stalinist kind of cooperative farming that have contributed to the changes in Soviet economy and society from the beginning of the industrialization drive up to the early 1950s. One may approve or disapprove of these changes, especially of the human suffering and economic setbacks which went with them; one may even argue--as did Lenin in his last article and Bukharin later on--that voluntariness is indispensable for the idea of cooperative farming to be realized, and that in the long run a more humane method might have achieved the same economic growth. Yet one cannot deny that in its own terms Stalin's collectivization was a success. But two things should be clear: first, Stalinist collectivization may or may not have been devised primarily as a transformation of the agrarian structure. In practice it turned out to be a means for imposing extreme forced saving for the sake of capital accumulation, on the vast majority of the population, including its urban segment. Second, if one suggests applying the modified type of Soviet cooperative farming of today, or a very similar one, to present-day developing countries, one has not only to start from a different setting, but one will also be applying the Soviet agricultural organization as it has developed only after the main thrust of Soviet industrialization was over. Historical experience is no clue to the applicability of such a changed model.

NOTES

1. There is an ample and valuable literature on the subject, enumeration of which would require a bibliographical reference volume of its own. I have tried to summarize the historical development of the agrarian question in Marxist thinking as well as in Soviet practice, and its spreading out to Eastern Europe after 1945, in a recent book on *Sozialistische Agrarpolitik in Osteuropa. I: Von Marx bis zur Vollkollektivierung* (Berlin: Duncker and Humblot, 1974). In the following paragraphs I draw largely on this book and its source references.

2. One of the few specialists on Communist agriculture who analyzed its implications as a model for development strategies was Otto Schiller. See his "The Agrarian Question: Communist Experience and Its Implication for Developing Countries," *Agrarian Policies and Problems in Communist and Non-Communist Countries*, ed. W. A. Douglas Jackson (Seattle and London: University of Washington Press, 1971), pp. 231-44. But Schiller, too, did not go beyond a few remarks on the question of application proper. At about the same time, Z. M. Fallenbuchl ("Collectivization and Economic Development," *Canadian Journal of Economics and Political Science* 33 [February 1967]: 1-15) dealt with the subject, but only incidentally with a view to the applicability to the less developed countries.

A book which deals extensively with the implications of Soviet and Chinese collective farming as a possible model for agricultural change in less developed countries is that by Heinrich Schweizer, *Sozialistische Agrartheorie und Landwirtschaftspolitik in China und der Sowjetunion: ein Modell fur Entwicklungsländer?* (Bern und Frankfurt: Verlag Herbert Lang, 1972), 266 pp. As to the Soviet collective farm, Schweizer rightly concentrates on its early form (up to the mid-1950s). For the rest, he is more in favor of the Chinese model, for a number of convincing and less convincing reasons. Thus, most of his book is only in part relevant to the present paper. Moreover, unlike Otto Schiller, he does not know thoroughly enough the Soviet *kolkhoz* in its reality as distinct from what it pretends to be, and he does not always make a clear distinction between Soviet collective farming as such and Soviet agrarian policies in general. Still, his is an interesting book, and the only Western one on the subject. The present writer's objections are directed mainly at Schweizer's belief in the function of collectivization as an indispensable means for the release of labor from agriculture and for capital transfer to industry, and against his deemphasizing the social and economic cost of Soviet collectivization. On more general grounds, Schweizer seems to underestimate the problems of transferring parts of a model and of neglecting the latter's overall social and political context.

3. See the testimony, among others, of the competent contemporary observer Otto Auhagen, "Russlands Wirtschaft unter dem Bolschewimus," *Osteuropa* 3 (1927-28): 125.

4. Derived from the more detailed data given by O. Anweiler, "Erziehungs- und Bildungspolitik," *Die Kulturpolitik der Sowjetunion*, ed. O. Anweiler, K.-E. Ruffmann (Stuttgart: Kroener, 1973), pp. 7-8.

5. On these questions, see J. F. Karcz, "Thoughts on the Grain Problem," *Soviet Studies* 18 (April 1967): 399-434; R. W. Davies, "A Note on Grain Statistics," ibid. 21 (January 1970): 314-29; and Karcz's rejoinder, "Back on the Grain Front," ibid. 22 (October 1970): 262-94. As early as 1926 it was reported that the Soviet statistical authorities had overestimated the grain stocks of the peasants, possibly in good faith, and that therefore peasant marketings necessarily remained below government expectations; the same Western observer suspected in early 1928 that the total grain crop and the peasants' stocks were again being overestimated. See O. Auhagen, "Wirtschaftsumschau," *Osteuropa* 1 (1925-26): 275, and 3 (1927-28): 443. Cf. the official correction of Soviet estimates reproduced by J. F. Karcz, "From Stalin to Brezhnev: Soviet Agriculture in Perspective," *The Soviet Rural Community*, ed. J. F. Millar (Urbana, Ill.: University of Illinois Press, 1971), p. 40.

6. Soviet statistics do not give data for the agricultural gross product during 1929-1935. Recent Soviet data (in prices of 1963) indicate 27.4 billion rubles worth for the annual average 1924-1928 and 29.8 billion for the average 1936-1940 (*Narodnoe khozyaystvo SSSR v 1972 g.* [Moscow, 1973], p. 284). The latter quinquennium not only includes the record harvest year of 1937 but also the output of the greatly expanded Soviet territory of 1940. Moreover, 1924 output was much below that of 1925-1928, so that the last pre-collectivization triennium (1926-1928) would show much better in comparison to, say, 1934-1936 or 1938-1940. But even the above averages for five years, calculated on a per capita basis for the growing total population (neglecting the population of territories annexed in 1939-40) imply, in constant prices of 1965, 180-190 rubles worth per head during 1924-1928, and 170-180 rubles during 1936-1940.

7. As an example of these, see V. A. Golikov, *Vazhneyshiy etap razvitiya sel'skokhozyaystvennoy kooperatsii v SSSR (1921-1929 gg.)* (Moscow, 1963), pp. 192-94.

8. For a most penetrating recent overview, putting many questions into new perspective, see Karcz, "From Stalin to Brezhnev: Soviet Agriculture in Perspective," pp. 36-63.

9. For the details, see K.-E. Wädekin, *Führungskräfte im sowjetischen Dorf* (Berlin: Duncker and Humblot, 1969), pp. 45-57.

10. The standard work on this older system is H. Wronski, *Rémunération et Niveau de Vie dans les Kolkhoz. Le Troudoden* (Paris: Société d'Edition d'Enseignement Superieur, 1957). For one of the numerous later Soviet criticisms of the disincentives associated with this remuneration system, see *Khozraschet i stimulirovanie v sel'skom khozyaystve*, ed. A. M. Yemel'yanov (Moscow, 1968), p. 86.

11. A detailed description is that of R. C. Stuart, *The Collective Farm in Soviet Agriculture* (Lexington, Mass.: Lexington Books, D. C. Heath and Company, 1972), pp. 11-29; cf. R. D. Laird, *Collective Farming in Russia* (Lawrence, Kans.: University of Kansas Publications, Social Science Studies, 1958). For the sociological aspects, see Wädekin, *Führungskräfte im sowjetischen Dorf*, pp. 155-300.

12. S. G. Stolyarov, *O tsenakh i tsenoobrazovanii v SSSR*, 3rd. ed. (Moscow, 1969), p. 119.

13. There are exceptions, of course, for very prosperous kolkhozes which could sell much of their above-plan production at higher official prices on the free market.

14. *Chislennost' skota v SSSR* (Moscow, 1957), pp. 6, 95, 102, 110, 116.

15. Cf. E. Schinke, "Some Peculiarities of the Employment of Factors in Soviet Agriculture," in Jackson, ed., *Agrarian Policies and Problems in Communist and Non-Communist Countries*, p. 145. Schinke's calculations may overestimate the coefficients of animal draft power in comparison to mechanical h/p units, but he is right in his contention that Soviet statistics grossly exaggerate the performance of mechanical power in farming operations by equating only 0.75 mechanical h/p to one horse, as the physicists do for quite different purposes. Surely, a 30 h/p tractor in farm practice is not a draft power equivalent to 40 horses with their many potential uses, and Schinke's equivalent of four horses is closer to reality.

16. K.-E. Wädekin, *The Private Sector in Soviet Agriculture* (Berkeley, Cal.: Univ. of California Press, 1973), chapter 7 and *passim*.

17. The competent, and well-documented, account of the Russian peasantry and its low degree of social differentiation on the eve of collectivization is that of T. Shanin, *The Awkward Class. Political Sociology in a Developing Society: Russia 1910-1925* (Oxford: Clarendon Press, Oxford Univ. Press, 1972), esp. chapters 9 and 10.

18. On the interesting Rumanian approach, see M. Cernea, "The Large Scale Formal Organization and the Family Primary Group," *Revue Roumaine des Sciences Sociales--Série de Sociologie* (Bucharest) 18 (1974): 85-98.

19. Cf. W. T. Shinn, "The Law of the Russian Peasant Household," *Slavic Review* 20 (December 1961): 601-621.

20. N. P. Voloshin, *Pravo lichnoy sobstvennosti kolkhoznogo dvora* (Moscow, 1961), p. 74.

21. B. A. Liskovets, *Imushchestvennye razdely i vydely v kolkhoznom dvore* (Moscow, 1963), p. 30; Voloshin, *Pravo lichnoy sobstvennosti kolkhoznogo dvora*, pp. 80-82.

22. See the decree of the Central Party Committee and Council of People's Commissars of May 27, 1939, *Direktivy KPSS i Sovetskogo pravitel'stva po khozyaystvennym voprosam*, vol. II (Moscow, 1957), p. 589; A. G. Kharchev, *Brak i sem'ya v SSSR* (Moscow, 1964), p. 216; U. Ilebaev, *Priusadebnoe zemlepol'zovanie v Kirgizskoy SSR* (Frunze, 1965), p. 18; cf. St. P. and E. Dunn, *The Peasants of Central Russia* (New York, 1967), p. 75.

23. Among the very few who did not commit this error were O. Schiller and Z. M. Fallenbuchl (see footnote 2 above).

24. As indicated above, capital inputs would have been less inadequate if the collectivization turmoil had not reduced the numbers of draft animals by half. On the other hand, the shrinking of livestock herds enabled the state to extract more grain, as less was needed for feed.

25. Lazar Volin, *A Century of Russian Agriculture* (Cambridge, Mass.: Harvard University Press, 1970), p. 257.

26. See J. R. Millar, "Soviet Rapid Development and the Agricultural Surplus Hypothesis," *Soviet Studies* 22 (July 1970): 77-93; and A. Nove, "The Agricultural Surplus Hypothesis: A Comment on James R. Millar's Article," ibid. (January 1971): 394-401; cf. both discussants, ibid. 23 (October 1971): 302-308. See also J. F. Karcz, "From Stalin to Brezhnev: Soviet Agriculture in Perspective," pp. 48-51.

27. J. R. Millar, "Mass Collectivization and the Contribution of Soviet Agriculture to the First Five-Year Plan: A Review Article," *Slavic Review* 33 (December 1974): 750-66; Millar relies on a recent Soviet reevaluation by A. A. Barsov, *Balans stoimostnykh obmenov mezhdu gorodom i derevney* (Moscow, 1969), and, by the same Soviet author, "Sel'skoe khozyaystvo i istochniki sotsialisticheskogo nakopleniya v gody pervoy pyatiletki (1928-1933)," *Istoriya SSSR*, 1968, no. 3.

28. J. F. Karcz, "From Stalin to Brezhnev: Soviet Agriculture in Perspective," pp. 68-69.

29. This point has been made by O. Schiller, "The Agrarian Question: Communist Experience and Its Implication for Developing Countries," p. 235.

30. See K.-E. Wädekin, "Auswirkungen und Konzeptionen sowjetischer Agrarpolitik für die Entwicklungsländer," *Agrarpolitik im Spannungsfeld der internationalen Entwicklungspolitik* (München, Bern, Wien: BLV Verlag, 1974), pp. 156-162.

31. For a recent Soviet (unofficial) statement, see A. Romanchenko, "Sel'skokhozyaystvennaya kooperatisiya razvivayushchikhsya stran," *Ekonomika sel'skogo khozyaystva*, no. 3 (1975), pp. 111-14; this author points out with satisfaction that in Algeria one of the forms of agricultural cooperatives is that "of the type of Soviet kolkhozes," and quite generally emphasizes, as all Soviet authors do, "Lenin's Cooperative Plan" as a guideline for developing countries. Yet the latter is not a precise collectivization program, and surely not identical with Stalin's kolkhoz model. Cf. the report, on a kolkhoz organized with Soviet assistance in India, in *Sel'skaya zhizn'*, 14 May 1974, p. 3.

5. Communization of Peasant Agriculture: China's Organizational Strategy for Agricultural Development

John Wong

INTRODUCTION

Chinese agriculture prior to the Communist revolution in 1949 was essentially of the pre-industrial, peasant farming type, which is a typology of underdeveloped agricultures reflecting wide institutional differences. In the case of China it was mainly characterized by small scale of operations, reliance on labor-intensive and traditional (but not inefficient) production techniques, and to a large extent (but not always) individual management of production. Like other orthodoz Marxists, the Chinese Communists from the outset had a strong ideological antipathy toward small-scale peasant farming.[1] Hence immediately after their accession to power they took steps to reorganize the institutional structure of agriculture which led to the final socialization of agriculture. The whole process of institutional (officially "socialist") transformation of agriculture in China comprised three spurts of revolutionary agrarian changes in the countryside--land reform, collectivization, and communization--all falling within the first turbulent decade of the People's Republic (the 1950s). In retrospect, the Chinese authorities have indeed achieved their objective of institutional revolution in agriculture with remarkable efficiency and smoothness, unmatched by any earlier efforts of socialist countries.

By virtue of its being the world's first radical rural experiment on a massive scale, the communized form of farming in China during the early years of its formation evoked great controversy and reaction from hostile commentators, some of whom tended to treat it as a "passing" phenomenon.[2] Communes have continued to exist as a viable rural institution and have in fact become a permanent feature of Chinese rural life. They are the mainstay of China's organizational strategy for agricultural development as well as the organizational structure within which agricultural production is planned and carried out to feed the world's most populous nation.

In this paper I shall recapture the main background of China's communization, and highlight both its long-term problems and its impact on agricultural development. The ultimate objective is to evaluate the Chinese organizational strategy with a view to drawing some deliberately provocative generalizations in order to stimulate new trends of thinking, especially in regard to the transferability of the Chinese experience to the Third World.

PROCESS OF INSTITUTIONAL TRANSFORMATION

From a purely technical standpoint, agricultural backwardness in traditional China has been the consequence of centuries-long technological stagnation, aggravated by the rapidly worsening man-land ratio and incessant political instability. To the Chinese Communists, however, Chinese agriculture was underdeveloped fundamentally because its institutional set-up was "feudalistic" or, to put it in Marxian terms, because "production relations" were impeding the development of "production forces." Such a contradiction had to be resolved before agricultural development could take place, which would mean the substitution of new production relations based on socialist organization for the old production relations long embedded in the feudal or semi-feudal agrarian society. The ideological rhetoric having thus been axiomatically established, all other issues became immaterial and irrelevant.

The Chinese Government could, of course, rationalize on cogent grounds its policy of depending on organizational strategy, rather than on technological progress, as the initial source of agricultural growth. *Pari passu* with the institutional revolution in agriculture was the launching of China's First Five Year Plan, 1953-57, which bore great resemblances to the classical "Stalin Model" of development requiring agriculture to serve industrial growth by contributing the maximum surplus. Furthermore, the agricultural surplus was to be generated with as little resource feedback from industry as possible.[3] Thus the agricultural sector was virtually starved of State development funds as was evident in the Plan's investment allocation pattern, which earmarked only 9.9 percent for agriculture as opposed to 64.5 percent for industry.[4] Growth of productivity in agriculture was to rely on the mobilization of traditional inputs until such time as industry began to supply agriculture with modern inputs. And the socialist transformation of agriculture would provide the most effective framework not only for the deployment of traditional resources but also for channeling the marketed surplus for industrial development.[5]

The first step in the institutional revolution was land reform. In terms of economic revolution, it was actually rather moderate in scope. It only sought to expropriate the landlords as a class while at the same time instituting measures to protect the rich peasants (as opposed to the Russian liquidation of the Kulaks in the Soviet land reform). Consequently, less than 50 percent of the land, and considerably lower proportions of other means of production such as draft animals and farm implements, were actually redistributed. Despite all its socialist trappings, the Chinese land reform shared with the conventional type of distributionist land reform in other developing countries the shortcoming of not having been able to transfer sufficient resources to make the acquisitions of the newly created owners economically viable. But the Chinese land reform did stand out for its extremely high rate of implementation, resulting in a drastic redistribution of political and social power in the villages and clearing the way for a more thorough institutional revolution in the future.[6]

After the land reform, the Government encouraged the peasants to form themselves into Mutual Aid Teams (MATs) along the traditional line of agricultural co-operation (see below). MATs were organized by a few households pooling their labor and animals for joint farming operations during the peak seasons. Meanwhile, the Government was making separate efforts to convert some of the MATs into agricultural producers' cooperatives (APCs) which were three or four times larger and were more elaborate in organization, based as they were on the motto of "centralized management but private ownership." They were considered as "semi-socialist" in nature because members contributed their land as capital shares to the cooperatives and received, after reduction for wages and costs, cooperative earnings as returns to their land shares.

As the Government's objective was to form full-fledged socialistic collectives, it pressed ahead for further socialist transformation. The main impetus for collectivization, however, seems to have come from political and ideological rather than economic considerations. The Cultural Revolution has disclosed that an intensive debate on the time schedule of collectivization took place among the top Chinese leadership. Liu Shao-ch'i, the dismissed President of the People's Republic, was reported to have opposed, based on the Soviet experience, the "adventurous" policy of moving toward full collectivization before China had built up a sound technical foundation in agriculture or had reached the stage of agricultural mechanization.[7] However, Mao Tse-tung took the opposite view that collectivization would better resolve the contradiction of the ever-increasing demand for marketable agricultural surplus and the low productivity inherent in small-scale peasant farming. He further asserted that Chinese agriculture could realize collectivization without prior mechanization.[8] Mao's endorsement actually touched off a "rural upsurge" in which more than 120 million of China's peasant households were reorganized into collectives (officially known as advanced APCs) in a single year--1956--with remarkably little disruption.[9]

In less than two years after collectivization, China launched the Great Leap Forward, in which 750,000 collectives were merged and regrouped into 26,000 people's communes, a truly Chinese innovation with a still higher form of socialistic organization. The first wave of communes attempted to experiment with a new system of production and consumption, including collective living arrangements. These communes soon ran into many "teething" problems, and their organizational excesses were widely reported in the foreign press.

After the initial collective euphoria subsided, the communes were streamlined. They underwent further reorganization in 1960 when the agricultural crisis set in, leading to the evolution of a three-level system: the commune, the production brigades, and the production teams. Consequently, the patterns of production and distribution virtually reverted back to the former collective stage, with the communes providing only an overall administrative framework.

With the formation of communes in 1958, the momentum of the institutional transformation sparked off by the land reform had spent itself. The agricultural crisis following the Great Leap Forward brought to the fore the vulnerable features of a peasant agriculture which relied on social organization as the main source of agricultural growth. After 1962 the Government resolved: (1) to place high priority on agriculture ("take agriculture as the foundation, and industry as the leading sector"); and (2) to step up the technological transformation of agriculture. The new emphasis was put on modern dynamic inputs. In sharp contrast to the uprooting nature of the institutional transformation in the 1950s, the technological transformation in the 1960s--which has lasted well into the 1970s--was a gradual process more accurately described as a "silent revolution." China has finally started her "Green Revolution."[10] The progress of institutional revolution is summarized in Table 1.

CHINESE ROAD TO COMMUNIZED FARMING

It is beyond doubt that the Chinese record of institutional reorganization in agriculture was most impressive in terms of speed, momentum, and smoothness of progress. Agricultural collectivization, which met with so much disruption and difficulty in the Soviet Union under Stalin, was achieved in China only four years after the land reform. The Chinese success has therefore long intrigued both Chinese and Soviet scholars. Among the generally accepted explanations, one attributed the Chinese achievement to the rural origin of the Chinese communist elite, who had a thorough understanding of the agrarian problem through their long contact with the peasants in the process of struggle for power. Others placed emphasis on the Chinese tactics of implementing the changes by stages so as to overcome initial obstacles. In my view, the intermediate stage of cooperativization played a vital role in preparing the peasants for collective life on the one hand and in enriching the organizational experience of the rural leadership on the other. Specifically, the formation of MATs was a crucial link in the whole chain of institutional change.

TABLE 1. THE PROGRESS OF INSTITUTIONAL TRANSFORMATION IN CHINESE AGRICULTURE (EXPRESSED AS % OF TOTAL RURAL HOUSEHOLDS INVOLVED IN EACH PROCESS)

	Land Reform	Mutual Aid Teams			APCs (Cooperatives)	Advanced APCs (Collectives)	People's Communes
		Simple MATs	Permanent MATs	Total			
1950	32	-	-	10.7	-	-	-
1951	58	-	-	19.2	-	-	-
1952	88	29.8	10.1	39.9	0.1	-	-
1953	-	27.8	11.5	39.3	0.2	-	-
1954	-	32.2	26.2	58.4	1.9	-	-
1955	-	23.1	27.6	50.7	14.2	-	-
1956	-	-	-	-	8.5	87.8	-
1957	-	-	-	-	-	-	-
1958 July	-	-	-	-	-	-	-
August							30.4
Early September							48.1
Mid September							65.3
Late September							98.0
December							99.1

SOURCE: State Statistical Bureau, *Ten Great Years* (Peking: Foreign Languages Press, 1960), pp. 35 and 43.

I have stressed elsewhere the evolutionary nature of China's communization by pointing out its historical connection in that group farming was actually not alien to the Chinese peasants and that the traditional agricultural structure of China contained elements which actually fostered cooperativization. The widespread existence of mutual-aid cooperation in many parts of China long before the advent of communism--a kind of traditional MATs--is the case in point.[11]

Traditional Group Farming in Chinese Agriculture

It is a myth that small peasant farming invariably means highly independent operations based on the individualistic family farm. While it is true that resources may be owned on the basis of the family farm, for economic and technological reasons the actual operational unit for resource utilization is often much larger. Because of the unfavorable man-land ratio, the ownership size is usually small, giving rise to potential underemployment of labor and capital. If the peasants are to optimize they must increase the operational size beyond the ownership size through some form of cooperative arrangements. In this way, the peasants can bring the otherwise slack resources into more efficient use as well as yield higher economies of scale without violating the existing patterns of landownership. Such is the analytical basis for the numerous types of agricultural cooperative activities found in traditional China.

In manv parts of China the seasonal character of agricultural production enforced an extremely tight and rigid farming schedule, resulting in a fluctuating demand for labor. Instead of having disguised unemployment, many peasants actually faced labor shortages during the harvesting and plowing periods, particularly in regions under double or triple cropping.[12] For lack of labor-saving capital equipment, most peasants responded to seasonal labor demands by acquiring additional labor through exchange. Hence the fairly ubiquitous existence of traditional cooperative teams under various local names, the most popular of which was the *jen-kung pien-kung* (labor service exchange). They operated by staggering the dates of planting to avoid simultaneous harvesting among the members. Some units chose their members on a variety of considerations including complementary economic interests and equal social standing, while others were created informally on the basis of kinship and friendship ties. In North China, it was also common to find labor gangs called *cha-kung* (collective hiring of labor) and *ho-chung ti* (collective cultivating).

Successful operation of those cooperative teams rests on two basic economic factors. First, the teams can better utilize all the available resources during the peak periods than can individual peasants, because group farming makes it possible to operate a tighter schedule of activities by requiring members to contribute to greater efforts. Whether or not peasants actually work more intensively as a group under pressure depends on a number of noneconomic--social and psychological--factors. The Chinese evidence seems to be on the positive side. Second, the teams can raise the efficiency of members by division of labor, and collective activities can also release women from household work. Here the constraint comes from the limitation of exploiting the economies of scale without basic changes in the input structure of production.

Cooperative activities are also very much in evidence in other agrarian societies in Asia, even though their agronomic and economic conditions may not be so favorable for such a widespread existence of cooperative teams as in China.[13] On account of their subsistence character, most traditional cooperative activities have generally failed to produce a dynamic impact on farming in terms of any induced change of the production structure. Their economic value comes from the survival role they play rather than from a developmental role, in which they are very limited. Only in the case of China have the cooperative teams carried a historically significant function as the vehicle of China's subsequent rapid and smooth socialization of agriculture.

During the war with Japan, the Chinese Communist leaders had a chance to work with the peasants in the cooperative teams in their wartime bases in Northwest China. Because of that experience the Communist leaders not only grasped the economic importance of the indigenous cooperative teams in overcoming agricultural crises but also their potential as the spearhead of China's future collectivization drive. In fact, Mao Tse-tung envisaged this as early as 1943![14]

After 1950, the Chinese Government promoted the formation of MATs based on traditional cooperative teams in areas where land reform had already been completed. Since the land reform did not basically alter the microeconomic structure of agriculture, it had not disrupted the original agronomic and economic forces that had fostered the development of the indigenous cooperative activities. If anything, land reform had increased the need for agricultural cooperation as resources were redistributed in such a way that the new owners became more interdependent. The MATs were, of course, different from the traditional cooperative teams because the former were more institutionalized, better organized, and led by more motivated leadership.

Clearly, a strong element of collectivism had existed in the traditional background of Chinese agriculture.[15] Such existence points to the grass-roots origin of China's massive collectivization and also explains its smooth progress.

CHANGING COMMUNE ECONOMICS

This section will be devoted to an economic analysis of the internal operational mechanisms of the communes, focusing on how production and distribution are actually managed in the communes.

Organization of Production

Apart from facilitating central planning by providing a framework for integrating the heterogeneous agricultural sector into the national economic plan, communized farming is designed to fulfill two major objectives: (1) to increase output and productivity through large-scale production, which in turn permits greater rationalization of land use and specialization of labor; and (2) to increase the level of resource utilization through mobilizing underemployed rural labor for extra-agricultural activities. The first aims essentially at maximizing output at the micro-level and the second at maximizing employment with macroeconomic implications. The two may not always work together, but may well conflict with each other.

A strong economic case could be made for the formation of MATs because, as we have seen earlier, it was the team and not the family unit that approximated the level of operation and decision-making in traditional agriculture. The same holds for the APCs, because the structure of production and the size of operation in an APC were not much different from those in an MAT. In fact, the lower APCs actually embodied many of the economic and social advantages long advocated by the protagonists of agricultural cooperation for the developing countries.[16]

Since economic benefits from the cooperatives are the net difference between the advantages of large-scale farming and the diseconomies of resource mismanagement arising from overexpansion in the size of production units, the APCs appeared to have reached the limit of an optimum size for maximizing output. Any higher form of organization would, therefore, only violate the microeconomic advantages of the cooperatives.

If so, the decision to collectivize could then be rationalized only on macroeconomic grounds. This was indeed the case. Increasingly the Government came to view the collectives as units capable of providing a larger base for more diversified agricultural activities as well as for utilizing the underemployed rural labor for the labor-intensive public work program. The early communes carried the official obsession with large agricultural planning units to an irrational degree, sharpening the inherent conflict between output maximization and employment maximization. A huge, centralized commune organization might well facilitate the mobilization of labor for turning labor into capital, but it was clearly unfit for peasant farming with rigid production sequences, which demands effective decision-making and personal initiative from farm operators at the lower level. Worse still, there was no synchronization between agricultural production and extra-agricultural activities in the early communes. A corvée type of labor was made to undertake capital construction works on a grand scale to the neglect of agricultural production. Even the management of labor mobilization at that time was poor. In short, one finds in the first wave of the communes many characteristics close to those of the so-called "command farming" model constructed by the Sovietologists.[17]

For a time, the initial organizational blunders of the early communes created doubts among many foreign observers as to whether such a radical social experiment could in fact be viable. However, the Chinese Government was prompt to check the initial excesses of the early communes through a series of decentralization measures, which eventually hammered the whole commune system into a realistic and flexible form. While the communes could have been a contributing cause to China's agricultural calamities in 1960, they had in fact provided an efficient organizational framework for the country to weather the ensuing agricultural crisis--the worst in decades--with minimum social and economic disruptions. Ironically, the communes first proved their worth by their ability to manage crises. Ever since, the "self-survival" function of the communes has been stressed and is now formulated into a policy of "self-reliance" for every commune.

Within the commune organization, the three-tier structural reorganization led to the evolution of optimal units for various specialized functions. While the commune as a central organization continued to handle the administrative responsibilities and policy-making concerning political, social, and cultural matters, it only sought to provide a broad framework to coordinate agricultural production, the direct control of which was now decentralized down to the lower units--the brigades and teams. The day-to-day farming operations were in the hands of the teams, which also served as accounting units for distribution. The size of teams varied between communes, and within each commune it could also vary according to economic and geographical conditions. Above all, peasants were no longer deployed for extra-agricultural activities at the expense of direct agricultural work, thus minimizing the potential conflict between agricultural production and agricultural capital formation (turning labor into capital).

Over the years, the communes have also managed to smooth out the frictions between collectivized interests and private interests, conflicts that party ideologues at first tended to play down but which have continued to pose a problem in almost all the socialist countries. Communization does not mean complete socialized ownership of all resources belonging to the peasants. Some parts of resources and activities--private plots, labor time after collective labor, and all those nonfarm activities managed by the peasant himself at his free time--constitute the private sector. The threat to the private sector became real the moment the peasants participated in any form of group farming, be it an MAT or APC. For a long time, the Chinese authorities assumed an ambivalent attitude toward the existence of the private sector as reflected in their fluctuating policies, which attacked it during a radical movement and restored it at the time of political relaxation.[18]

In recent years, the private sector has become less problematic to the Government. As the socialized sector has grown stronger and more diversified due to the technological transformation, the private sector has become correspondingly weaker.[19]

While communized farming is taking deeper roots in all aspects of rural life, the peasants have become used to the *importance* of the socialized sector from which, and *not* from the private sector, they derive their subsistence. The incessant political campaigns and the colossal government effort to educate the peasants to sacrifice self-interest for the public good must have also done their work in eroding away such petty selfishness as is so often found among the small peasants. The private sector has thus been assimilated into the communized system, within which the socialized and private sectors have finally worked out their mutually beneficial complementary existence.[20]

Organization for Distribution

Distribution is a fairly simple process in peasant agriculture. In socialized agriculture, however, it is both important and complicated, involving two basic decisions: (1) what is the proportion of the total product to be distributed among the members? and (2) how is the distributable portion actually distributed? A commune is not a state farm but is owned by its members. From the total proceeds, deductions have to be made for government levies, welfare, capital accumulation, etc., and the total size of all these obligations determines the distributable proportion.

The actual distributable proportion has changed over the years, reflecting changes in the government's rural policy. For instance, the average government levy increased from 5.4 percent for the cooperatives to 10.0 percent for the collectives, while accumulation funds increased correspondingly from 5.4 percent to 9.2 percent, resulting in the reduction of the distributed proportion from 87.8 percent to 79.2 percent.[21] This merely supports the earlier contention that one of the motivations for the Government to collectivize agriculture was to increase the transfer of surplus.

Some of the early communes maintained an extremely high proportion of their incomes for accumulation and reserves, thereby further reducing the distributable proportion to 50-60 percent of the total output. As the radical commune movement was in retreat, the Government urged the communes to be more realistic in putting aside the share for reserves (now standing at 5-7 percent), while at the same time the Government tax was fixed at a rather low level (now averaging about 8 percent). With the nondistributable proportions having thus been stabilized, the actual amount available for distribution would increase with increasing output.

Once the proportion of the product is earmarked for distribution among members, some means have to be found to divide it up in a manner consistent with the basic policy requirements. Group farming requires a group decision to apportion the product which derives from collective effort. A formal system of distribution came into being with the introduction of the APCs (cooperatives). The general distributive device was known as the *kung fen chi* (the work-point system), with local variations in application.[22] It was essentially a kind of piece-rate system. This system was later refined by incorporating a norm system in order to differentiate the quality of labor. Such a work-point cum-norm system was also adopted by all the collectives and has since become a standard distributive mechanism in China.

When the communes were first formed, an attempt was made to do away with the distributive system by introducing free supplies based on the communistic principle of "to each according to his need," and then a "part wage and part supply" system. But the experiment was quickly abandoned and the former distributive device restored. In the post-Great Leap years, the general distributive device was strengthened through the enforcement of stricter financial controls.[23] Today, China appears to have finally developed an effective distributive mechanism which strikes a balance between the ideological requirement for greater egalitarianism, and the economic reality requiring some amount of differentiation in rewards for the sake of maintaining work incentives.

For socialized agriculture, organizing a workable remuneration system can pose greater problems than organizing collective production, because satisfactory distribution rests on proper job evaluation and strict financial and accounting management. But the Chinese experience has shown that the problems are not insurmountable. First, the present distribution system was not imposed overnight from above but has evolved over the years, providing members with a sufficient learning period. As soon as a departure from individual farming was made--back in the early 1950s--the peasants began to learn concepts like labor days and adapt themselves accordingly. Few communes today still lack experienced cadres versed in the necessary accounting and bookkeeping practices after well over two decades of group farming. Second, the Government's pragmatic policy in allowing the communes to retain some incentive scheme within their distributive mechanism and in checking any unrestrained form of complete egalitarianism has greatly facilitated the smooth operation of the present distributive device. Third, constant indoctrination and political exhortation have prepared the peasants to accept, within some reasonable limits, "moral incentives," the presence of which has undoubtedly greatly simplified the knotty problems in the remuneration system.[24] There is no need for skeptics who are unfamiliar with the inner mechanisms of the Maoist economic ideology to dismiss outright the ethical elements of economic development and the working of nonpecuniary incentives. The Chinese experiment, in addition to Israel's kibbutz and moshav, should serve as a reminder to the conventional economists that there is a possibility of their misconceiving the structure of material incentive and of overemphasizing its role in the economic mechanism.[25]

While glaring income disparities are a distressing scene in the rural areas common to many developing countries today, their absence in China should show how the communes' distribution works as a whole. The Chinese authorities thus far have only tackled the microeconomic aspect of income distribution within the communes. But the inter-commune differentials still stand, and to even them out would require much greater macroeconomic effort and regional economic planning.

Potential Problems

The Soviet and Eastern European experience has clearly shown that agriculture does not take kindly to centralized planning. Their socialized agriculture compares quite miserably with the western capitalist agriculture in terms of efficiency in resource utilization and in the level of labor productivity. The communized agriculture of China at the present stage is, of course, structurally quite dissimilar to the Soviet collectivized agriculture or its East European counterpart, not to mention the difference between China and other socialist countries in their overall government policies toward agriculture. So far, the Chinese have been strikingly successful in mobilizing their organizational resources to overcome many structural rigidities (e.g., bureaucratic problems) inherent in their super-collectives--the communes--that would otherwise have been inimical to agricultural progress. Will the many problems that have long plagued agriculture in other socialist countries be brought to the fore in China, once her communized peasant agriculture grows beyond the "peasant subsistence" stage?

In China today, technological progress is the mainspring of agricultural growth. Within the general sweep of technological progress, some economic problems are easily glossed over. Also, success in agriculture tends to reinforce itself--higher productivity results in higher income and then higher incentive and so on--so that many basic problems arising from organizational failure may not immediately surface. However, as the communes continue to grow and become more diversifeid in their economic undertakings, not only will the managerial complexities multiply and the problems of organizing production and remuneration become more complicated, but many microeconomic issues regarding resource allocation will arise. Clearly a well developed commune has to utilize a more precise economic calculus with respect to the allocation of labor and capital, as the opportunity costs of their misuse will rise sharply due to the increase in per capita income and in capital intensity. Likewise, as the communes become more affluent, many more problems of choice will loom larger. With the range of economic life in the communes growing more sophisticated, the commune organization simply has to adapt and change, perhaps in the direction of further decentralization.[26]

GROWTH PERFORMANCE AND PATTERNS

China's agricultural performance over the last two and a half decades will be analyzed here on the basis of her grain production. National aggregate grain output figures are among the few statistics that are available outside China for almost the entire period. There are three main objectives for this exercise: (1) to evaluate the short-term impact of institutional changes, from land reform to communization, on agricultural production; (2) to analyze the pattern of structural change in production; and (3) to examine China's long-term agricultural performance in relation to other Asian countries.

In Table 2, I have constructed a grain output series for 1949-1974 based on official and "semi-official" figures which appear, on the whole, reliable and consistent, with the exception of the Great Leap Forward years 1958 and 1959.[27] Most agricultural production series should display a relatively smooth, slow-rising trend reflecting the small incremental nature of agricultural production and some amount of random fluctuations due to such factors as climatic influence. The Chinese grain output series, however, is most unusual in that it displays violent fluctuations for the given time span. As shown in Figure 1, fluctuations are found in the beginning period as well as in the middle section. The amplitude of the fluctuations is so great as to practically render the whole series unsuitable for the fitting of any single trend. The dip in the early years represents the recovery period, while the middle section of the series clearly reflects the disturbance created by the Great Leap Forward. To analyze the size and patterns of fluctuations caused by the institutional transformation, we need to establish, first of all, a hypothetical trend showing the average movement of the "normal years" to be the base line. Hence a "potential trend" for 1952-1974, excluding the intermediate period 1958-1963, is fitted, which yields a very good fit: $Y = 156.0 + 4.48\ X$; $r = 0.98$. Since the years in the two periods 1949-51 and 1958-63 are obviously the "statistical outliers," their removal from the regression would certainly produce a trend which can best describe the movement of the series in a stable manner. The existence of our "potential trend" 1952-74 (which actually comprises two subperiods: 1952-57 and 1964-74), can in fact be supported by a more sophisticated statistical analysis. Performing Chow's Test, which yields the F-ratio of 4.7, we can establish that our "potential trend" itself is statistically consistent and that the two disjointed parts of the series actually do share a common trend.[28]

Next, we extrapolate the "potential trend" to the early period (1949-52) and interpolate the intermediate period (1958-63), and then work out the average deviations from the trend. To analyze the pattern of fluctuations, we divide up the whole series into five periods. We can then draw the following conclusions: (1) land reform had not disrupted production but, if anything, actually facilitated recovery; (2) nor had cooperativization and collectivization disrupted agricultural production; (3) the early communes together with the Great Leap Forward had thrown the growth trend out of gear, first by shooting it up very high and then precipitating an abrupt downswing; (4) production throughout the 1960s was once again back to the stable path (see Tables 3 and 4).

Is the price for the commune movement, as expressed in the 18.5 percent downward swing away from the trend, too high? Officially, the downturn was explained away as solely the consequence

TABLE 2. GRAIN PRODUCTION IN CHINA, 1949-1974
(Million Tons)

Year	Output	Indices				
1949	108	100	70	54	72	82[a]
1950	125	116	81	63	83	95
1951	135	125	88	68	90	103
1952	154	143	100	77	103	118
1953	157	146	102	79	105	120
1954	160	149	104	80	107	122
1955	175	163	114	88	117	134
1956	183	170	119	92	122	140
1957	185	172	120	93	123	141
1958	250	230	162	125	167	191
1959	270	251	175	135	180	206
1960	150	139	97	75	100	115
1961	162	151	105	81	108	124
1962	174	162	113	87	116	133
1963	184	171	119	92	123	140
1964	200	186	130	100	133	153
1965	200	186	130	100	133	153
1966	220	205	143	110	147	168
1967	230	214	149	115	153	176
1968	-	-	-	-	-	-
1969	-	-	-	-	-	-
1970	240	223	156	120	160	183
1971	246	229	160	123	164	188
1972	240	223	156	120	160	183
1973	250	232	162	125	167	191
1974	258	240	168	129	172	197

[a]Average of 1949-52 = 100 (which yields 131 million tons, and the pre-war peak was 139 million tons).

SOURCES: For 1949-58: direct official sources, as contained in State Statistical Bureau, *Ten Great Years* (Peking: Foreign Languages Press, 1960), p. 119. For 1959: Li Fu-ch'un's report in *Peking Review*, 5 April 1960, p. 6. For 1960-65: the semi-official data in Robert Michael Field, "How Much Grain Does Communist China Produce?" *China Quarterly*, no. 35 (January-March 1968), pp. 98-110. For 1966: Han Suyin, *China in the Year 2001* (New York: Basic Books, 1967), p. 54. For 1967: Anna Louise Strong, *Letter from China*, no. 55, 15 January 1968. For 1970-71: *Far Eastern Economic Review*, 8 January 1972, p. 27. For 1972: *Peking Review*, no. 1, 5 January 1973, p. 12. For 1973: ibid., 4 January 1974, p. 8. For 1974: Chou En-lai's report to the 4th National People's Congress in January 1975, as reported in *Straits Times* (Singapore), 21 January 1975.

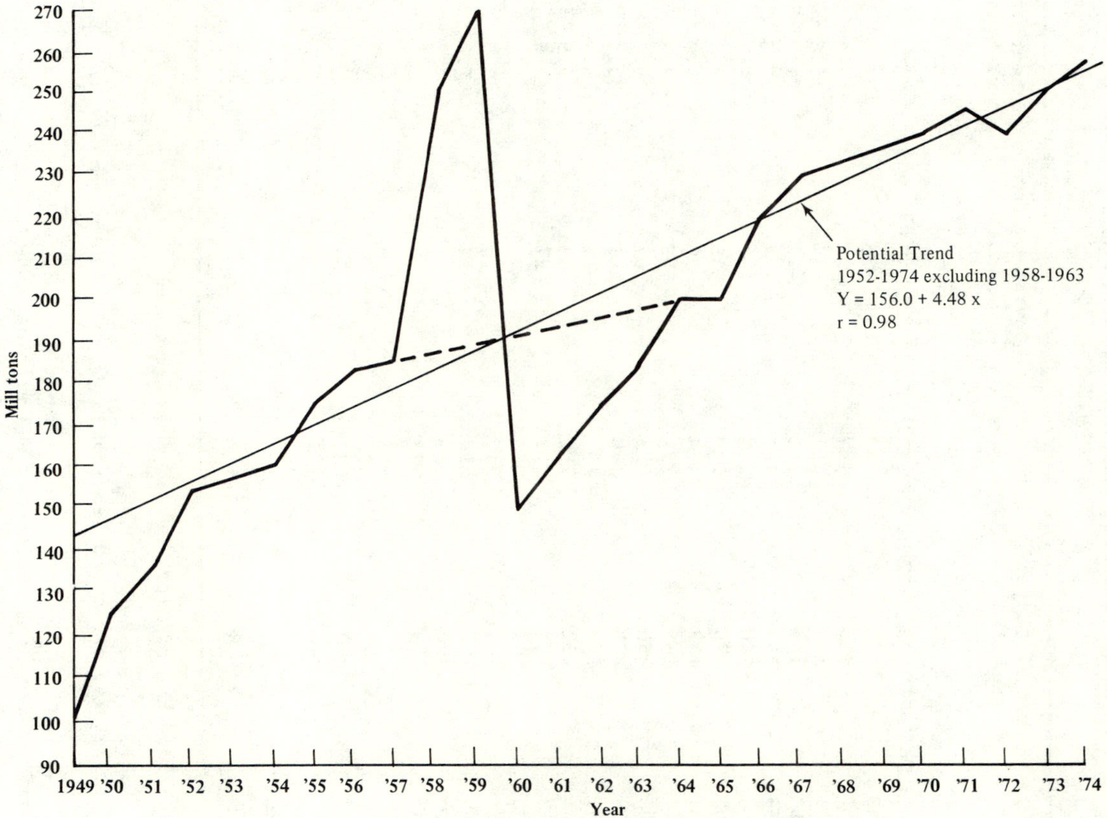

Figure 1. Grain Output Trend in China, 1949-1974

TABLE 3. ESTIMATION OF CHINA'S GRAIN OUTPUT TREND BY REGRESSION, 1949-1974

Year	Actual Output Reported Y	Estimated Output $\hat{Y}$	Deviation from Actual $Y - \hat{Y}$	Extent of Deviation $\frac{Y - \hat{Y}}{Y}$ %
1949	108	143	-35	32.4
1950	125	147	-22	17.6
1951	135	151	-16	11.8
1952	154	156	- 2	1.3
1953	157	161	- 4	2.5
1954	160	165	- 5	3.1
1955	175	170	5	2.8
1956	183	174	9	4.9
1957	185	179	6	3.2
1958[a]	250	183	67	26.8
1959[a]	270	188	82	30.4
1960[a]	150	192	-42	28.0
1961[a]	162	196	-34	20.9
1962[a]	174	201	-27	15.5
1963[a]	184	205	-21	11.4
1964	200	210	-10	5.0
1965	200	214	-14	7.0
1966	220	219	1	4.5
1967	230	223	7	3.0
1968	-	229	0	0
1969	-	232	0	0
1970	240	237	3	1.3
1971	246	241	5	2.0
1972	240	246	- 6	2.5
1973	250	250	0	0
1974	258	255	3	1.2

[a]1958-1963 are not included in the regression.

$Y = 156.0 + 4.48\ X\ (r = 0.98)$
1952-1974, excluding 1958-1963

of bad weather; but admittedly the mismanagement of early communes must have also been a major contributing factor to the agricultural crisis. On the other hand, such a judgment should be balanced against the positive role played by the communes in bringing about the structural change in agricultural production. In the long run, the commune movement acted as a catalyst for the technological transformation of Chinese agriculture. First, as we have seen, the setback of the early communes made the Chinese authorities realize

TABLE 4. PATTERNS OF AVERAGE DEVIATION FROM THE POTENTIAL TREND

	Percentage
1949-1952 (Recovery)	15.8
1953-1957 (Institutional Transformation)	3.3
1958-1959 (Great Leap Forward and Initial Commune Movement)	28.6
1960-1963 (Post-Leap Aftermath)	18.5
1964-1974 (Technological Transformation)	3.3

Computed from the last column of Table 3.

that they had reached the apex in their exploitation of the institutional forces for agricultural growth. Second, the whole Great Leap episode led to the evolution of a new generation of more manageable and practicable communes on the one hand, and the emergence of a pro-agricultural development policy on the other, eventually combining to give birth to China's equivalent of the green revolution.[29] Third, once the new strategy of relying on technological progress got underway, the communes furnished a viable organizational framework to step it up and consolidate it. China gives an example of employing institutional revolution to lead the technological transformation in agriculture, and such characteristics are reflected in the N-shape curve of the original time series. (See Figure 2.)

How successful has China's long-term strategy of agricultural development been in terms of growth? Growth series with different base levels have been worked out in Table 2. Compared with many other developing countries in Asia, China's stand in agricultural growth is actually quite creditable, as can be seen in Table 5. In Table 6, linear growth trends for selected Asian countries based on the production series in Table 5 have been worked out. The results in Table 6, with a higher b-value indicating a steeper growth rate, *ceteris paribus* tend to support the conclusion in Table 5. As a mature agrarian economy with a traditionally efficient agriculture and a relatively high level of productivity to start with, China is really not doing badly in terms of growth performance for food grain production. There are, however, signs that a slower growth trend has set in during the recent years, implying that it could be harder for China in the future to push ahead further productivity

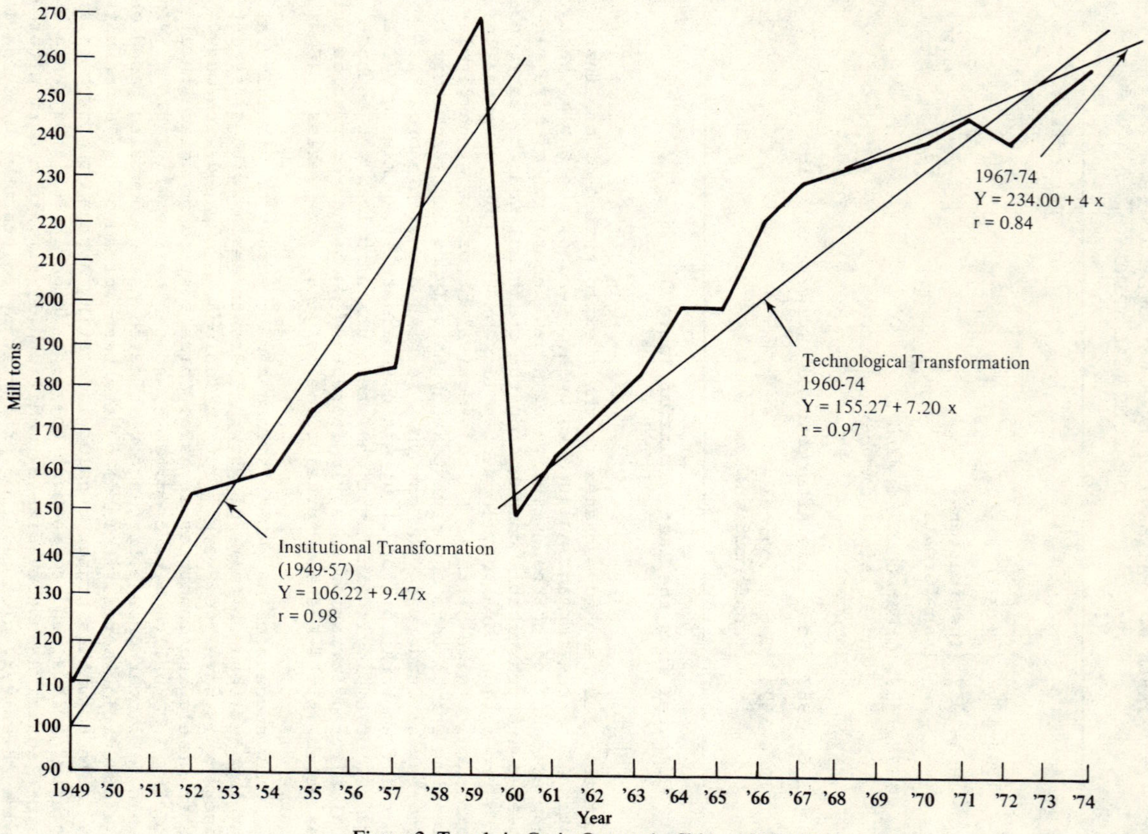

Figure 2. Trends in Grain Output in China, 1949-1974

TABLE 5. INDEX NUMBERS OF FOOD PRODUCTION IN SELECTED ASIAN COUNTRIES
(1961-1965 = 100)

Year	China	India	Bangla-desh	Pakistan	Sri Lanka	Burma	Indonesia	Philippines	Thailand	South Korea	Japan	Taiwan
1961	88	100	95	91	93	91	96	95	90	94	94	93
1962	95	99	91	96	98	99	102	98	98	86	100	93
1963	100	101	104	101	102	99	96	99	104	96	99	95
1964	109	104	104	103	107	107	104	102	103	112	104	105
1965	109	96	106	109	101	103	103	105	105	112	103	114
1966	120	97	102	109	102	91	107	112	122	121	107	118
1967	125	105	118	117	107	103	102	115	106	113	118	124
1968	124	111	117	130	114	106	111	114	120	114	124	128
1969	126	116	122	131	112	106	112	121	131	132	121	126
1970	130	123	118	145	118	110	127	127	133	128	119	131
1971	134	123	109	144	115	110	131	128	140	129	113	137
1972	130	118	105	147	114	106	129	132	127	128	117	141
1970-1972	131	121	111	145	116	109	129	129	133	129	116	136

SOURCES: For China: computed from figures in Table 2 infra. For Taiwan, 1961-1971: *FAO Production Yearbook*, vol. 25 (Rome, 1971); for 1972: Economic Planning Council, *Industry of Free China* (Taipei, 1972), p. 58. For all other countries: *FAO Production Yearbook*, vol. 26 (Rome, 1972), p. 38.

TABLE 6. TREND VALUES OF GRAIN OUTPUT IN SELECTED ASIAN COUNTRIES, 1961-1972
(y = a + bX)

	a	b[a]	Index	r[b]	Index
China	89.52	4.05	100	0.96	100
India	92.27	2.38	59	0.86	89
Bangladesh	96.02	1.78	44	0.67	70
Pakistan	82.97	5.48	135	0.98	102
Sri Lanka	94.08	1.98	49	0.91	95
Burma	94.47	1.25	31	0.70	73
Indonesia	89.55	3.15	78	0.91	95
Philippines	89.65	3.45	85	0.99	103
Thailand	88.62	4.05	100	0.92	96
South Korea	89.14	3.79	93	0.90	94
Japan	94.71	2.34	58	0.85	89
Taiwan	86.88	4.65	115	0.98	102

[a]High b-value indicates a steeper slope for the trend.
[b]Coefficient of correlation.

increases at the same rate as before, as reflected in the third trend line (1967-74) with a more gentle slope (Figure 2).

SOME IMPLICATIONS OF CHINA'S AGRICULTURAL DEVELOPMENT EXPERIENCE

In conclusion, rather than summarizing China's past and current organizational strategies for agricultural development, we would attempt to assess the relevance of the Chinese experience for Third World countries, bearing in mind that factors and conditions that succeeded in one country could be hard to reproduce in another. Can other developing countries learn anything from the Chinese experience if the Chinese model is stripped of its intrinsic social and ideological assumptions?

There seem to be too many unique factors in the structure of Chinese development experience that will pose a transferability problem. To begin with, even at the peak of the Cold War when the Chinese Government was zealous about the "export potential" of its revolutionary model to the Third World, it held no such vision for its agricultural development model. Perhaps the Chinese knew only too well that their communes and other institutional innovations had evolved through experience and practice based on China's special agronomical environment and socio-economic structure. Ironically, the type of agriculture which bears the greatest similarity to Chinese agriculture is to be found in Korea, Taiwan, or Japan, all in East Asia; but these are precisely the countries with successful agricultural development under a different system. Nor should the historical condition be ignored. The Chinese success in the institutional revolution owed a great deal to the effective implementation of a thorough-going land reform at the beginning, which in turn owed its success to a revolutionary government with all the political will to institute radical changes. It is actually rare for a government in a developing country to be able to implement thorough institutional changes today under stable conditions: President Marcos of The Philippines had to declare martial law in order to re-introduce land reform into his country.

What then are the transferable elements in the Chinese experience? To the extent that the major innovative features in the Chinese organizational strategy are compressed in the communes, must the communes be reproduced *in toto* before the full attributes of the Chinese model can be made to work? Can the Chinese-style commune organization actually be transplanted into a different social and ideological setting? Once again, the whole issue boils down to whether a developing country can actually emulate the Chinese organizational strategy without first undergoing a thorough social and political revolution beforehand.

Without falling into the fallacy of eclecticism, one can nonetheless generalize that the Chinese experience not only holds valuable lessons for many Third World countries but also serves to highlight the problematic areas in their own agricultural development policies. The main thrust of the bold commune experiment in China has forcibly demonstrated that organized farming in the context of peasant agriculture can be a viable alternative to the time-honored system of individualistic small peasant farming, provided that group activities are properly organized with sufficient flexibility and pragmatism--this organizational hurdle in itself being likely to be too formidable for many developing countries.[30] Group farming is also compatible with technological transformation of peasant agriculture: it can just as well initiate, accelerate, and sustain the so-called green revolution as it has done in China. In fact, the present green revolution could well be stepped up or consolidated in some developing countries if their small, poor peasants could organize themselves into some form of cooperatives which could help spread out the risks arising from initial technological changes as well as increase the market power of the peasants in their purchase of modern inputs.[31] Furthermore, group farming structures can act as a bridge for the poorer section of the peasantry to be integrated into the modern segment of the village economy.

With regard to the macroeconomic concern over reducing the extent of rural unemployment (be it Ricardian unemployment or disguised unemployment) and mobilizing the underemployed labor for extra-agricultural activities, it seems beyond doubt that group farming provides a more effective organizational basis than small peasant farming. Much of the current setback of the green revolution can be traced to the failure in water control and irrigation improvement. As most governments in the Third World cannot undertake extensive rural public works on account of financial constraints, peasants would have to organize themselves on a cooperative basis to bring about the vital infrastructural development in their villages.[32] In the event of an outbreak of calamities, peasants as an organized group can better manage crises by pooling their resources rather than operating as unorganized individuals. A viable cooperative structure can serve as an ad hoc body for organizing relief efforts and, above all, getting production started in the immediate aftermath by preventing the peasants from being dispossessed of their land and drifting to the cities.

The Chinese model has also served to show that organizational strategy alone is not sufficient to achieve the development goal. It has to be bolstered by appropriate government policies which are not urban-biased in nature as well as by a government leadership not alienated from the peasantry. Over the long run, the organizational strategy has to be sustained by real progress in technological transformation. Communes can furnish the structural framework for the mobilization of human and material resources but they are not meant to be a substitute for agro-techniques or agro-chemicals.

China's agricultural development is not a sparkling example of success if one measures it by the conventional economic yardsticks or if one is used to the kind of spectacular performance that is seen, for example, in Japan's postwar economic growth. In terms of production increases, what China has so far done is rather modest--merely doing a little better than holding her own--and the Chinese Government has claimed just that.[33] It may, of course, be argued that what is considered to be a modest achievement is in fact a considerable step forward by a large, overpopulated country toward solving the so-called Malthusian dilemma. Where the Chinese performance stands out is in comparison to what other Third World countries have achieved, countries not few in number today which cannot even hold their own ground.

NOTES

1. Most economists in the Classical period viewed the peasants as indolent and economically inefficient, as opposed to the progressive and economizing urban capitalists. Marx showed similar scorn for the traditional peasants and treated peasant agriculture with "the townsman's contempt for all things rural and the economist's disapproval of small scale production." D. Mitrany, *Marx Against the Peasant* (Chapel Hill: University of North Carolina Press, 1951), p. 6. See my "Socialization of Agriculture and Peasant Response to Economic Incentives in China," *Review of Southeast Asian Studies* 2 (December 1972): 14-26.

2. At the time of its inception, communization was widely misinterpreted in the West as almost a process of mass enslavement of the peasants. The misinterpretation was perhaps typically characterized by the observations from one of the leading China experts, Professor A. Doak Barnett, who wrote:

> The decision to embark upon the communization program is perhaps the Chinese Communists' biggest gamble to date. In treating the Chinese

> people callously, impersonally, and ruthlessly, as raw material to be organized and manipulated by the state for its own purposes, they may be going too far, even for a totalitarian regime . . . . It is not possible at present to predict the long-term reactions of the Chinese people to the communes. Nor can we know how the leaders of Communist China may react to pressures from the people. . . .

Communist China and Asia: A Challenge to American Policy (New York: Vintage Books, 1960), p. 25.

3. Anthony M. Tang has analyzed the Stalinist development strategy as applied to China. "Agriculture in the Industrialization of Communist China and the Soviet Union," *Journal of Farm Economics* 40 (December 1967).

According to some Western critics, there is an inherent contradiction in an agricultural development policy so heavily dependent on institutional reorganization, which, while facilitating the fulfillment of the "extractive function" required of the Stalinist model by maximizing "surplus," may come into conflict with the "developmental function" in the sense of maximizing "output." In actual fact, whether or not such a conflict would arise depends on the ideological premises of the argument. According to the Chinese Government, if the production relations and the production forces were made compatible, the developmental and extractive objectives would not be conflicting. For further discussion of this topic, see K. K. Fung, "Output vs. 'Surplus' Maximization: The Conflicts Between the Socialized and the Private Sector in Chinese Collectivized Agriculture," *The Developing Economies* 12 (March 1974): 41-55.

4. For an official statistical analysis of the Plan, see State Statistical Bureau, *Ten Great Years* (Peking: Foreign Languages Press, 1959), p. 59.

5. John Macrae has estimated that during the Plan period the Government had extracted between 30 and 40 percent of the market surplus from agriculture for development in other sectors, as compared with 5 percent for India for the same period. "Mobilization of the Agricultural Surplus in China for Rapid Economic Development, 1952-57," *The Developing Economies* 8 (March 1970). See also the excellent study by Shigeru Ishikawa, "Resource Flow between Agriculture and Industry: The Chinese Experience," ibid., 5 (March 1967).

6. For a more detailed discussion of the land reform policy and implementation, see John Wong, *Land Reform in the People's Republic of China: Institutional Transformation in Agriculture* (New York: Praeger Special Studies, 1973).

7. See "Outline of the Struggle Between the Two Lines from the Eve of the People's Republic of China through the 11th plenum of the 8th CCP Central Committee," *Current Background*, no. 884, 18 July 1969, American Consulate General, Hong Kong. The debate was reported to have continued and become intensified after the formation of the communes in 1958. See "The Conflict Between Mao Tse-tung and Liu Shao-ch'i over Agricultural Mechanization in Communist China," *Current Scene* (Hong Kong) 6, no. 17, 1 October 1968.

8. Mao Tse-tung, "On the Question of Agricultural Cooperation" (July 31, 1955), *Selected Readings from the Works of Mao Tse-tung* (Peking: Foreign Languages Press, 1967), pp. 316-39.

9. For the official versions of the collectivization movement, see Tung Ta-lin, *Agricultural Co-operation in China* (Peking: Foreign Languages Press, 1958), and *Socialist Upsurge in Chinese Countryside*, Preface by Mao Tse-tung (Peking: Foreign Languages Press, 1957). For a western analysis, see Kenneth Walker, "Collectivization in Retrospect," *China Quarterly* 26 (April/June 1966), pp. 1-43.

10. The 10th Plenary Session of the 8th Central Committee of the Chinese Communist party has been identified as a crucial watershed in the agricultural history of China, the point at which the Chinese Government made a firm decision to change the basic agricultural development strategy. It gave birth to the Green Revolution in China. See the excellent study by Benedict Stavis, *Making Green Revolution: The Politics of Agricultural Development in China*, Cornell University, Rural Development Committee Monograph Series no. 1 (1974).

11. See John Wong, "Peasant Economic Behaviour: The Case of Traditional Agricultural Co-operation," *The Developing Economies* 9 (September 1971).

12. See Yong Sam Cho, *Disguised Unemployment in Underdeveloped Areas, with Special Reference to South Korean Agriculture* (Berkeley, Cal.: University of California Press, 1963). For labor shortage in pre-war China, see J. L. Buck, *Land Utilization in China* (Shanghai, 1937), p. 301.

13. For example, the *oyakatanago* cooperation in traditional Japan was very much akin to the Chinese *pien-kung*. (Thomas C. Smith, *The Agrarian Origins of Modern Japan* [Stanford, Cal.: Stanford University Press, 1959], chapter 3.) In the rice-growing regions of Burma, there were cooperative bodies similar to the Chinese *cha-kung* (David E. Pfanner, "A Semi-Subsistence Village Economy in Lower Burma," in Clifton R. Wharton, Jr., ed. *Subsistence Agriculture and Economic Development* [Chicago: Aldine, 1969]). Elementary cooperatives also thrived in India, e.g., the *virhi*, the *mang*, and the *sanjn* in rural Punjab. (S. K. Goya, *Some Aspects of Cooperative Farming in India* [Asia Publishing House, 1966].)

14. Mao stated in his address, "Get Organized," in November 1943:

> Among the peasant masses a system of individual economy had prevailed for thousands of years, with each family or household forming a productive unit. This scattered, individual form of production is the economic foundation of feudal rule and keeps the peasants in perpetual poverty. The only way to change it is gradual collectivization, and the only way to bring about collectivization, according to Lenin, is through cooperatives. We have already organized many peasant cooperatives in the border region, but at present they are only of a rudimentary type and must go through several stages of development before they can become cooperatives of the Soviet type known as collective farms.

Selected Works of Mao Tse-tung (Peking: Foreign Languages Press), 3: 156.

15. Joseph Needham finds in China "a very old tradition not only of great public works, but also of nationalized production." Similarly for agricultural collectivism. "Another aspect of Chinese life which should not be underestimated is the great extent to which mutual aid took place among the peasant farmers. Throughout the ages there was cooperation at the village level, sometimes more, no doubt, and sometimes less. Mutual aid teams were not something absolutely new and unheard of when they were encouraged at the beginning of the present government." "The Past in China's Present," *Pacific Viewpoint* 4 (September 1968), p. 133.

16. It is to be remembered that the APCs actually earned considerable admiration from the visiting Indian Delegation of Agricultural Cooperation to China in 1956. See Government of India, *The Report of the Indian Delegation to China on Agrarian Cooperatives* (New Delhi, 1957).

17. See, e.g., Jerzy F. Karcz, ed., *Soviet and East European Agriculture* (Berkeley, Cal.: University of California Press, 1967).

18. Kenneth Walker has focused attention on the economic importance of the private sector in terms of food and cash to the peasants. He estimates that in 1956 the private plots, which largely ranged from 2.5 percent to 3 percent of the cultivated land, supplied 14 percent of the calorie-intake to the peasants, while the private sector as a whole provided from 14 percent to 34 percent of the peasants' total income. *Planning in Chinese Agriculture: Socialization and the Private Sector* (London: Frank Cass, 1965), p. 32 and p. 34.

19. While the private sector has certainly exerted great psychological effect on the peasants, especially in the initial period of collectivization, its overall economic significance should not be over-stressed: at all times the basic means of subsistence of the members in the collectives or communes, and the bulk of the marketable surplus were coming from the collective sector. Even Kenneth Walker has conceded that the increasing supply of chemical fertilizers, while raising the output of the public sector, will also reduce the Government's need for the dependence on the private sector for the supply of animal fertilizer so that the significance of the private sector will begin to disappear. Ibid., p. 98.

20. The new Constitution, adopted by the Fourth National People's Congress in January 1975, stipulates that the peasants "may farm small plots for their personal needs." This should not be construed as the revival of the economic importance of the private sector; rather, it implies that the private sector no longer constitutes a threat to the ever-growing, wide-ranging socialized sector.

21. This is based on the analysis of the 1955 official survey of national incomes and expenditures of 26,935 cooperatives and collectives. See my "Economic Organization of the Collectives in China," in *Selected Seminar Papers on Contemporary China*, edited by Steve S. K. Chin and Frank H. H. King (Center of Asian Studies, University of Hong Kong, 1971), pp. 113-118.

22. Ten work-points form one labor day, and payment to members is based on the number of labor days each has earned. In its simplest form, each member is assigned a definite number of work-points based on his known skill and capacity for work. The quantity is then subject to revision after a discussion of the work the member has actually done for the day. For greater details, see *Model Regulations for an Agricultural Producers' Co-operative* (Peking: Foreign Languages Press, 1956), pp. 33-34.

23. For further discussion of the distribution problem, see Charles Hoffmann, *Work Incentive Practices and Policies in the People's Republic of China, 1953-1965* (Albany, N.Y.: State Univ. of New York Press, 1967).

24. For further discussion of the "moral" or ideological elements of the Chinese economy, see E. L. Wheelwright and Bruce McFarlane, *The Chinese Road to Socialism* (England: Pelican, 1970), Chapter 8; and John G. Gurley, "Capitalist and Maoist Economic Development," in Edward Friedman and Mark Seldon, eds., *America's Asia: Dissenting Essays on Asian American Relations* (New York: Vintage Books, 1971). These authors understand enough orthodox economics to see Maoist economics in a western setting.

25. Hence writes an Israel expert on kibbutz and moshav: "Modern economic theory occasionally tends to overstress the importance of the profit motive in farming. It can be shown easily by the example of Kibbutz and Moshav in Israel that there are other no less decisive factors for making a choice." Yehuda Lowe, "Kibbutz and Moshav in Israel: An Economic Study," in Rodger N. Dixey, ed., *International Explorations of Agricultural Economics* (Ames, Iowa: Iowa State University Press, 1964), p. 127.

Professor James Tobin of Yale, as one of the group of American academics visiting China in 1972, writes: "China really is at the beginning of an experiment to see if nonpecuniary incentives can be substituted for substantial income differences as inducements for high-quality professional, scientific and administrative performance. In conversation with our Chinese companions, I was surprised at how easily and cheerfully they accepted this fact of their lives, and how little value they placed on the freedom of choice they lack." "The Economy of China: A Tourist View," *Current Scene* 12 (May 1974): 11.

26. Perhaps a good example of how inefficiency in central planning can be tolerated in the subsistence-type economy can be illustrated by James Tobin's observation: "most of China's consumers are peasants who can find a way to consume whatever supplies of many products are not purchased for urban consumption. If they have a big harvest of apples, and prices are held stable or reduced insufficiently to attract extra urban consumption to match, the peasants can simply eat the apples or preserve them. If the apple harvest is poor while urban demand stays high, the peasants will simply have fewer apples to consume or preserve. The elasticity of their demand with available supplies provides a buffer that makes controlled prices workable." Ibid., p. 10. For a more developed economy, the penalty for the wrong decisions leading to either shortage or overproduction is clearly much higher.

27. For further discussion on this subject, see my "Grain Output in China: Some Statistical Implications," *Current Scene* 11 (February 1973).

28. Chow's Test is commonly used to ascertain whether the economic relationship remains stable in two periods of time, or whether the same relationship holds for two different groups of economic units; or, statistically, whether two sets of observations can be regarded as belonging to the same regression model. Having estimated a linear regression with *p* coefficients, one then goes ahead to test whether *m* additional observations belong to the same regression:

$$F\ (p,\ m + n - 2P) = \frac{Q_3/P}{Q_2\ (m + n - 2p)}$$

where $Q_3 = Q_1 - Q_2$, and Q_1 being the sum of squared residuals of the overall regression (m + n) while Q_2 and Q_3 being that of the sub-series.

See Gregory C. Chow, "Tests of Equality between Sets of Coefficients in Two Linear Regressions," *Economietrica* 28 (July 1960): 591-605.

29. For the technological change in Chinese agriculture in the early 1960s, see Kang Chao, *Agricultural Production in Communist China, 1949-1964* (Madison, Wis.: University of Wisconsin Press, 1970). For an overall discussion of China's green revolution, see Benedict Stavis, *Making Green Revolution*.

30. It is not just a tautology to suggest that much of the Chinese success in their organizational strategy depends on their organizational resources. Such is the impression that struck John Kenneth Galbraith during his visit to

China in 1972. "That the Chinese have a genius for organization seems to have been the revelation of an infinity of scholars." *A China Passage* (Boston: Houghton Mifflin Co., 1973), pp. 120-21.

31. For an analysis of how cooperatives can increase the bargaining power of the peasants, see Sidney Hoos, "Joint Decision-Making Processes in Present-day Agriculture," in Ugo Papi and Charles Nunn, eds., *Economic Problems of Agriculture in Industrial Societies* (London: Macmillan, 1969), pp. 379-89.

32. See Keith Griffin, "Policy Options for Rural Development," *Oxford Bulletin of Economics and Statistics* 35 (November 1973).

In China, the massive waterwork programs undertaken by the communes over the years have started to produce effects. In the last few years cultivation conditions had been among the worst for a century, and yet grain production could hold its own. "Chinese agriculture [in 1974] was hit for the fourth year in succession by highly unfavourable weather and the overall impression which Peking sought to give in releasing crop statistics was that the farm situation was much better than most observers might have guessed." *Far Eastern Economic Review: Asia Yearbook, 1975*, p. 158.

33. Premier Chou En-lai in his report to the 4th National People's Congress in January 1975 remarked: "While China's population has increased 60% since Liberation, grain output has increased by 140% and cotton 470%"; and he went on, "In a country like ours with a population of nearly 800 million, we have succeeded in ensuring the people their basic needs in food and clothing." *Straits Times* (Singapore), 21 January 1975.

6. China's Agriculture: Present Situation and Prospects

Gilbert Etienne

Dr. John Wong has provided us with a clear presentation of the main trends in China's agricultural strategy. Although I shall have some specific comments on Dr. Wong's analysis, I propose to concentrate on an extension of his remarks to cover some aspects of the agricultural situation in China and future prospects which Wong does not cover.[1]

THE EARLY PHASES

The heavy industry bias of the first five-year plan is well known, yet the data given by Dr. Wong can be misleading. In fact, the imbalance against agriculture was not so large as figures in the plan would indicate, since, unlike industry, agriculture was benefiting from expenditure categories in the plan that were not specifically identified with agriculture--relief funds, land-clearing projects undertaken by the army, State loans and grants--as well as by investments made by farmers themselves. These categories total Y 18.4 billion, which should be added to the Y 3.26 billion specifically identified with agriculture in the plan. (Industrial development investments for capital construction total Y 24.85 billion.)[2]

With respect to land reform one could also be more specific. There are countries where land redistribution makes sense because there is enough surplus land available to benefit most of the very small landowners and landless laborers (e.g., in parts of Latin America). In China, as in India, Bangladesh, or the Island of Java in Indonesia, *there is simply not enough land to accommodate all people who need land in order to reach a minimum economic holding.* That is why a drastic land redistribution, as in China, can only lead to an "atomization" of holdings, which necessarily must be followed by a regrouping of holdings in order to have economic units of production. Thus at times ideology and economic practice and necessity are closely connected. Those who would advocate a drastic redistribution of land in India, for example, should recognize that such a measure either would help only a very small number of the have-not population or would have to be followed by some kind of collectivization.

The Number and Size of Communes

The introduction of communes discussed by Wong involved more than "teething" problems, and excesses were not only reported in the foreign press: Chinese newspapers were filled with such reports in the early 1960s and all of them were not inspired by Liu Shaochi! In fact, it was no mean achievement to gradually put things in order and to arrive at a proper organization at the three levels: the commune, the brigade, and the team. This pattern of organization has not been altered in any major way since that time.[3]

One question that remains a puzzle is the number of communes. Wong mentions 26,000 in 1958, a figure which may have come down to 24,000 in 1959. In the next several years some of these communes must have been split, since Audrey Donnithorne reports Chinese sources showing 74,000 communes in 1961-1962.[4] For about ten years this figure was not questioned, but "a recent Chinese press report puts the number of communes at 50,000."[5]

It would be interesting to know why this consolidation of communes has taken place. So far, however, explanations are lacking. In any case, the size of brigades and teams does not seem to have been altered. The former range from less than 1,000 to about 1,500-2,000 persons, the latter from 50 to 200-300, as I observed in different provinces in the summer of 1961 and the fall of 1972. The population of the communes varies greatly, however, from around 10,000 to as large as 50,000. Available data do not permit determination as to whether the size of communes varies from province to province.

Agricultural Taxes and Delivery Quotas

More details could also have been presented by Wong in his discussion of the "government levy." The agricultural tax, which amounted to 12 percent of production in 1952, had fallen to 5 percent as a national average by 1974.[6]

In addition, of course, there are compulsory quotas delivered to the State for which the State makes payments at fixed prices. Since 1971 such quotas have been fixed for a period of five years (previously, they were set for three years), after which they are reassessed. Both taxes and quotas are usually paid in kind: cereals and other major agricultural commodities such as cotton, depending upon the locality.

Production teams are encouraged to sell more than their quota to the State. In a commune near Peking, above-quota grain deliveries were purchased at a price 50 percent higher than the price for quota deliveries. In Honan it was 20 percent; in Hopei, Hupeh, and Kwangtung, 30 percent (observations made in 1972).

SOME CONSIDERATIONS OF POPULATION AND AGRICULTURE IN CHINA

I wish now to turn to some aspects of the agricultural situation in China which are not covered in any detail in Dr. Wong's paper. Certainly in the great concern today over world population and food supplies, the importance of China is unmistakable. China has succeeded in eradicating the most acute poverty. But although economic growth continues at a good pace, the entire growth process is becoming increasingly complex. China also has a substantial influence on international trade in grains since she is a sizable net importer. For these and other reasons, it will be useful to take a closer look at present and prospective trends in population growth and agricultural development before commenting briefly on those aspects of China's experience that might be found useful by other developing countries.

Population Growth and Prospects

The actual number of people in China is not precisely known. The 1953 Census reported a population of 582 million people on the mainland of China. Taking an average rate of 2 percent annual increase, which may be more an under- than an over-estimate in view of the progress in public health, one arrives at 870 million in 1973, or 900 million (in round figures) in 1975.

For unexplained reasons, the Chinese authorities have stopped referring to the 1953 Census in the last ten years or so. They now mention a population of more than 500 million in 1949, increasing at a rate of about 2 percent, and a current population of 800 million.

The problems associated with family planning and reduced births are far from being resolved. Only some naïve foreign visitors can claim that China has controlled its population. In fact, several of its top officials are much more reserved on the subject (in particular Chairman Mao Tse-tung, the late Premier Chou En-lai, and Vice-Premier Li Hsien-nien). I got similar expressions of concern from a number of Chinese cadres in 1972. *Pekin Information (Peking Review)* summarized the current status of family planning in its issue of November 11, 1974: "Family planning is now widespread in towns and gradually introduced in the countryside," where, one might add, the whole problem is much more complex than in the cities. In 1975, several radio broadcasts emphasized the need of sustained propaganda in the countryside to overcome practical as well as psychological difficulties.[7] One can readily appreciate the magnitude of the task of spreading birth control information and practices in more than 1 million villages.

The Chinese are also making serious efforts to postpone marriages until after age 25. Such restraints are again much more readily implemented in the towns than in the villages.

When one realizes that almost 85 percent of the population lives in the countryside, one can imagine that the birth rate cannot fall appreciably in a few years' time. Demographic trends cannot be rapidly altered. Thus it is not likely that the rate of population increase will decline sharply in the immediate future. For the next 10-15 years, population may increase by 20-35 percent, taking different combinations of rates between 1.8 percent and 2.0 percent per year.[8]

Grain Production Trends

Chinese agriculture has entered a phase of increasingly sophisticated development, including the growing use of such modern technical inputs as mechanized irrigation, complex major flood-control works, chemical fertilizers, pesticides, and, to some extent, tractors and power tillers. These trends explain the achievements of the 1960s and early 1970s. In the 1950s Chinese agriculture grew mainly through the more intensive use of traditional inputs and techniques, but these more or less reached the limit of their capacity by the end of the decade.[9] The only way to move ahead was through the introduction of a *package of new inputs*.

In spite of determined efforts and substantial achievements, growth has not been all that rapid: around 60 percent between 1952 and 1971. There has been an acceleration in 1974 and 1975.[10]

While the strategy of growth seems quite sound and very cleverly combines human labor with modern and semi-modern techniques, progress so far has been uneven. It has been remarkable in the field of water works: flood control; land reclamation from waterlogging and salinity; major, medium, and minor irrigation works; etc. There has also been tremendous progress in small pumping sets and large pumping stations using surface water, and tubewells pumping groundwater, as I was able to observe and compare during my different visits to China in 1958, 1961, 1965, and 1972. However, the total irrigated area is difficult to assess because even the gross and net cultivated area is not well known. In recent years, several official statements have used 107 million hectares as the net cultivated area. This is probably an underestimation, since official data show 108 million hectares in 1952 and 112 million hectares in 1957. On the other hand, some foreign visitors obtained in China the figure of 127 million hectares, excluding multiple cropping. I would accept a figure of around 120 million hectares, including several land reclamation schemes completed in the last few years.

As to irrigation, it seems to cover 50-60 percent of the cultivated area. It is necessary, however, to differentiate between safe irrigation and inadequate irrigation. The latter often depends on reservoirs which may not fill in years of low rainfall, a rather common occurrence, especially in the Northern Plain. For these reasons, the Chinese have in recent years introduced a concept of hectareage with safe or stable yields--13 million hectares in 1971, and increasing to 33 million in 1974.[11]

Much progress has also been achieved in multiple cropping. We can assume that around 50 percent of the land bears more than one crop a year or perhaps somewhat more.

As is well known, the amount of new land to be brought into cultivation is very limited, except in remote areas where investments may be very heavy. According to some Chinese officials from the Ministry of Agriculture, 14-20 million hectares may become available. It is thus quite obvious that the most promising development lies with increasing yields and additional multiple cropping.

If water is the prerequisite to most other improvements, a close second in terms of priorities is basic agricultural science and research. This is so far one of the relatively weak links in China's agricultural strategy. First, we should not confuse improved seeds, where substantial progress has been achieved since 1949 as well as in earlier times, with high-yielding varieties (HYV). The former may boost yields by 10-50 percent; the latter by 100-300 percent, provided they are accompanied by appropriate amounts of water, chemical fertilizers, and pest control.

During my visit in 1972, some hybrid sorghum and hybrid corn could be found in North China. But high-yielding varieties of wheat were rare. Experiments were going on in research institutes. Some new varieties such as AFU (a Chinese-Albanian crossed variety) were giving good results, but they were still used on a limited basis, particularly in parts of Honan and Shensi. Research was more advanced in new paddy varieties. In Hunan, these new varieties were grown on 20 percent of the paddy fields with average yields of 5,250 kg/hectare for the early crop and 3,750 for the second crop. Further progress must have been achieved since then. In 1974, yields per hectare of grain for the entire province averaged 5,250 kg/ha., including all crops (either 2 crops of paddy, or 1 crop of paddy and 1 of wheat).[12] The U.S. Plant Studies Delegation mentions in its report that land sown to HYV of paddy in Southern areas has increased from 3.3 million hectares in 1965 to 6.7 million hectares in 1973.[13]

In any case, one thing is clear--HYV of wheat are definitely lagging behind. This is confirmed by the data on yields which I gathered in 1972 in several areas of Hopei and Honan,[14] and it is at least implied by the sizable imports of wheat which are required by the large cities of the North.

Observations made by various teams of foreign agronomists, including the one referred to above, indicate that much remains to be done in order to speed up and strengthen fundamental research. This, of course, is a complex process. The Chinese fully admit it and are showing a growing interest in cooperating with foreign research specialists. The Chinese no doubt have good scientists, but as is evident from the experience of a number of other Asian countries, such cooperation may provide a precious short cut in plant breeding, seed multiplication, etc. However, here too, things are not simple. Some foreign varieties may prove well adapted to local Chinese conditions, others not. In 1972, I was told in several research institutes that seeds of IR 8, one of the varieties of the International Rice Research Institute of the Philippines, and of Mexican wheat had not proved successful. One has also to take into account the diversity of climates. The same HYV of wheat may not be satisfactory in both Hopei and Hupeh. The diversity is even greater in the case of paddy. Some new seeds developed in Kwangtung, for instance, would not do well in Hunan or Hupeh.

The Chinese are aware of such shortcomings, which explains the renewed emphasis on basic research. *The People's Daily*, in an article on Shanghai' Futan University, mentioned that the university "pays great attention to the study of basic theory of natural sciences."[15] However, in 1976, the whole issue is again the object of strong criticisms in the renewed struggle against revisionism.

On the other hand, great progress in applied research has already been made at the local level. Often communes or brigades have experimental plots for trying improved seeds, better seed selection, and seed multiplication. Such experimentation will help considerably to spread the HYV when these become available on a larger scale. Proper seed multiplication, often a difficult process, should be easier than in some other Asian countries thanks to China's organization of agriculture.

After research, chemical fertilizers are a top priority.[16] Since the early 1950s, national production and imports of chemical fertilizers have very much increased. In addition to large and medium factories, since 1965 an increasing emphasis has been given to local indigenous factories based on a specially designed intermediate technology. For nitrogen the standard pattern is a plant of 200-300 workers producing 3,000-5,000 tons of synthetic ammonia giving 12,000-20,000 tons of ammonium bicarbonate or, less frequently, aqua ammonia. Such factories supply 54 percent of the total production of nitrogen fertilizers.[17] Thanks to the increasing number of phosphate mines discovered in the South and the Center-South, medium and small plants of single superphosphate are also expanding and now represent around 75 percent of production.

Such intermediate technology is quite appropriate for the production of superphosphate, but the same cannot be said of the small nitrogen plants. The fertilizer produced by these plants is not giving very positive results, though it evidently is better than nothing. Ammonium bicarbonate cakes and evaporates rapidly, while aqua ammonia is very bulky. The former contains 16-17 percent of N, the latter 14 percent.[18]

For these reasons, by the end of 1972 the Chinese began again to encourage the erection of large plants for the production of nitrogen. No less than 13 full-scale plants,[19] each including one unit for the production of synthetic ammonia and another unit with a capacity of 500,000 tons of urea per year were negotiated on a turn-key basis: 2 plants with Japanese firms; 8 plants with Kellogg Continental; and 3 plants with a French group. The original schedule has been delayed somewhat following further discussions on the terms of several of the agreements. These difficulties were apparently overcome by early 1975, with the result that by 1980-1981 the entire construction program should be completed, with the new plants reaching their full production capacity. In 1975, the Chinese planned to order another three combined plants of the same size from Kellogg in Texas. However, the agreement was still not signed by January 1976. The Chinese have sent 250 engineers and technicians to Houston for training, who could, in any case, be used in the plants built by Kellogg Continental.

In spite of these orders for large plants, the Chinese, for the time being at least, continue to build additional small plants. (It seems that these small factories may be gradually phased out after 1979.) In the first ten months of 1974, 100 new units entered production, so that the total of such plants by the end of 1974 was around 1,000.[20]

Finally, according to recent reports, the Chinese are now constructing 5 or 6 medium-size plants with a capacity of 100,000 to 150,000 tons of urea per year. For these they need practically no foreign know-how, and import requirements are very limited.

TABLE 1. PRODUCTION AND IMPORTS OF CHEMICAL FERTILIZERS IN TERMS OF N, P_2O_5, AND K_2O[a]

	(July-June)	1972/73	1973/74	1974/75	1975/76[b]
		Production (million tons)			
N		2.01	2.40	2.50	
P		0.95	1.20	1.30	
K		0.30	0.30	0.50	
Total		3.26	3.90	4.30	4.70
		Imports (million tons)			
N		1.40	1.20	0.80[c]	1.30[c]
P		0.10	0.10	0.40	0.30-0.40
K		0.08	0.01	0.04	0.06
Total		1.58	1.31	1.24	1.66-1.76

[a]A large part of the increase for N and P, if not most of it, came from the indigenous factories.

[b]Provisional estimates for this year. Total increase may be around 10 percent.

[c]The sharp fall in imports of N in 1974/75 is due, to a large extent, to the increase in world trade prices. The increase for 1975/76 is due partly to the decrease in world prices.

It is clear that the situation will much improve when the new nitrogen plants enter production. At present, most chemical nitrogen applications in terms of N are in the range of only 30-50 kg/hectare, if not less. Besides, especially in North China, which is far removed from the phosphate mines localized in South and South-Central China, fertilizer applications do not include enough phosphatic chemical fertilizers. Such an imbalance with phosphate deficiencies may result in soil deterioration.

Finally, the Chinese are trying to compensate for the present shortage of chemical fertilizers by an ever-increasing use of farm manures.[21] In research institutes and in advanced communes, I gathered data on yields of paddy of 5,000-7,500 kg/hectare on lands fertilized with 45-50 kg/hectare of N and 50 of P_2O_5 plus 50 to 75 tons/hectare of manure. As Claude Aubert, a French agronomist and Sinologist rightly points out, such large amounts of organic manure could obviously not be collected and applied on a large scale all over China.[22]

To sum up, and in addition to variations in the weather, when assessing the growth of agriculture in the last five years one should take into account the insufficient emphasis given so far to fundamental research, and the difficulties associated with the production of chemical fertilizers such as the low quality of ammonium bicarbonate.

There is a final question to which a foreign observer can hardly provide an answer: how do the various institutions, agencies, district and provincial administrations concerned with agricultural development work? Here and there, when in the field, I was told of delays in getting fertilizers, of lack of power for pumps, etc. The Chinese newspapers often mention defects and shortcomings in the management and planning of agriculture, or cadres of uneven talent or uneven dedication to work, etc.

The month-long agricultural conference which took place in the fall of 1975 threw some more light on these problems. In learning from Tachai, both practical and ideological problems were strongly underlined. In his report, Hua Kuo-feng (then Minister of Public Security and now Prime Minister) expressed his concern over county party committees and other cadres. He criticized "softness, looseness and laziness" as well as "bourgeois factionalism." Speculation and excesses of private rural trade were also criticized.[23] That such defects exist is nothing unusual, especially in a vast country such as China. But are they rare or widespread? So far it seems the Chinese have done comparatively well, though accurate confirmation could come only through detailed field studies. However, it is obvious that such management problems will become increasingly important in the future. A growing dependence on modern inputs will require more planning than before--at the national and the provincial, as well as at the district levels.

If management problems do not become too severe, one can expect a more rapid growth of agricultural production as the HYV become more widely adopted and as the new chemical nitrogen factories come into operation. However, additional efforts for the production of phosphatic fertilizers must also be made, a subject about which little information is available, at least to my knowledge. The exact requirements of potash are not too clear, but in the last few years the Chinese have also imported some amounts of that fertilizer.

SOME TENTATIVE PROSPECTS FOR AGRICULTURAL GROWTH

The future growth and development of agriculture in China is a crucial question, given the large and growing population. As noted above, it is quite clear that within the next 15-20 years the rate of population growth will not fall sharply even if the birth rate continues to decline. In so far as foodgrains are concerned, an increase of about 40 percent within the next 10-15 years does not seem impossible (see appendix). In this period population is likely to grow by 20-35 percent. In normal years, China may be able to get by without grain imports while maintaining some exports of rice (see appendix).

The sharp increases in foodgrain production in 1974 and 1975:[24] 275 and 285 million tons, respectively, versus 250-257 million in 1973, explain the important fall in imports for 1975/76. However, unfavorable weather may still have a sizable impact on production and imports, at least in the immediate future.

PRIVATE VERSUS COLLECTIVE INTERESTS: LOCAL VERSUS NATIONAL GOALS

Ideological factors in China are too intimately connected with agricultural organization and production to be omitted in any analysis. Campaigns against egotistic aims and the bourgeois spirit have continued even after the cultural revolution.[25] Individual interests are still so strong that it would likely prove impossible to take the brigade as the basic unit of production and accounting since within the brigade it would lead to an egalitarianism in favor of poor teams which the richer production teams would not be prepared to accept.[26]

The same can be said of private plots enabling the farmers to raise vegetables, pigs, and poultry. Such private income may represent 20-35 percent of the collective income.

It would be difficult to achieve a better balance than that which now exists between private and collective interests, both from an ideological and an economic point of view. More radical measures would upset the farmers and affect both collective and private production.[27] That is why the production teams and private plots are specifically mentioned in the new constitution released in 1975.

There are also contradictions between collective and national interests. Especially in the richer areas of Central and South China, collective units are tempted to emphasize crops and activities other than rice production since the former yield a higher collective income. That is why Chinese authorities continue to insist on the priority to be given cereals and more recently cotton production--commodities which are badly needed by the urban population.

To maintain both types of balance (private versus collective interests and collective versus national interests) represents a very complex task and requires constant vigilance to avoid possible deviations toward the right or the left. This is of course complicated by the difficulty of closely supervising the 5 million production teams, often far removed from roads or railway lines, especially in view of the limited number of vehicles available at district headquarters.[28]

A factor that will have a growing influence on the organization of agriculture is the increased use of modern inputs. This will require more local planning and thus involve wider responsibilities of district and commune authorities. One can thus envisage possible contradictions between the autonomy of the team on the one hand, and a greater integration of the teams to brigades and communes under the compulsion of technical and economic factors on the other.

LESSONS FROM THE CHINESE EXPERIENCE

In trying to glean from the recent past what China's experience has to offer other countries, one must first consider the impact of China's history and civilization on contemporary developments: the traditional labor-intensive techniques and hard work of the farmers; the ability to mobilize the masses for large and small rural works; the sophistication of the bureaucracy; etc. One must also emphasize the outstanding talent of China's revolutionary leaders.

In other developing countries, situations vary greatly. The inefficiency of a number of revolutionary movements has been striking. The traditional background of most nations differs widely from that of China. Thus the commune experiment is likely to be of limited value for many countries, at least for those that have not achieved a successful revolution.

As to the promotion of cooperative efforts among peasants, there are many examples in Latin America, Africa, and Asia showing that unless they are introduced through a real revolutionary process, they rarely help the have-nots. The main beneficiaries are the large- and/or medium-sized farmers.[29]

However, there are some useful lessons to be drawn from China's experience. The need to curb the urban bias of the elites as it now exists in many developing countries is certainly one of them. There are also a number of practical lessons which might be learned and used by policy-makers under almost any type of political system. The importance of industries supporting agriculture, the combined use of advanced techniques and intermediate technologies, large- and small-scale plants within the same industries are all striking features of China's development strategy.

The Chinese have understood well the global process of rural development: in addition to agriculture, emphasis is placed on the creation of small indigenous industries (plants of 50-300 workers) based mostly at the level of the district towns and rarely in villages where local infrastructure is lacking; repair and maintenance workshops are encouraged and located primarily at the village level; the expansion of handicrafts is promoted; finally, the active encouragement and promotion of rural trade is correctly used as a powerful tool of development. The latter is of particular importance, and the basic idea is to stimulate farmers to sell more (collectively and privately) on the one hand, and in turn to sell them more goods--agricultural inputs as well as consumer goods. Such a process stimulates economic growth, helps to diversify the economy, and increases job opportunities in the secondary and tertiary sectors of the economy.

The promotion of rural trade is facilitated through the creation of shops (cooperative- or State-owned) all the way down to the village level. For instance, mini-department stores selling processed food, beverages, textiles, plastic goods, bicycles, clocks and watches, etc., are now to be found in the larger villages of at least the most advanced districts.

This process is not totally unique to China. It can be found in certain other parts of Asia, such as in the more advanced districts of India and Pakistan. Yet it could be much more systematically encouraged and organized through the promotion of intermediate technologies, standardized models of indigenous factories as in China,[30] more encouragement and dispersion of repair and maintenance workshops, etc. As to the encouragement of rural trade, it is striking to see in most of Asia today how rural markets remain under-exploited due to a lack of proper trade channels and facilities extending down to the villages. Actually active trade is usually restricted to the larger villages and the smaller towns along the main roads. The processes for expanding and encouraging more widespread rural trade could work under very different economic and political systems: a mixed economy, semi-capitalist, or semi-socialist. It would help to promote economic growth as well as to create more employment--especially for the poor, very small landowners, and landless laborers.

APPENDIX

Unless otherwise noted, the data herein on acreages, yields, and production are from FAO, *Production Yearbook 1972*, and U.S. Department of Agriculture, *The Agricultural Situation in the People's Republic of China* (Washington, D.C., 1974). See also the reports of the Joint Economic Committee of the U.S. Congress.

TOTAL PRODUCTION OF FOODGRAINS
(million tons)

1952	154	(good weather)
1970	240	
1971	250	(good weather)
1972	240	(bad weather)
1973	250-257	
1974	275	(good weather)
1975	285	

Includes cereals and tubers. Rice is accounted in terms of paddy; tubers (mostly sweet potatoes) are taken at one-fourth of their gross weight. It is uncertain whether soybeans are included in the 1974-75 estimates.[31]

IMPORTS, MOSTLY OF WHEAT AND SOME CORN (in million tons)

1962/63	1971/72	1972/73	1973/74	1974/75	1975/76 (July-June)
6.0	3.4	6.1	7.7	5.5	3.5 (preliminary estimate)

EXPORTS OF RICE (in million tons)

1968	1970	1972	1974
.886	.885	.811	1.4

Excluding exports to North Korea and North Viet Nam.

VALUE OF IMPORTS AND EXPORTS OF GRAIN (US $ millions)

	1971	1972	1973
Imports	205	345	840
Exports	95	155	535

Some Tentative Prospects for Agricultural Growth in the Next 10-15 Years

Paddy: 34 million ha., 1971/73, average yield of 3,000 kg/ha. of paddy, production 108 million tons. It is probable that through additional multicropping and/or more reclaimed land, around 40 million ha. could be achieved, i.e., an increase of 6 million ha. Average yields could, with HYV and other inputs, reach at least 4,000 kg/ha. These two influences combined could result in an increase of 55 million tons.

Wheat: 28.7 million ha., 1,200 kg/ha., 34 million tons average 1971/73. It is not sure that the acreage can be substantially increased, but average yields of 2,000 kg/ha. belong to the art of the possible; around 2,500 in well-irrigated areas, 1,000-1,500 in nonirrigated areas or poorly irrigated areas. We assume that irrigated wheat would cover a wider area than dry wheat. We would obtain an increase of 25 million tons.

Barley: 13.5 million ha., 1,400 kg/ha., 19.5 million tons in 1973. With yields of 2,000 kg., as for wheat, 8 million extra tons would be realized.

Corn: 10.5 million ha., 2,800 kg/ha., 29.8 million tons in 1973. Such yields look doubtful. Chinese officials provided me with estimates of 1,900-2,250 which even appear too high as averages. In several districts in 1972 I gathered average yields of 1,500-2,000-2,400 kg., with the latter having been grown with the use of some chemical fertilizers. With more chemical fertilizers and HYV one could reach the average figure of 2,800. That is why we keep the total figure of production as it is said to be now.

Millet and Sorghum: 29 million ha., 760 kg/ha., 22 million tons in 1972. Increases will probably not be too great, considering the low and erratic rainfall of areas which grow these cereals and the fact that several of these areas cannot be irrigated. Thus, it may be difficult to go beyond a 33 percent increase leading to rough average yields of 1,000 kg/ha., i.e., an increase of 7 million tons.

Sweet Potatoes and Other Tuber Roots: A forecast is difficult in view of the paucity of data.

Summing Up: The total increase of 95 million tons arrived at may be rounded to 100 million tons allowing for some progress in sweet potatoes, corn, etc. The increase totals approximately 40 percent. All the above targets do not represent the upper limits which could be reached.

Other Crops

Soybeans: 14.3 million ha., 810 kg/ha., 11.6 million tons (average 1971-1973). The latter figure seems high. D. L. Dawson gives 10 million tons for 1957, 8-9 million for 1967.[32] For 1970 to 1973 various estimates range between 7 and 9.5 million tons.

Sugarcane: For 1973, 700,000 ha. produced 34-36 million tons of cane, giving 2-2.5 million tons of sugar. Imports reached 0.5 million tons in 1973. Sugar beets play a very minor role.

Groundnuts: Production of 2.6 million tons (in shell) in 1957. After a sharp fall in the early 1960s, production recovered and reached 2.9 million tons by 1967;[33] we would have 2.6 million tons and 2.7 million tons in 1973 and 1974, respectively.

The concentration on grain versus cash crops since the 1960s may explain the relative stagnation in the production of soybeans and groundnuts.

Cotton: Production seemed to have reached 1.7 million tons (lint) in both 1973 and 1974 (1.3 million tons in 1967). Some advanced communes have reached yields of 700 kg/ha. (lint), sometimes even 1,000, but national average yields may be below 500. On the other hand, imports of cotton lint have increased considerably: 149,000 tons in 1971/72, 418,000 tons in 1972/73, and 452,800 tons in 1973/74.

It is clear that in all of these crops a considerable increase in yields could be achieved.

Sideline Activities: Vegetables, poultry, pigs, and fish are old and familiar features of China's civilization. Their production has expanded rapidly on the private plots and on some of the collective lands, especially in suburban vegetable communes. Further progress can be achieved, but data are not available.

NOTES

1. My comments are unavoidably brief. More details can be found in my book, *La voie chinoise, la longue marche de l'économie 1949-1974* (Paris: Presses Universitaires de France, 1974), and in *China's Agricultural Development*, ed. G. Etienne (Geneva: Asian Documentation and Research Centre, Graduate Institute of International Studies, 1974).
2. *First Five-Year Plan for the Development of the National Economy of the People's Republic of China in 1953-57* (Peking: Foreign Languages Press, 1956), pp. 29-33.
3. If we except some attempts during the Cultural Revolution to revert to the brigade as the basic unit.
4. *China's Economic System* (London: Allen and Unwin, 1967), p. 49.
5. U.S. Department of Agriculture, *The Agricultural Situation in the People's Republic of China* (Washington, D.C., 1974), p. 6. The same figure is also reported by the U.S. Plant Studies Delegation to the People's Republic of China, *Trip Report*, Draft (New York: Rockefeller Foundation, April 1975), p. 21.
6. It seems that a readjustment of the tax was made in 1958. *Pekin Information*, 5 September 1975.
7. Broadcasts from Canton, Kweiyang, Nanning, Hofei, Chengchow, Tsinan. See U.S. Department of Commerce, Foreign Broadcast Information Service (cited hereafter as FBIS), 1975: 21 January; 3, 10, 11 February; 5 March; 2, 3 April.
8. These projections seem rather conservative, since, as mentioned above, rates of that magnitude may be on the low side.
9. These developments are consistent with the major thesis developed by T. W. Schultz in *Transforming Traditional Agriculture* (New Haven: Yale University Press, 1964).
10. Growth rates and trends are always difficult to assess when taking into account the weather variable. Both 1952 and 1971 enjoyed favorable conditions;

1972 was a bad year, whereas 1974 was good. In 1975, natural conditions seemed, on the whole, rather favorable.

11. Joint Economic Committee of the U.S. Congress, *People's Republic of China, an Economic Assessment* (Washington, D.C., 1972), and New China News Agency, 26 December 1974 (FBIS, 3 March 1975). Additional evidence of irrigation expansion can be deduced from figures on the growing use of power for irrigation and/or drainage: 1964, 6 million HP; 1972, 20 million; 1974, 30 million. See L. T. Kuo, *The Technical Transformation of Agriculture in Communist China* (New York: Praeger, 1972), pp. 81-82; New China News Agency, 28 September 1972; Cheng Shih, *A Glance at China's Economy* (Peking: Foreign Languages Press, 1974); *Pekin Information*, 9 December 1974. The 1974 area covered by this power irrigation and/or drainage may total 33 million hectares. Joint Economic Committee of the U.S. Congress, *China: A Reassessment of the Economy* (GPO: Washington, D.C., 1975), p. 336.

12. New China News Agency, 22 January 1975 (FBIS, 23 January 1975).

13. The figure for 1965 is surprising. Were the Chinese actually referring to improved varieties? If the figure actually refers to HYV, it seems quite high. U.S. Plant Studies Delegation to the People's Republic of China, *Trip Report*, Draft, p. 25.

14. 750-1,000 kg/ha. in several districts, elsewhere 1,000-1,500, more rarely 2,000 as a district average.

15. *Current Scene*, no. 12 (1975).

16. I wish to thank Marcel L'Eplattenier, agricultural economist, Zurich, who helped me greatly on the subject of chemical fertilizers.

17. New China News Agency, 20 March 1975 (FBIS, 20 March 1975).

18. For Hunan, 82 percent of nitrogen chemical fertilizers used in 1973, at an average rate of 67 kg/ha. in terms of N, came from indigenous factories. New China News Agency, 30 December 1974 (quoted in FBIS, 3 January 1975).

19. The 13 plants will be located in 12 provinces, between Yunnan in the extreme South-West and Heilungkiang in the North. They will be near to gas and/or oil fields, or near to refineries where they can use refinery gas as a raw material.

20. Shih, *A Glance at China's Economy*, p. 25. It seems that such plants have a rather low rate of utilization, perhaps around 50 percent of their capacity, due to technical problems at least partly inherent in the intermediate technologies which they employ.

21. It is well to remember that China belongs mostly to the temperate world where farm manure has much more lasting effects than in tropical climates.

22. See *France Asie*, no. 4 (1974), p. 106. On a large scale it would be difficult to use more than 20-30 tons/ha. of manure. Many radio broadcasts in early 1975 insisted on greater efforts to collect manure. Kweichow Radio set a target of 22-37 tons/ha. for the province on 14 March 1975 (see FBIS, 19 March 1975). Harbin radio mentioned 30 tons/ha. on 30 January 1975 (FBIS, 4 February 1975).

23. *Current Scene*, no. 12 (1975), as well as a number of radio broadcasts on these issues in January-April 1975.

24. *Far Eastern Economic Review*, 19 December 1975, quoting Chinese sources.

25. See the very clear analysis by Chang Chun-chiao, "On Exercising All Round Dictatorship over the Bourgeoisie," *Peking Review*, 4 April 1975, and the overall campaign against bourgeois attitudes which followed the anti-Confucius-Lin Piao campaign.

26. See particularly on this point the examples given in chapter 12 of my book, *La voie chinoise*.

27. As far as the latter is concerned, it remains important since it supplies a large part of the pork, poultry, and eggs sold in towns or exported.

28. In the rich districts of Hupeh and Kwangtung, district authorities have only 2-3 cars or jeeps at their disposal. The continuous emphasis on sending upper cadres down to villages helps in checking for such dangers of deviation, but the problem remains far from simple and is not easily resolved.

29. See for example the eight volumes of studies, dealing mainly with cooperatives in Latin America, Africa, and Asia, by the U.N. Research Institute for Social Development, *Rural Institutions and Planned Change* (Geneva, 1969-1975).

30. One has to be fairly selective in that field. Small nitrogen factories are not satisfactory. But for superphosphate, cement, pumps and engines, power

tillers, other agricultural implements, small-scale factories using intermediate technologies do make sense.

31. See *Current Scene*, no. 6 (1976).
32. *Communist China's Agriculture* (New York: Praeger, 1970).
33. Ibid., p. 262.

Part IV: Mixed Collective/Individual Farming Systems

Editor's Introduction

The country experiences included under this category are those of Yugoslavia, Chile, Peru, Tanzania, Dahomey, and Sri Lanka. Most of these would come under Galeski's Type Two. However, the correspondence here is not as clear as in those discussed in the preceding sections. Yugoslavia presents a particularly ambiguous case with respect to Galeski's classification. Galeski himself suggests that Yugoslavia might be considered in his fourth type, along with other intermediate forms. In any event, government leadership was central to the organization of all the group farms discussed in this section. Thus they have characteristics of Galeski's Types Two and Three.

All cases other than Yugoslavia represent countries from the developing continents of Latin America, Africa, and Asia. Physical and climatic conditions vary greatly; these countries have had different historical experiences with colonialism; current governments range from democratic socialism to military rule; and the countries are at various stages of industrialization with the proportion of the population in agriculture ranging from less than 25 to over 75 percent.

The experience (and experimentation) with group farming is also quite unique in each country: an announced policy of rural villagization and group farming to eventually encompass all of agriculture in Tanzania; a relatively small (at least to date) program of settlement and group farming on some large estates and new land in conjunction with land reform legislation in Sri Lanka; a massive land reform in both Chile and Peru, establishing group farming enterprises on the reformed land but thus far at least retaining pre-reform small- and medium-sized farmers on their individual farms; and a relatively limited, though with possibilities of expansion, experience with group farming in areas of Dahomey devoted principally to the production of oil palm. In all these cases (excepting again Yugoslavia) the experience is quite new--mostly within the past 10 years. Thus it should not be too surprising that these group farming enterprises have thus far not met all the expectations that were expressed by governments when they were established.

There are unique situations and innovations in each country, and it would be difficult to maintain that these countries have followed any "outside" model. One common issue faced by all is the coexistence of a private farm sector (in most cases still dominant) and a cooperative, group farming sector. This results in some ambivalence in government policy, and also in the expectations of farmers in both sectors. In some cases (Peru) the private farmers complain that they are not receiving sufficient government support and that all government effort is devoted to the newly established production cooperatives. In Sri Lanka, the lure of private property and the expectation of becoming private landowners in the future is still much on the minds of cooperative members. In Chile, the entire experience and its future have been clouded since the military take-over in September 1973.

One of the issues that both Chile and Peru have struggled with has been the incorporation of landless or near-landless workers into the production cooperatives. The tendency has been for resident workers on the former estates to resist the incorporation of outsiders.

Yugoslavia, as we have said, presents quite a special case. There have been a number of basic changes in policy with respect to agriculture. Collectivization during the late 1940s and early 1950s attempted to establish kolkhoz-like units. Although the ruthless character of Soviet collectivization was avoided, the Yugoslav effort at establishing collective farms was abandoned. General farmer cooperatives were then introduced to organize and eventually socialize the peasant sector. These were more or less traditional-type, multipurpose cooperatives. Farmer cooperatives along with worker-managed, agro-industrial combines and agricultural estates comprise the socialist sector of Yugoslav agriculture. But even today, individual peasant producers comprise almost 40 percent of the total population and own and operate about 85 percent of the agricultural land.

7. Group Farming Practices in Yugoslavia

Robert F. Miller

INTRODUCTION

Even the casual visitor to Yugoslavia is struck by the great variety of social, cultural, demographic, economic, and topographical characteristics of this dynamic Balkan country of 21 million inhabitants. In the words of Branko Horvat, the well-known economist, "Yugoslavia has been described as one country with two alphabets, three religions, four languages, five nations, and six federal states called republics."[1] This underlying complexity is, if anything, magnified by the propensity of the country's leaders continually to experiment with and alter the major political and economic institutions. The resulting systemic changes have affected agriculture perhaps more than any other sector; and the country has experienced one of the most dramatic shifts from a predominantly rural to an urban socio-economic orientation in the modern world.[2]

From the standpoint of initial economic backwardness and cultural heterogeneity, often aggravated by traditional ethnic antagonisms, Yugoslavia represents a classic case of the problems of modernization. The obvious, if incomplete, successes of the developmental effort are in no small measure a tribute to the political sensitivity and relative pragmatism of President Josip Broz Tito and his closest colleagues. As Marxists, the Yugoslav leaders have an almost built-in tendency to conceptualize and theorize about the processes of socio-economic development. This ideological orientation, substantially modified from the Soviet model by Yugoslav theory and practice into what is sometimes called the "Titoist" variant of socialism, has provided them with an ideal model of the kind of social and economic order they wish to see ultimately realized in Yugoslavia. It also suggests some of the institutional means for getting there. Whether this combination of ends and means has been an appropriate vehicle for Yugoslav agricultural development under the prevailing conditions in the village is an open question, one which constitutes one of the central themes of this chapter.

An important feature of the role of ideology in Yugoslav policy-making has been the variability of its effect over time. This fact must be understood if we are to make any sense at all of the twists and turns of agricultural policy. The swings in intensity and salience of the ideological factor have followed a more or less cyclical pattern since World War II in response to major foreign and domestic crises. In greatly oversimplified terms the pattern is as follows: during periods of serious military or political danger from the Soviet Union or of domestic economic difficulties--either of which has customarily led to greater reliance on the West--ideological considerations have tended to be relegated to the background; in periods of closer relations with Moscow and/or of social and economic malaise at home resulting, according to Tito, from too great an involvement with the West, ideological considerations become much more salient in the policy-making process.

For agriculture these shifts have often been of crucial significance. Given the traditional dislike of Marxian socialists for the peasant, for example, the whole question of agricultural cooperation acquires a deeply ideological coloration. Cooperatives are seen as a way of weaning the peasant away from his individualistic mode of production and his private plot of land prior to transforming him into a member of the proletariat--that is, into a rural counterpart of the urban industrial working class. During phases of intensified ideological concern pressures for the expansion of cooperative forms of agricultural organization have markedly increased. The actual performance of the cooperatives and the attitudes of the peasants toward them became almost irrelevant during these phases--almost, but not entirely; for the Yugoslav leaders have remained committed to real economic development and the maintenance of peace and a modicum of harmony in the countryside. The record of Yugoslav agricultural cooperation is thus the story of an ongoing search for an accommodation between long-term ideological goals and the tangible requirements of an effective economic and social policy.

EXTERNAL INFLUENCES AND CONDITIONS

As they undertook the liberation of their country from Axis occupation and the conquest of power for their own movement, Tito and his fellow leaders of the Communist Party of Yugoslavia (CPY) had in mind a quite explicit model of agricultural organization for a future socialist Yugoslavia. That model was, of course, the Soviet system of collective and state farms. To be sure, CPY leaders expected that the path to collectivization would be a long one, with many difficulties to be overcome. For one thing, the Yugoslav peasants were no more likely to welcome collectivization than had their Soviet counterparts. Moreover, the Yugoslav peasants had been heavily involved in the national liberation struggle, comprising the bulk of the Partisan forces. Indeed, during the war the peasants had accounted for more than 50 percent of the total membership of the CPY itself.[3]

Consequently, unlike their Bolshevik political forbears, the Yugoslav leaders could not automatically assume general peasant hostility as a justification for repressive measures. Peasant interests had to be taken into consideration. This constraint was reflected in the moderation of postwar agrarian legislation. The law on Agrarian Reform and Colonization of 23 August 1945, set a limit of 25-35 hectares of arable land under individual peasant ownership and established a state land reserve for colonization and redistribution out of the resulting surplus and out of land confiscated from enemy aliens and collaborators. Special enlarged land grants were provided for groups of peasants who agreed to engage in collective cultivation (Art. 23), but the primary orientation of the law was toward individual peasant proprietorship.[4]

In any case, the total amount of land thus available for redistribution was not great.[5] Probably the most important reason for the general moderateness of the reform was the relative egalitarianism in landholding before the war. Tables 1a and 1b show the changing patterns of agricultural landholdings in Yugoslavia for selected years.

From the table it is clear that neither the 1945 land reform nor the 1953 reform which succeeded it greatly changed the distribution of peasant land. The natural processes of subdivision by inheritance and parcellation, which had been the bane of prewar East European agriculture, continued apace, with the result that although polarization between large and small farms was significantly reduced, there was a growing concentration of farms at the small end of the range (by 1969 almost 75 percent of all farms were less than 5 hectares). In 1969 the average Yugoslav peasant farm had only 2.57 hectares of arable land. Table 2 shows the changing regional and national patterns of land distribution during the 1960s.

Given the traditional, extensive methods of grain and livestock production which still prevailed in most regions, it was evident that the vast majority of farms were too small and too poor for modern commercial agriculture. Some form of producer cooperation seemed to be indicated. Yugoslavia's new leaders had come to this conclusion well before the patterns described in Tables 1 and 2 were manifest; but they did so for questionable ideological reasons, and their consequent choice of institutions and methods often proved to be counterproductive.

Agricultural cooperation had been fairly well developed in prewar Yugoslavia. In 1938 there were some 9,100 primary agricultural cooperatives throughout the country with 1,183,000 members, or almost 10 percent of the agricultural population.[6] The great majority were credit and supply cooperatives, but about 20 percent engaged in specialized production operations.[7] Despite the formation of a national Main Cooperative Alliance in 1919, regional cooperative movements operated autonomously with their own distinc-

TABLE 1. CHANGING DISTRIBUTION OF INDIVIDUAL PEASANT FARMS BY SIZE FOR SELECTED YEARS (1931-1969)

Size of Farm (hectares)	Year 1931	1941	1951	1953	1955	1960	1969
	a. Number of Farms (in thousands)						
Up to 1 ha.	710[a]	1,221[a]	276	253	950[b]	470	549
1-2 ha.			344	358		446	465
2-3 ha.	698[a]	650[a]	322	333	947[b]	393	400
3-5 ha.			452	464		559	526
5-8 ha.	420[a]	524[a]	322	328	391	422	384
8-10 ha.			103	103	321[b]	142	129
Over 10 ha.	241	241	186	189		187	147
TOTAL	2,069	2,636	2,005	2,028	2,609	2,619	2,600
	b. Relative Distribution (in %)						
Up to 1 ha.	34.3[a]	47[a]	13.8	12.5	36.4[b]	18.0	21.1
1-2 ha.			17.2	17.6		17.1	17.9
2-3 ha.	33.6[a]	24[a]	16.1	16.4	36.3[b]	15.0	15.4
3-5 ha.			22.5	22.9		21.3	20.2
5-8 ha.	20.4[a]	20[a]	16.0	16.2	15.0	16.1	14.8
8-10 ha.			5.1	5.1	12.3[b]	5.4	5.0
Over 10 ha.	11.7	9	9.3	9.3		7.1	5.6
TOTAL	100	100	100	100	100	100	100

[a]Data for 1931 and 1941 are for the following farm size groups: up to 2 ha.; 2-5 ha.; 5-10 ha.; over 10 ha.
[b]Data for 1955 are for the following size groups: up to 2 ha.; 2-5 ha.; 5-8 ha.; over 8 ha.
SOURCES: Petar Marković, *Strukturne promene na selu kao rezultat ekonomskog razvitka, period 1900-1960* (Belgrade: Zadružna knjiga, 1963), pp. 22, 25, 27; *Naše teme*, no. 10-11 (October-November 1970), Table 20, p. 2025.

TABLE 2. CHANGES IN AVERAGE PEASANT FARM SIZE, 1960-1969
(in hectares)

Region	Average Total Land		Average Arable Land	
	1960	1969	1960	1969
Bosnia & Hercegovina	3.86	3.32	2.49	2.19
Montenegro	5.02	4.42	2.04	1.83
Croatia	3.60	3.35	2.43	2.27
Macedonia	3.55	2.57	2.79	2.00
Slovenia	6.76	6.39	2.54	2.54
Serbia	4.37	4.18	3.18	3.04
Serbia proper	4.78	4.63	3.17	3.06
Vojvodina	3.51	3.37	3.34	3.22
Kosovo	4.23	3.49	2.76	2.37
YUGOSLAVIA	4.41	3.97	2.75	2.56

SOURCE: Edhem Dilić, et al., eds., *Društvene promene u selu* (Belgrade: Biblioteka sociologije sela, 1974), p. 23.

tive legal and functional characteristics, reflecting the ethnic heterogeneity of the country. The relatively strong cooperative movements in Slovenia and Croatia, for example, were closely tied to the Roman Catholic Church in those republics. The various sectional cooperative establishments were actively involved in the unhappy political life of prewar Yugoslavia. And as might be expected, cooperative politics tended to attract the more prosperous, politically conservative strata of the peasantry. These characteristics rendered the existing cooperative institutions suspect in the eyes of the new Communist leaders, who openly criticized them as potential centers of a bourgeois-capitalist restoration.[8] Hence they were considered unsuitable as a vehicle for the kind of socioeconomic transformation that underlay Tito's plans for the future.[9]

By the fall of 1946 it was already clear that the officially favored type of cooperative was the "peasant working cooperative" (SRZ), essentially a carbon-copy of the Soviet *kolkhoz*. However, collectivization--the mass formation of SRZs--began in earnest only in late 1948. The immediate cause of this drastic step was a series of internal and external crises. Domestically, the severe drought of 1947 and growing peasant opposition to governmental control over trade in farm products (compulsory deliveries and restrictions on private trading) created an atmosphere of political and economic tension. This feeling was intensified by Yugoslavia's sudden explusion from the Cominform and the ensuing efforts by Moscow to undermine Tito at home. The Yugoslav leaders responded by tightening political and social control throughout the country.

Tito was determined to rebut Cominform charges that his policy had been too lenient toward the peasantry. Collectivization would thus demonstrate Yugoslavia's credentials as a Marxist-Leninist state.

The pace of collectivization was rapid, reaching a maximum of 6,972 SRZs in April 1951, when about 17 percent of all peasant households, containing approximately 20 percent of the total arable land supply, were in collective farms.[10] Pressures to enroll were, on the whole, probably less intense than during Soviet forced collectivization. Peasants were given positive incentives to sign up their land, labor, draft animals, and implements for a standard three-year trial period; there were promises of modern agro-technical services, higher yields, and higher family incomes. But more direct forms of pressure were used as well. Highly discriminatory income tax rates--a flat 3 percent rate for SRZ members as against a steeply progressive rate (up to 90 percent in 1949) for non-members[11]--were bound to stimulate peasant resentment. Political and social coercion were not uncommon. Those who did join saw little, if any, improvement in their lives. Yields were at best only marginally better than on private farms. Technical supplies and services were woefully inadequate, an especially serious problem in the case of draft power, since many peasants had slaughtered their draft animals before joining the SRZs. Finally, managerial ineptitude within the SRZs demoralized the members and made them all the more conscious of their loss of economic independence.

Meanwhile, the situational factors which had seemed to require collectivization had been subtly changing. As confidence in their ability to withstand the Cominform danger mounted, Tito and his colleagues became more critical of the Stalinist model and began to consider alternatives. The most outstanding manifestation of this process was the introduction of Workers' Councils and the system of worker self-management in 1950. Steps were taken toward dismantling the machinery of centralized state direction of the economy. Change came more slowly in agriculture, but after the Sixth Party Congress in 1952 the new atmosphere of domestic detente had become irresistible. Although the growing demands of the peasants to leave the SRZ once their three-year contracts were up were strongly resisted, and new inducements were offered to make them reconsider (elimination of compulsory deliveries, promises of more generous funding and more machinery, etc.),[12] it was increasingly obvious that the SRZ experiment had been a costly, unnecessary, and politically dangerous mistake. The ensuing decline in the number of SRZs was as rapid as their growth had been three years earlier. By the end of 1953, only 1,165 SRZs remained; by the end of 1956 there were but 575.[13]

The Yugoslav leaders only grudgingly accepted the failure of collectivization. A new Regulation on Property Relations and Reorganization of Peasant Working Cooperatives, promulgated on 30 March 1953, further reduced the limits on individual property holding to 10 hectares of total land (15 hectares in certain cases) and placed other restrictions on the reestablishment of private farms.[14] Clearly the intention was to demonstrate to the peasants the lack of perspective for private farming and to pave the way for a further attempt at socializing the rural economy at some future date.

Therefore, de-collectivization did not signify the abandonment of socialist goals in agriculture. While legalizing the dismantling of the SRZ, the 1953 Regulation encouraged the peasants to reorganize themselves into general farmer cooperatives (OZZ). These rather amorphous cooperatives, both the form and content of which remained to be worked out, now became the major hope for undertaking the eventual socialist transformation of the peasantry.[15]

During the 1954-1956 period, however, the official policy was one of relative toleration toward the individual peasant farm, while the wounds of collectivization were allowed to heal. Some economists began to argue that peasant agriculture, under free-market conditions, could move the economy out of the doldrums left by collectivization, even without massive state investments.[16] These laissez faire expectations were disappointed. Political tensions did abate, but agricultural output did not significantly increase, and a dangerous rise in prices began.

Thus, the stage was set for another concentrated effort at socialization. This time the approach was much more gradual and flexible. A new Regulation on Farmer Cooperatives enacted in the spring of 1954 provided for the establishment of more or less traditional type multi-purpose cooperatives by ten or more peasants on the basis of limited liability (up to a maximum of ten times the value of shares held).[17] Basic capital was to be supplied by the state, for which the cooperatives were to pay specified interest charges. In the fall of 1955 cooperatives were assigned a central role in state crop procurements, replacing the network of state procurement stations as the main channel of centralized commodity purchases.[18] Private trade was not officially proscribed; however, through the cooperatives the state was attempting to strengthen its control over major commodity marketings.

At a special session of the Federal Parliament devoted exclusively to agriculture in April 1957, a major campaign of cooperative construction was announced as the centerpiece of a new agricultural policy. Past shortcomings and future goals were put into general perspective, and the commitment to agricultural modernization and social transformation in the village was reaffirmed.[19] The basic institutional framework of farm cooperation laid down at this time has remained in effect to the present time, even if the vigor of its implementation has fluctuated widely with the twists and turns of Yugoslav politics. Before we turn to a discussion of the institutional and operational patterns of the cooperative system, a few comments are in order on the application of worker self-management to agriculture.

The system of worker self-management is the hallmark of the Yugoslav version of socialism. It is based on the concept of "social" ownership of the means of production, as opposed to state ownership which is characteristic of Soviet-type systems. Under social ownership the labor force of a given enterprise manages it (by means of an elected Workers' Council, a management board, and a director elected by the council) as the steward for society as a whole, the ultimate proprietor, who "rents" the given property to the labor force. Somehow the workers reconcile their collective interest with that of society at large, for example, in making their decisions on production development and income distribution. Needless to say, there exist powerful institutions and procedures to ensure that the collective observes broad social objectives.[20]

Every citizen is theoretically enlisted into active participation in the self-management system in at least two ways: as a producer--through membership in a work collective, called a basic organization of associated labor (OOUR); and as a resident of a given commune, the basic local territorial-political unit. The Marxian socialist orientation of the system is reflected in the special emphasis placed on the former type of involvement. Indeed, legislative bodies at all three political levels--commune, republic, and Federation--contain special chambers for producer representation.

Given their proclaimed commitment to self-management as the essence of Yugoslav socialism, it is perhaps easy to understand why political leaders (and others) consider the situation of the private peasant highly anomalous. Still comprising almost 40 percent of the population and owning 85 percent of the agricultural land, the "individual peasant producer" has remained essentially outside the self-management system. The peasant has usually been underrepresented both as a producer and as a citizen of a territorial community. This civic disability applies to membership in social and political organizations such as the LCY as well (peasants make up only about 5.6 percent of LCY membership at present).[21] How much this isolation is disturbing to the peasant himself is difficult to determine. It is clearly a matter of grave concern to the regime. Discussing the agricultural scene in late 1955, Tito asserted, "A socialist country cannot have two systems of production--socialist and capitalist--if it wishes to create socialism in full measure."[22] Recent statements by Yugoslav leaders indicate that this viewpoint is held no less strongly today.[23]

Ideally, agriculture was to be reorganized along industrial lines in enterprises where the worker has no independent property claims. As will be seen later, the agro-industrial combines, which precisely fulfill these conditions, have received special support since the early 1960s. Nevertheless, for a number of reasons, combines have been unsuccessful in providing a "final solution" to the peasant question; nor is it expected that they will be able to do so in the foreseeable future. Thus, agricultural cooperatives have remained an important instrument for integrating the peasants into the system.

INTERNAL OPERATING MECHANISMS AND PROCEDURES

Both the operating mechanisms and the general orientation of the OZZ have varied considerably over time, even if their basic rationale has remained constant: to provide the nuclei for the association of peasant labor and its integration into the socialist self-management system. Farmer cooperatives, together with agro-industrial combines and agricultural estates and farms (*farmi*), comprise the socialist sector of Yugoslav agriculture. Since the early 1960s the combines have increasingly tended to dominate the socialist sector, chiefly by the absorption of cooperatives and estates. Table 3 shows the changes in the mix of the various types of farms in the socialist sector from 1955 to 1974.

TABLE 3. CHANGES IN THE COMPOSITION OF THE SOCIALIST SECTOR, 1955-1974

Year	Total[a]	No. of SRZs	No. of OZZ	No. of Combines, Estates, and Farms
1955	8,366	688	6,006	914
1957	7,586	507	5,472	776
1959	6,087	232	4,817	559
1961	4,133	127	3,228	469
1963	3,156	39	2,438	348
1965	2,559	10	1,937	282
1967	2,327	--	1,712	278
1969	2,073	--	1,279	271
1970	1,925	--	1,102	269
1971	1,813	--	993	270
1972	1,772	--	906	293
1973	1,745	--	832	306
1974	1,944	--	813	486

[a]The totals in columns 3, 4, 5 do not add up to the figures in column 2, which includes a residual category--"other socialist farms" of unspecified types.
SOURCE: Dilić, et al., eds., *Drustvene promene u selu* (Belgrade: Biblioteka sociologije sela, 1974), p. 36; *Statistički Godišnjak Jugoslavije 1975*, pp. 432, 433, 434.

The relative stability in recent years, and the sharp increase in 1974, in the number of combines, estates, and farms reflects the increasing concentration within the socialist sector around the combines, which represent the regime's ideal of the large-scale, vertically integrated socialist agribusiness enterprise. But to a great extent all three major types of socialist farms (combines, estates, and farmer cooperatives) have been performing the same kinds of functions: (a) direct agricultural production on their own land, (b) procurement and processing of agricultural products, (c) wholesale (and often retail) marketing of agricultural products, (d) providing various technical and financial services to individual peasant farms. This fourth function was originally envisaged as the exclusive mission of the farmer cooperatives. Under the quasi-free-market conditions prevailing in the Yugoslav economy, especially since the reforms of 1965, the first three of these functions proved to be most efficiently performed by the combines.[24] Meanwhile, economic conditions have tended to dampen enthusiasm in the socialist sector for the fourth function--by the OZZ no less than the combines and estates--thus further attenuating the claim of the cooperatives to a special relationship with the peasants.

These phenomena are only dimly reflected in Table 3. In total numbers the OZZs still appear comparatively strong; that is until one considers that the average OZZ now has about 250 hectares of its own land, as compared with almost 4,000 hectares in the average combine, and that in terms of large-scale farm machinery the average OZZ is only about 15 percent as well equipped as the combine.[25]

On a per hectare basis the average OZZ did not appear badly off in terms of equipment (over eight tractors and almost four combines per cooperative in 1969). This observation is relevant, however, for only one of the specific areas of activity of the OZZ--production on its own landholdings, its so-called *ekonomija*. The cooperatives also had broad service responsibilities toward the surrounding peasant farms. Under the increased emphasis on the cooperatives announced in 1957, the OZZ were charged with the introduction of modern mechanized farming methods in the peasant economy. Peasants were to be given access to newly increased capital investments in agriculture but only by tying their activities to the cooperatives. It was expected that the growing material dependence of the peasants on the socialist sector would lead to their gradual integration into the socialist economy.

In order to facilitate this process a new form of relationship with the socialist sector, less binding than full membership in the cooperatives, was introduced. This was "socialist cooperation" (*kooperacija*, as opposed to *zadrugarstvo*--cooperation as commonly understood in the West), a formal contractual relationship between the individual peasant producer and the cooperative for a specified period, usually one to three years.[26] In return for a commitment to sell his produce to the cooperative, the peasant received credits, farm supplies, agronomic counseling, and modern cultivation services. Peasants who signed such agreements were to be considered already part way toward inclusion in the socialist sector.[27] The arrangement promised to guarantee state influence on crop orientation and cultivation practices in the private sector, and it was expected that favorable economic results would induce the peasants to increase their involvement and eventually join the cooperatives as full members.[28] Table 4 shows the changes in the number of cooperative members and participants in kooperacija in recent years.

As already noted, in addition to their service functions for members and kooperacija contractees, farmer cooperatives customarily maintained demonstration farms (ekonomijas), originally to

TABLE 4. CHANGES IN THE NUMBER OF FARMER COOPERATIVES, COOPERATIVE MEMBERS, AND KOOPERACIJA PARTICIPANTS FOR SELECTED YEARS (1955-1974)

Year	No. of Farmer Coops.	Coop. Members (in thousands)	Participants in Kooperacijas (in thousands)
1955	6,006	1,285.8	--[a]
1957	5,472	1,518.2	55.1
1960	4,086	1,506.9	801.2
1962	2,816	1,397.5	876.8
1964	2,096	1,519.8	1,260.9
1966	1,712	1,345.0	1,240.8
1968	1,403	775.1	1,082.0
1970	1,102	840.8	928.8
1971	993	--[a]	863.5
1973	832	--[a]	712.0[b]
1974	813	--[a]	700.0[b]

[a]Data unavailable.
[b]Author's estimate, correcting for double counting.
SOURCES: *Statistički godišnjak FNRJ and SFRJ*, 1962, 1963, 1964, 1965, 1972, 1974, and 1975.

teach the peasants modern methods of production. The ekonomijas consisted of land supplied from state land reserves, supplemented by acreage rented from members and leased or purchased from local peasants unable to farm it themselves.[29] In time the ekonomijas became important production undertakings in their own right. For a while the expansion of their acreage was a major target of agricultural policy. Indeed, statistics on the growth of OZZ landholding were considered a measure of progress toward the socialization of agriculture.

The Regulation of Farmer Cooperatives of 1954 had provided for active participation by members in OZZ decision-making on a one-man one-vote basis. Questions of basic policy had to be decided by a full assembly of the members, while current business was entrusted to a management committee of elected members and an elected director.[30] Hired specialists and workers were eligible for membership upon purchase of shares and the payment of an entrance fee. Under the principles of worker self-management and the working class bias of the system, it was probably inevitable that these "outsiders" would eventually come to dominate the decision-making processes of the cooperatives.

For as the cooperatives became increasingly oriented toward profitability and toward the more easily controllable production activities of their ekonomijas, two things happened, both of which were detrimental to participation by the peasant members. First, there was a steady increase in the permanent labor force of the cooperatives. Second, there was a change in the basic character of the cooperatives themselves: they became self-contained business enterprises, almost indistinguishable from the agricultural estates and the newly emerging agro-industrial combines. The second phenomenon led increasingly to conflicts of interest between the cooperatives and their peasant members, placing the latter and the permanent work force in a condition of fundamental antagonism. Savo Cukanovic dates the beginning of the dominance of the workers in cooperative affairs to 1961, when mandatory share-purchase and entrance-fee payments and the principle of collective responsibility for debts were all abolished.[31] From then on, the workers tended to dominate the cooperative councils (the counterpart of the workers' councils in industry).

TABLE 5. REPRESENTATION OF WORKERS AND COOPERATIVE MEMBERS IN AGRICULTURAL COOPERATIVE COUNCILS FOR SELECTED YEARS (in %)

	1958/59	1961	1964	1970
Workers	28.4	47	51.3	53
Peasant members	71.6	53	48.7	47

SOURCES: Čukanović, *Zemljoradničko zadrugarstvo u agrarnoj politici Jugoslavije* (Belgrade: Mala poljoprivredna biblioteka, 1971), p. 82; *Statistički godišnjak Jugoslavije, 1971*, p. 66.

This situation has evidently had a demoralizing influence on peasant membership[32] and goes part way in explaining the decline in the number of members noted in Table 4. But the primary explanation must be sought in the realm of economics. Given what the farmer cooperatives had become by the early 1970s, neither the peasants nor the cooperatives were much interested in maintaining or expanding the membership.

If the efficacy of peasant members' participation in cooperative decision-making has been seriously diluted by the requirements of worker self-management, the participatory status of the peasants involved in kooperacija has been even less satisfactory. Participating peasants are organized into councils of cooperators to facilitate the coordination of production activities and provide a conduit for the dissemination of information from cooperative management. The councils also serve as a sounding board for peasant opinions and grievances, providing in theory a kind of organized forum for the articulation of peasant interests to the cooperative. It should be stressed that the councils have only consultative status and are in no way a true equivalent of the cooperative councils of members and permanent workers. When in the early 1960s agro-industrial combines and estates were encouraged to involve themselves directly in kooperacija with private peasants, they were expected to set up special units (*pogoni*) for dealing with the peasants. Councils of cooperators were also supposed to be established to advise these units. In practice, however, these councils appear to have been even less vigorous than their counterparts attached to the farmer cooperatives. Indeed, the majority of combines and estates neglected even to create special units for kooperacija, preferring to sign up the peasants on short-term contracts on a purely individual basis.

The stronger financial position of the combines allowed them to pay higher prices and extend more generous credit terms to the peasants, making them in the short run more attractive partners for kooperacija than the OZZ. However, with the establishment in 1965 of fixed markups for produce and livestock procured from individual producers, many combines lost interest in formal contracts with the peasants. Once again the cooperatives were left in a dominant position as the socialist organization of last resort for the private peasants.[33] By this time, unfortunately, the OZZ were in considerably weakened economic circumstances and were unable and unwilling to involve themselves in anything but purely commercial relations with the peasants. Under the harsh realities of economic life brought about by the economic reform of 1965, which seriously emasculated the socializing role of the cooperatives, many cooperatives in outlying regions were compelled to close up shop, leaving the local peasants to fend for themselves.

Since kooperacija played a central role in the development of the relationship between the individual peasant producers and the socialist sector for much of the post-collectivization period and was a major focus of official socio-economic policy in the village, it is important to have some idea of who the main cooperators were and the kinds of operations for which they sought assistance from the socialist sector. Table 6 shows the distribution by size of peasant farms involved in kooperacija with the socialist sector in 1962. That year was prior to the peak in popularity of kooperacija, but there is no reason to believe that the proportions significantly changed.

The data in Table 6 suggest some interesting patterns. First of all, it was the middle-sized farm (3-8 ha.) that was most likely to be involved in some form of kooperacija, although the larger farms were also slightly overrepresented (17.7 percent as compared with the average of 16.6 percent of all farms participating in that year). Secondly, except for the smallest farms, which were not self-supporting in any case, the distribution of participation among groups was fairly close to the distribution of farms by size (see Table 1b for 1960 for comparison). However, when it came to the kind of close involvement in production service activities that the regime favored most, the data in columns 5 and 6 of Table 6 suggest that only the smallest farms were very receptive. The

TABLE 6. DISTRIBUTION OF PEASANT FARMS IN KOOPERACIJA WITH SOCIALIST SECTOR FARMS BY SIZE OF LANDHOLDINGS IN 1962

Groups of Farms	No. of Farms in Kooperacija	Percent of Total No. of Farms in Group	Percent of Total No. of Farms in Koop.	Contracts for Production Services on More Than 70% of Area	
				No.	%
Up to 1 ha.	49,336	10.4	11.4	35,942	97.3
1-3 ha.	149,658	16.6	35.2	56,231	37.6
3-5 ha.	102,939	18.4	24.0	19,518	19.0
5-8 ha.	75,569	17.8	17.6	6,352	11.0
8-10 ha.	25,392	17.7	5.9	1,965	7.7
Over 10 ha.	25,382	11.8	5.9	1,697	7.6
TOTALS AND AVERAGES	428,276[a]	16.6 (15.45)	100.0	121,705	31.2 (30.03)

[a]These data are from a survey of approximately half the total number of farms in kooperacija in 1962.

SOURCE: Savezna prevredna komora, Glavni savez Zemljoradničkih zadruga Jugoslavije, *Razvitak i problemi kooperacije zemljoradničkih zadruga i individualnih proizvodjača* (Belgrade, June 1964), p. 23.

larger farms, which were best equipped with implements and draft power, were progressively less likely to seek production assistance from the cooperatives. This suggests that the peasants preferred to avoid close reliance on the socialist sector if they could. A high-level study of kooperacija in 1964 noted a tendency for a decline over time in the intensity of peasant interest in complete mechanical services by the socialist sector.[34]

In the middle 1960s a few of the medium- and large-sized peasant farms began to acquire obsolescent tractors from the socialist sector. In February 1967 it became legally possible for the first time for private peasants to purchase new tractors and other large farm machines.[35] By 1968, private peasants were already accounting for 75 percent of all domestically produced medium-sized tractors sold and a large share of imported machines as well. At the end of 1973 they owned almost 80 percent of all tractors in Yugoslavia.[36] Obviously there is little resistance to modernization among the Yugoslav peasants. In fact, many peasant households have been sending their members to work in Western Europe for extended periods to obtain hard currency to buy farm machinery and other farm and domestic appliances.

Further evidence of peasant acceptance of modern farming practices is their continuing interest in kooperacija for operations which they cannot efficiently perform with their own equipment.

Table 7 shows the changes in peasant reliance on kooperacija for various kinds of cropping operations.

TABLE 7. KINDS AND VOLUME OF KOOPERACIJA IN FIELD CROP OPERATION

Year	Volume of Operations on Individual Farms with Socialist Sector Machinery			
	Plowing (000 ha.)	Sowing (000 ha.)	Harvesting (000 ha.)	Threshing (000 tons)
1958	397	56	68	874
1959	560	137	213	1,356
1960	627	116	194	1,424
1961	674	129	130	784
1962	937	222	150	859
1963	1,154	251	379	1,454
1964	969	211	303	1,301
1965	1,058	226	230	1,538
1966	1,041	258	294	1,632
1967	931	282	305	1,687
1968	774	325	316	1,377
1969	703	351	331	1,435
1970	661	342	299	1,021
1971	627	372	308	1,188
1972	390	307	316	977
1973	435	308	269	836
1974	339	236	276	996

SOURCES: Edhem Dilić, et al., eds., *Društvene promene u selu* (Belgrade: Biblioteka sociologije sela, 1974), p. 43, table 24; *Statistički Godišnjak Jugoslavije 1975*, p. 173.

The general decline in the volume of plowing reflects the growing availability of tractors and simple hitch equipment in private hands. The more complicated machines are not yet widely owned by the peasants, and given their small landholdings there is not much prospect that they will seek them, but will continue to rely on the socialist sector so long as the price for machinery services in comparison with commodity prices makes this an economic proposition. The most recent data available (1974) suggest the effects of these price relativities.

Summing up the experience of kooperacija, we might do well to quote the conclusions of a recent Yugoslav study:

> Viewed as a whole, one cannot speak of a crisis of *kooperacija*. *On the basis of the data available one can conclude that kooperacija is stagnating or declining in crop farming*

> *but increasing in animal husbandry.* The decline in the volume of *kooperacija* in crop farming has been influenced by various factors, among which the following must certainly be mentioned:
>
> - the improved equipment of the individual farms with modern means of production (they possess today about 100,000 tractors),
>
> - the more favorable material position and greater ability of the individual farms to pay cash for farm inputs,
>
> - a decline in the interest of socialist subjects [farms] in *kooperacija* (the machinery pool of farmer cooperatives has significantly decreased).
>
> In animal husbandry the situation is different. The individual farms do not have the conditions for independence and are more directly led to cooperation with socialist subjects.
>
> But regardless of these data, *kooperacija* has played, and is playing today, a significant role. *It has contributed to the spread of modern technological processes in agricultural production and to the raising of the level of production know-how of the individual farmers. Kooperacija* has also created the conditions here and there for the speedier development of the village and the standard of living therein.[37]

Thus far the discussion has focused on operating mechanisms and procedures at the lowest level--in the cooperatives and in relations between the individual peasants and the cooperatives and other socialist-sector farms. To a certain extent this picture is misleading. For in fact the self-management bodies at this level are not at liberty to make economic decisions entirely as they wish. In addition to general economic and financial conditions which constrain the cooperatives' freedom of action there are several higher level institutions which set the parameters within which decisions are made.

The most general of these is the national economic plan. In Yugoslav practice the plan does not have the directive force that it does in Soviet-type economies. The "social plan" merely establishes basic goals of sectoral and regional development, which may or may not be observed. On the other hand, fundamental proportions for resource allocation--e.g., the share of funds to be made available for agricultural development--can affect basic decisions at all levels. In any case, the depressed state of agricultural finances in recent years has robbed the plan of much of its relevance for positively directing the policies of individual cooperatives. I shall have more to say on this subject in the following section.

Structurally, the cooperatives, like other socialist enterprises, are enmeshed in a web of formal governmental and economic institutions. They are subject to general agricultural enactments by federal, republic, and communal legislatures and to overall administrative direction by federal and republic secretariats for agriculture and communal secretaries (or communal economic and labor relations departments).[38] The cooperatives are also formally incorporated into the system of economic chambers (*komore*), which formulate general business recommendations. Finally, they are subject to the controls of the system of central and local agricultural, sanitary, veterinary, and plant protection inspectorates.[39]

Another important control agency which affects the cooperatives along with all other enterprises is the Social Accounting Service (SDK). The SDK monitors all financial transactions of enterprises through a nationwide system of registered accounts (*žiro računi*); fund transfers are debited and credited to the respective accounts of the parties.[40] When an enterprise is in deficit, the SDK has the power to declare its account "blocked" and restrict further transactions. Unless funds or credits are obtained, receivership and, ultimately, bankruptcy proceedings may be instituted. These procedures have been invoked against a number of farmer cooperatives in recent years and are frequently the stimulus for mergers with large combines.

The system of economic chambers is an interesting example of the effort to reconcile the desire for higher level coordination of economic activity without formally impinging on the rights of enterprise self-management. The system was introduced in 1962 in line with the policy of economic decentralization.[41] As of the beginning of 1969 the cooperative movement has been represented in the federal economic chamber by a Council for Agricultural Cooperation and Kooperacija. Similar councils exist in the economic chambers of the republics and communes. Membership in the system of chambers is obligatory for all cooperatives.[42] The formal list of functions of the cooperative councils includes the strengthening of cooperation; the development of self-management relations in agriculture; the integration of agricultural production in the private and socialist sectors; the implementation of federal regulations on agriculture and the formulation of recommendations for agricultural development; and the facilitation of linkages with other economic sectors.[43] In the new integrative atmosphere reflected in the 1974 Constitution, the chambers appear to be playing the role of a forum for exerting pressure on agricultural organizations and republican and local agricultural officials to commit themselves to joint financial and production undertakings.[44]

In short, self-management affords numerous opportunities for the cooperatives to make their own decisions, but there are many consensus-forming institutions in the system to insure that these decisions will conform to official policy.

MEASURES OF SOCIO-ECONOMIC PERFORMANCE

Before I attempt to assess the performance of the Yugoslav farmer cooperatives, it might be well to summarize briefly what these organizations are and what their basic objectives have been. From the descriptions presented in the preceding sections it is clear that they differ considerably from agricultural cooperatives as commonly understood in the West and in many of the developing countries.[45] One of the frankest descriptions of the farmer cooperatives has recently been given by Svetolik Popović, a prominent Belgrade agricultural economist closely connected with top policy-making circles:

> Our cooperative has never been an organization of peasants, nor has it had all the elements of association. The basic motive for the organization of the cooperative is the advancement of production on the basis of certain social relations. The public interest is emphasized as primary. For that reason the goal of the cooperative has been income and the strengthening of socialist reproduction, and not the reproductive capabilities of individual peasant farms. The focus of reproduction is the expansion of the resources of the public community, and the form of ownership is public property, which is connected with a definite territory, where the accumulated surplus value is created. These are public organizations for the development of the material base of socialism in the village.
>
> Thus, in its organizational structure and its operating principles the general farmer cooperative has had more the characteristics of an enterprise than an association. It has differed in many elements from the well known forms of cooperatives.[46]

The present agricultural establishment, headed by Ivan Kuštrak, is completely committed to the large-scale socialist sector as the only progressive element in Yugoslav agriculture. For Kuštrak and his colleagues the private peasant sector has no future in commodity production except to the extent that it is totally submerged in the socialist sector.[47] Not all Yugoslav agricultural spokesmen are so firmly convinced of the validity of this formula, especially of the ability of the socialist sector to assume the full burden of agricultural production. Nor are persons of this persuasion so ready to accept Popović's unfortunately valid characterization of the cooperatives, quoted above, as necessarily desirable. They recognize the positive benefits that the cooperatives have conferred on the peasants and see a continuing need for some kind of cooperative assistance to the peasants. In the concluding section we shall see that several measures have been taken recently to meet their arguments.

Turning to the question of socio-economic performance, the cooperatives clearly did not fulfill their original ideological purpose of convincing peasants of the superiority of large-scale socialist methods of farming and inducing them to tie their fate to the socialist sector. To the extent that the cooperatives were also intended to aid in improving agricultural techniques and raising household incomes in the individual peasant farms, they undoubtedly did have an impact.

There is considerable evidence of specific practical benefits from linkage with the socialist sector. This is especially true of the less intensive forms of association represented by kooperacija. Table 8 shows the effect on yields through kooperacija for the two main peasant crops, wheat and corn.

Yields on participating peasant farms were thus on the average 50-100 percent greater than on the farms of their nonparticipating counterparts. The disparity with the socialist sector is even more striking. The reason for these differential yields is the greater availability of quality hybrid seed, agricultural chemicals, and agro-technical advice and services for the participating peasants--not so intensively available as on the socialist sector farms, but considerably more so than on the farms of the nonparticipants, largely because of the high retail prices of these factors relative to producer prices in recent years. The declining acreage figures for both crops in kooperacija in recent years reflect the effects of these price relativities, which have tended to dampen socialist sector interest in signing up the peasants for kooperacija. Still, there can be little doubt of the positive influence of kooperacija in raising peasant yields and improving the quality of peasant livestock. Even nonparticipating peasants have benefited from the demonstration effect of the innovations brought by kooperacija. Improvements in peasant yields in the 1970s have been quite respectable, as Table 8 shows. The main barrier to the fuller dissemination of these innovations has been their high costs relative to prices for farm products.

Where they have been able to earn the trust of the surrounding peasants--in effect by ignoring the official policy signals--the general farmer cooperatives have had their share of success, too, although such cases are unfortunately not numerous. Dragoljub Simonović describes one successful venture with cooperative grain combining in the hilly regions of southern Serbia, where joint efforts helped to generate the classic self-help spirit generally sought by cooperativist ideologists. It began with kooperacija:

> *Kooperacija* in combining, which many cooperatives in the hilly-mountainous regions have undertaken with caution because of the high cost of the combine and the tiny size of the parcels of the members, has shown in some parts of this region a relatively new and more direct relationship of the members toward the cooperative. This new relationship has shown once again just how much the cooperative as a collective organization has grown up alongside the individual

TABLE 8. HARVESTED ACREAGE AND YIELDS OF WHEAT AND CORN BY TYPES OF FARM, 1964-1974
(Acreage--000 ha.; yields--quintals/ha.)

Year	Socialist Sector Harvested Area	Socialist Sector Yield	Peasant Farms[a] Harvested Area	Peasant Farms[a] Yield	Peasants in Koop. Harvested Area	Peasants in Koop. Yield
			a. Wheat			
1964	387	25.7	1,734	16.1	499	19.4
1965	292	32.5	1,025	15.9	363	24.3
1966	368	38.1	1,082	19.3	380	29.2
1967	397	38.3	1,074	19.8	409	28.6
1968	490	33.6	1,163	16.0	357	23.8
1969	502	36.0	1,165	18.5	353	26.0
1970	488	29.1	1,023	16.6	320	21.1
1971	486	40.6	1,170	23.2	272	33.4
1972	499	36.5	1,182	19.9	243	27.8
1973	419	42.2	1,087	21.8	191	31.9
1974	496	48.8	1,137	27.0	209	38.0
			b. Corn			
1964	183	51.3	1,719	22.2	528	41.7
1965	201	44.9	1,763	17.1	586	34.2
1966	205	57.3	1,705	24.1	590	45.7
1967	241	52.8	1,713	20.5	556	43.6
1968	245	52.2	1,720	20.6	495	40.1
1969	213	59.1	1,708	25.9	476	46.8
1970	190	55.0	1,726	23.6	436	43.0
1971	246	53.8	1,723	23.8	453	44.7
1972	241	55.9	1,740	27.3	402	45.6
1973	258	52.8	1,789	29.7	330	47.9
1974	230	57.8	1,699	35.9	327	49.2

[a]Computed as a residual from total acreage and total production.
SOURCE: *Statistički godišnjak Jugoslavije, 1971*, tables 107-9, 107-44, 107-52; ibid., *1975*, tables 108-8, 108-43, 108-50.

farms and how much the latter farms feel linked to the cooperative. The innovation in this cooperative relationship is that many cooperatives have acquired combines on the initiative and with the financial assistance of the members. Individual members have loaned the cooperative from 20,000 to more than 200,000 [old, presumably, -R.F.M.] dinars at 6% interest. This is more favorable than bank credit. On the other hand, the creditors are cooperative

> members, potential *kooperacija* participants, and self-managers [*sic*] of the given cooperative. The cooperative, thus, although it operates like any other kind of enterprise, nevertheless operates in a special relationship with its members.[48]

Simonović cites this example to urge greater official support for the cooperative movement as the key to the kind of socio-economic transformation the regime has long sought. Unfortunately for his argument, his examples were a relatively rare exception to the prevailing pattern of cooperative relations with the peasantry.

In discussing peasant attitudes toward the cooperatives it is important to distinguish between their feelings toward cooperation as a concept and toward the specific institutions with which they are forced to deal. Under Yugoslav law the registration of a cooperative gives it a monopoly on cooperation in a specified region. This reflects the regime's determination to retain control over whatever organizational activity may emerge in the countryside. There is some evidence that the peasants resent this monopolistic practice, especially where the given cooperative does not serve their interests adequately. In 1971 and 1972 the author informally questioned a number of peasants in central Serbia and the Vojvodina about their relations with the local *zadrugas* (farmer cooperatives). Their answers were invariably to the effect that, "We don't have any zadruga." In their eyes the local cooperatives were merely commercial jobbing shops which bought their produce at the lowest possible price and gave them no special advantages in return. Their attitude was well illustrated by the events in the central Serbian town of Gornji Milanovac in the fall of 1971. A group of peasants there had sought to establish their own cooperative to transport and market agricultural and handicraft products in Belgrade. The local authorities used force to break up their operation on the grounds that a cooperative already existed in the area to perform these functions. The peasants disliked dealing with this cooperative and merely wished to associate and "self-manage" their efforts on their own.[49] The coercive aspects of the current cooperative system must obviously be listed among the factors that diminish its attractiveness to the potential peasant membership.

On the other hand, there are several published studies and surveys which show a continuing readiness to participate in cooperative ventures even with the existing arrangements, at least at the level of kooperacija. A recent survey of peasant youth in Croatia, for example, showed that a small but significant proportion of those who were optimistic about the economic prospects of their family farms in the immediate future (32 out of the 620 who were optimistic) saw improved cooperation with the OZZ as the primary reason.[50] Of the 1,144 who thought the situation of the peasant could be improved, 98 chose some reply connected with cooperatives as the main avenue for improvement.[51]

A broader study of nationwide peasant opinion, based on surveys of six communes in widely scattered parts of Yugoslavia, published in 1974, found some interesting distributions of attitudes toward cooperation, the socialist sector, collective farming, and other aspects of agricultural policy. The majority of those giving a definite answer thought that the socialization of agriculture on a voluntary basis was impossible.[52] Similarly, a majority of definite replies indicated doubt that self-management in a cooperative or combine could ever provide the kind of care for the land which the private peasant exhibited toward his own land.[53] On the question of the utility of cooperation in specific farm activities, joint cultivation of the land had the least amount of support (28 percent), followed closely by joint ventures in animal husbandry (33 percent). The two most favored forms of cooperation were joint acquisition and use of farm machinery (65 percent) and joint supply and marketing operations for field crops (64 percent).[54]

On the basis of these findings it is difficult to agree with the conclusions of the Yugoslav research team that today's Yugoslav peasants are more favorably disposed toward cooperation than ever before, at least without injecting qualifications as to the form of cooperation and the formal status of the participant, which they fail to do.[55] The receptivity of the Yugoslav peasant to traditional forms of cooperation can hardly be questioned; his attitude toward the quasi-cooperation of the OZZ and other governmental-type farming enterprises is another matter entirely.

An extremely important consequence of Yugoslav agricultural policy, although other factors have been involved as well, is the high rate of exodus from farming. Future policy toward cooperation as well as other agricultural institutions will have to take this phenomenon of "de-agrarization" into consideration.

De-agrarization, the rapid decline of the active agricultural population, is a composite phenomenon with the following major elements: the increasing age of heads of farm households;[56] a growing tendency of rural youth to continue their schooling and to choose nonagricultural employment after graduation; changes in the socioeconomic profile of individual peasant farms, particularly the emergence of so-called "mixed" farms (*mesovita gazdinstva*) in which members of the household are primarily employed in nonagricultural jobs. All of these phenomena are characteristic of developing economies. Yugoslav commentators are somewhat ambivalent about the very rapidity of the process in their country, however. On the one hand, they exhibit a certain pride in the fact, which seems to signal the success of their developmental efforts. On the other hand, they are uneasy over the possibility that the best part of the rural labor force will disappear before effective alternatives to the private farm can be introduced.[57]

Considerations of space preclude a thorough treatment of the demographic details of de-agrarization. A few general remarks will have to suffice. Table 9 shows the decline in the agricultural (not necessarily the rural) population of Yugoslavia over the last half century.

TABLE 9. CHANGES IN THE TOTAL AND AGRICULTURAL POPULATION OF YUGOSLAVIA, 1921-1971 (in thousands)

Year	Total Pop.	Agric. Pop.	Nonagric. Pop.	% Agric. Pop.
1921	12,545	9,885	2,660	78.8
1931	14,533	11,132	3,401	76.6
1938	16,057	12,027	4,030	74.9
1948	15,842	10,606	5,236	67.0
1953	16,991	10,316	6,675	60.7
1961	18,549	9,198	9,351	49.6
1971	20,523	7,515[a]	13,008	36.6

[a]The agricultural population for 1971 is reduced here for the household members in long-term employment abroad.
SOURCE: Edhem Dilić, et al., eds., *Društvene promene u selu* (Belgrade: Biblioteka sociologije sela, 1974), p. 48.

The present demographic situation in active agricultural households is not encouraging. A recent study of farm population trends in the Vojvodina and Slavonia, the two main agricultural production regions in Yugoslavia, showed that 31.3 percent of farm households had no heirs to continue the family farm; 47.0 percent were sending their children to school to prepare them for nonagricultural occupations; and only 21.7 percent had heirs prepared to continue farming.[58] According to Professor Marković's projections, over the two decades from 1964 through 1984, 470,000 persons will have left the farm in the Vojvodina (agricultural population 738,000 in 1971) as a result of finishing their education alone.[59]

The problem of aging is particularly serious in this respect. The agricultural population is substantially smaller proportionally in all age cohorts up to 49 years than the nonagricultural population. In the important 25-34 age group the proportions are 10.9 percent to 15.5 percent.[60] In the older age groups the relationships are reversed. Table 10 shows the age structure of heads of farm households by region in 1969.

The number of farms without heirs that are headed by old people is growing steadily. The fact that social security and health care provisions for private peasants are virtually nonexistent makes the situation of these persons extremely precarious.[61] Some Yugoslav specialists see their plight as a golden opportunity to expand the acreage in the socialist sector--by offering pensions and health insurance to aged peasants in return for the donation of their land.[62] This may prove to be an acceptable solution for some peasants, but suitable provisions have yet to be worked out on a large scale.

TABLE 10. AGE STRUCTURE OF HEADS OF PEASANT FARM HOUSEHOLDS BY REGION IN 1969

Region	Total No. of Hsehlds.	Age Groups (in %) Under 20	20-27	28-34	35-44	45-54	55-60	61-65	Over 65
Yugoslavia	100	0.1	2.2	8.6	22.5	19.2	16.7	12.0	18.7
Bosnia & Hercegov.	100	0.2	3.5	13.7	27.9	17.7	14.0	9.6	13.3
Montenegro	100	0.2	2.4	9.1	19.2	17.0	15.5	10.9	25.7
Croatia	100	0.0	1.5	6.9	20.6	19.4	18.0	13.5	20.1
Macedonia	100	0.1	3.0	9.0	22.1	21.6	17.4	9.6	17.3
Slovenia	100	0.0	1.3	6.2	18.5	18.8	17.4	14.2	23.6
Serbia	100	0.1	2.1	7.5	22.1	19.6	17.0	12.2	19.4
-Serbia proper	100	0.1	1.8	7.0	22.3	19.3	16.7	12.8	20.0
-Vojvodina	100	0.0	2.3	7.3	21.4	19.7	19.8	12.2	19.3
-Kosovo	100	0.4	0.4	10.6	22.2	20.7	17.3	8.4	16.4

SOURCE: "Popis individualnih poljoprivrednih gazinstava:Statistički bilten," br. 720, in Edhem Dilić, et al., eds., *Društvene promene u selu* (Belgrade: Biblioteka sociologije sela, 1974), p. 62.

An equally important, if somewhat more ambiguous, element of de-agrarization is the change in the occupational character and income orientation of the individual peasant farms, particularly the emergence of the "mixed" farms. It has become common in the growing number of studies of social change in the village to use a four-fold classification system: (1) the "purely" agricultural farm, where all active members participate in farming; (2) the "mixed" farm (sometimes further divided into "primarily agricultural" and "primarily nonagricultural"), where part of the household members are employed as wage earners outside the farm; (3) nonagricultural households, where all active members are permanently employed outside the farm; and (4) farms without labor, where there are no active members to farm the land (usually aged peasants without heirs). Table 11 shows the changing distribution of private farms among these categories in the decade 1960-1969.

The mixed farm has proved to be a much more stable component of the Yugoslav village scene than had originally been predicted. Rather than a transitional phase on the way to complete de-agrarization, the mixed farm household appears to be a continuing economic phenomenon; the agricultural component of family income usu-

TABLE 11. CHANGES IN THE SOCIO-ECONOMIC STRUCTURE OF PEASANT FARMS IN YUGOSLAVIA, 1960-1969

Year	Purely Agricultural Farms		Mixed Farms		Nonagricultural Farms		Farms Without Labor	
	No.	%	No.	%	No.	%	No.	%
1960	1,533,329	58.5	874,089	33.4	143,462	5.4	67,233	2.6
1969	1,403,587	54.0	987,103	38.0	150,605	5.8	56,651	2.2

SOURCE: Computed from Edhem Dilić, et al., eds., *Društvene promene u selu* (Belgrade: Biblioteka sociologije sela, 1974), p. 25, table 9.

ally comprises an important supplementary element. This is partly a result of the particular Yugoslav situation where many industrial workers (mainly unskilled and semi-skilled) continue to reside on the family farm and commute long distances to work. It is of interest to note, incidentally, that by the mid-1970s about 50 percent of family income of all farm households is derived from outside the farm. In recent years the off-farm share of income has been growing at more than twice the rate of farm sources of income.[63]

The purely agricultural farms tend to be larger in area and better equipped with farm implements. It is these farms that the cooperatives and other socialist farms hope to involve in joint commodity-production ventures. The continuing decline in their numbers is consequently a source of no little concern to agricultural policy-makers.

I began this section on the socio-economic performance of the cooperatives with a quotation by Svetolik Popović on the true nature of the OZZ. His conclusion in that 1970 article was that only those cooperatives that were economically viable would survive and deserved to survive.[64] The special functions and the socializing mission of the cooperatives, to the extent that they were still of any relevance in official policy at the time, were as likely to be carried out by the combines and estates as by the cooperatives. On the surface it appeared that the decline in the importance of the cooperatives was directly related to the decline in their external performance. As we have seen, the number of cooperatives was decreasing, and their financial and technical capacity to provide services to the peasants was substantially less than the other institutions of the socialist sector.

When one looks more closely at the performance record of the cooperatives relative to the other socialist farming enterprises, however, one finds some surprising facts. On the average for the 1965-1970 period it turns out that the less integrated types of enterprise, such as the OZZ, had better financial results per unit of invested capital than did the more highly integrated enterprises.[65] In terms of income per unit of labor, income per 100 dinars of expended funds, and financial results per 100 dinars of invested capital, the cooperatives did as well as, or better than, the combines and estates.[66] These surprising findings could to a certain extent be attributed to the difficult financial conditions under which combines were forced to operate.[67] But these conditions also affected the cooperatives to some extent. One is forced to the conclusion that a large part of the reason for the shift in emphasis from cooperatives to the large-scale, vertically integrated combines was ideological: the combines simply fit better the official image of what a modern socialist agricultural enterprise should be.

At the same time, agricultural policy-makers, despite their obvious preference for the combine, apparently see no immediate prospect for the total absorption of the cooperatives. The latter have been left a large share of the business of dealing with the individual peasant producers, especially in the area of commodity purchases. Table 12 shows the continuing high proportion of such purchases handled by the cooperatives through 1968.

TABLE 12. PERCENTAGE SHARE OF FARMER COOPERATIVES IN TOTAL COMMODITY PURCHASES FROM INDIVIDUAL PEASANT FARMS, 1959-1968

1959	1960	1961	1962	1963	1964	1965	1966	1967	1968
82.6	86.8	85.5	85.1	81.2	74.7	82.6	82.8	83.7	78.7

SOURCE: Savo Čukanović, *Zemljoradničko zadrugarstvo u agrarnoj politici Yugoslavije* (Belgrade: Mala poljoprivredna biblioteka, 1971), p. 158, table 59.

The substantial dip in the cooperative share of purchases in 1963 and 1964 reflects the abolition of the cooperative monopoly in purchases of peasant commodities in their respective territories in June 1963.[68] The fixing of profit margins on the resale of commodities from the peasants in 1965, noted above, dampened the interest of the combines in this trade; consequently, the cooperatives restored much of their prominence in subsequent years.

As far as providing employment in the countryside is concerned, the role of the cooperatives and other socialist sector institutions has been a modest one. After rising to a maximum of 284,000 persons employed in the socialist sector in 1965, the number declined steadily; in 1970 the total was 202,000.[69] The share of the cooperatives in socialist sector employment has decreased quite rapidly. In 1962 the cooperative share was 45.3 percent; by

1969 it had declined to 36.5 percent.[70] Personal incomes for OZZ workers were generally lower than for their counterparts in combines and estates.[71] Increasing concentration of farming enterprises in the socialist sector and the rising technical level of production have meant a certain rationalization in the use of human resources. There is an increasing demand for higher professional and technical skills in the socialist sector labor force, precisely the skills that are likely to be lacking in the surplus peasant population. Therefore, the socialist sector will probably not be an important source of employment for redundant peasant labor. This is especially true of the cooperatives, which have become progressively less involved in direct production operations.

THE NATURE OF CHANGE, TRANSITION, AND EVOLUTION

The details of the changes in individual aspects of the Yugoslav farmer cooperative system and its relations with the peasants have been discussed in the previous sections. In this section some conclusions will be presented on major implications of these changes for the future role of peasant farming and cooperation in the overall context of Yugoslav agriculture. Finally, attention will be given to recent governmental responses to the agricultural situation.

Until the mid-1960s and the rise of the agro-industrial combines, the farmer cooperatives appeared to have a secure future in the projected socialist agricultural system. Collective farming, as such, had been totally discredited by the SRZ experience. Although the goal of socialization remained, there was no inclination to return to collectivization. Instead, official hopes were placed on the general farmer cooperatives, which by 1957 were envisaged as a mediator between the individual peasant producer and the socialist economy (specifically, the processing industry). This mediatory role was to have a social as well as an economic content. The higher household incomes presumably generated by economic involvement with the cooperatives were expected eventually to convince the peasants to intensify their dependence until ultimately they came to identify their own interests with those of a given cooperative. In the end the cooperative and its network of surrounding peasant farms would come to constitute an integral agricultural enterprise, which, under the principles of worker self-management, would be essentially no different from other kinds of socialist enterprises. Meanwhile, the national economy would be enjoying the payoff of increased agricultural output at a higher, more dependable technical level of production.

The effectiveness of the socialization process was understood to depend very much on the quantity and quality of the services the cooperatives performed for their actual and prospective members. To facilitate their proselytizing mission the cooperatives were given a virtual monopoly of the supply of agro-technical materials and services to the peasants and on the purchase of commodities from the peasant farms in their respective regions.

Owing partly to this monopoly, the cooperatives were fairly successful in carrying out many aspects of their mission. There can be little doubt that the cooperatives played a major role in the dissemination of agro-technical innovations in the countryside and in orienting the previously subsistence economy of the peasantry toward production for the market. These achievements were greatly magnified by the introduction of kooperacija, which extended many of the benefits of cooperation to peasants who were unwilling to associate formally with the OZZ as full members. On the other hand, there can be just as little doubt that the cooperatives largely failed in their socializing mission. Although they were willing to enjoy the material benefits of association with the cooperatives, the peasants remained substantially less willing to commit their own resources to joint ventures. Whereas in prewar Yugoslavia peasant cooperatives were able to rely on member savings for almost 40 percent of their operating funds (in 1938), the postwar OZZ were never able to approach that level of participation.[72]

Certainly among the main reasons for this failure was the conviction among the peasants that the cooperatives were not really *their* organizations--that they had little say in cooperative decision-making. The proletarian bias of worker self-management led inevitably to the domination of the permanent labor force of the cooperatives over the members in collective decision-making bodies. The peasants had even less opportunity to influence decisions in other kinds of socialist farms. The prohibition against forming cooperatives of their own where an OZZ already existed further contributed to the peasants' sense of powerlessness and isolation.

The most important reason for the failure of the socializing mission of the cooperatives and, ultimately, their economic mission as well was the series of radical economic policy changes which began in the early 1960s. One of the hallmarks of development policy in Soviet-type systems has been the extraction of capital from agriculture to finance industrialization. Until the mid-1950s, Yugoslavia essentially followed this policy. A major departure of the 1957 agricultural policy shift was the decision to plow funds back into the agricultural sector. The agricultural share of investment in basic capital rose from a low of 5.7 percent in the 1947-56 period to a high of 16 percent in 1959; of this, the proportion going to the cooperatives rose to 42 percent in 1958.[73]

By the mid-1960s, however, state agricultural investment and the cooperative share of it were allowed to decline once again. Agricultural enterprises were forced to compete with industrial and commercial enterprises for basic capital and operating funds on a substantially free money market. The cooperatives were less able to stand the competition than were the big combines, and they were forced to pare away many of their service functions, the linchpin of their socializing mission. A significant number of cooperatives in inaccessible mountain regions simply closed up shop, leaving the surrounding peasants without farm supplies, credits, agronomic and technical assistance, and, worse still, without a reliable market for their produce. Under the new economic conditions the service functions of the cooperatives withered away. They were replaced by a preoccupation with profits and short-term liquidity, which often witnessed the most brazen violation of formal contractual obligations with the peasants on purely commercial grounds. By 1970 little remained of the social mission of the cooperatives.

This blatant neglect of peasant interests inevitably led to disaffection in the countryside. Prohibited by the 10-hectare limit from rationalizing their farming operations, deprived of reliable production and marketing assistance from the socialist sector, yet forbidden to organize their own joint ventures, seriously under-represented in the organs of political power, and virtually unprotected by social security and health coverage, the Yugoslav peasants came to consider themselves the pariah class of the social system--a system which many of them had fought hard during the war to bring into being. Some peasants reacted by abandoning their farms. The majority became embittered and cynical, suffering (but occasionally benefiting from) the fluctuations of the commodity market and the fickleness of their socialist-sector "partners."

By the end of the 1960s, the rise in domestic social and nationalistic tensions and the uncertainties of the international situation--the Soviet-led invasion of Czechoslovakia and friction on the borders with Italy, Austria, and Bulgaria--made the discontent of the peasantry a matter of serious concern. The Ninth Congress of the LCY in early 1970 formed a special Commission to study the agricultural question and suggest a program of remedial action.[74] Other public organizations, such as the Socialist Alliance, also carried out special studies of village problems during 1970.[75] The recommendations of these various forums were essentially for more of the same, as far as institutions were concerned. However, more attention was to be given to peasant grievances over representation and participation in local government and cooperative affairs. Also, improvements in the economic aspects of the problem were promised.

The turmoil which beset the Yugoslav polity over the nationalities question and its political and economic ramifications evoked a characteristic search for solutions through constitutional amendments. One of the new amendments debated throughout 1970 and promulgated in June 1971--Amendment XXIV--was dubbed the "peasant amendment." It called for freer combinations of the peasants among themselves into cooperatives or in association with existing socialist-sector farm agencies.[76] This amendment was further revised when it was decided to introduce an entirely new constitution. Articles 61-63 of the 1974 Constitution have extended the right to form cooperatives and guarantee members in cooperatives an equal voice with the permanent workers in all self-management bodies.[77] These provisions signaled a new campaign to sign up the peasants in various forms of association with the socialist sector. Significantly, more intensive forms of relationship than kooperacijas are now being stressed, as the ideological aspects of cooperation have been brought once again to the center of the stage. The emphasis now is on longer-term involvement of the peasants and their resources in joint ventures with socialist-sector partners.[78] On the whole, neither the peasants nor the larger socialist-sector enterprises seem to be in much of a rush to take advantage of the new constitutional provisions.[79] This is not surprising in view of the past record of disappointments. Nevertheless, if the political authorities have the staying power to enforce the new orientation and the will to commit the necessary financial resources, there is no reason to believe that the new program of association will not be fairly successful. Who will benefit most from the program is difficult to determine at this point. Some peasants are expressing the fear that another wave of forced collectivization is actually in progress,[80] a charge the regime is anxious to refute; but there is no doubt that pressures to associate are intensifying.

Probably more important in the short run are the measures that have been taken to improve the economic position of the individual peasant farm and the socialist sector. The main innovation was the forging, in June 1973, after six months of intense haggling among the republics, of a so-called "green plan" for the development of agriculture through 1975.[81] The "green plan" involved the allocation of almost a billion dinars for agricultural development and stabilization over the period. Almost half of these funds were to be spent on the private sector; *but* they were to be made available only through socialist-sector enterprises for approved developmental projects on the basis of formal agreements of association. The expansion of livestock production for the export trade was an object of special attention.

Unfortunately, in 1974 the bottom dropped out of the world livestock market. Peasants who had accepted credits to expand their stock-raising facilities found their creditors once again reneging on obligations to purchase the animals for which they had contracted. Once again the peasants had been let down. In the latter half of 1974 and early 1975 it became impossible for the socialist sector to interest the peasants to sign up for "green plan" credits. There is ample evidence that short of force there were limits to the extent of the government's ability to influence the peasants to turn to cooperation if economic conditions were not right.

CONCLUSIONS

A common element in the programs of Marxist-Leninist regimes has been the commitment to transform the peasant village along socialist lines. The practical impact of this ideological commitment has, naturally, varied from country to country. In Yugoslavia the pragmatic, relatively tolerant system of worker self-management has permitted the evolution of types of agricultural institutions and processes which may be more relevant to the needs of developing economies than are the more rigid patterns and practices of, say, the Soviet or Chinese models. Yet in Yugoslavia, too, the ideological commitment has remained; its costs and constraints cannot be ignored.

Agricultural cooperation, as a specific form of property relations and decision-making processes, has been no more trusted in Yugoslavia than in other Marxist-Leninist countries. The group-property interests of traditional cooperation are seen to be in inevitable conflict with the general social interest as determined by the monopolistic political leadership. Autonomous cooperatives, it is feared, would be more difficult to enlist in the pursuit of the regime's goals or to harness to national economic and social plans. By strengthening the economic position of their peasant members, moreover, cooperatives ipso facto impede progress toward the creation of a classless society based on collective ownership.

Thus, true cooperatives have not been allowed to exist in Yugoslavia. However, the close association of the official surrogate cooperatives with the state has had some positive benefits. Especially in those periods when the socializing aspects of OZZ functions have been emphasized, the government has been willing to expend substantial amounts of capital on agro-technical and other forms of assistance to the private peasants. The result has been a considerable improvement in the technical level of peasant farming and a general rise in village living standards. Had this mission been consistently pursued, the results might in time have conformed in large measure to the objectives of the regime. However, the periodic flirtations with market-type relationships forced the cooperatives to neglect their more popular service functions. The consequence has been a feeling of insecurity and cynicism by the peasants toward the official cooperatives.

The prevailing line from the mid-1960s to 1972 was increasingly to write off the individual peasant sector as a major agricultural factor. The actual performance of the socialist sector alternative, however, has made it clear that such a policy was at best premature. Whether recent decisions to re-emphasize cooperative ventures with the peasants will be able to remedy the deterioration of the psychological and economic climate in the village it is still too early to predict. Current reports are not encouraging.

If the overall performance of the Yugoslav cooperatives has been disappointing in terms of the goals of the regime, specific forms and procedures developed to attain these goals have proved their vitality, at least from an economic standpoint. Kooperacija is probably the best example. The enlistment of peasant participation with state-sponsored agricultural enterprises in joint production ventures on a limited contractual basis is certainly a viable alternative to collectivization or total state control over agricultural production. There will undoubtedly be costs under such an arrangement, but from a political, social, and economic point of view these costs will probably be less than under a more rigid collectivist format. To be sure, the utopian goal of a total socialist transformation of the peasantry cannot be attained through kooperacija. But that goal has remained elusive even under the more "progressive" systems tried thus far in the socialist countries.

NOTES

1. Branko Horvat, "Yugoslav Economic Policy in the Post-War Period: Problems, Ideas, Institutional Developments," *The American Economic Review* 61 (June 1971), Part 2, Supplement, p. 71

2. Petar Marković, "Starenje sela," *Politika*, 27 February 1975, p. 13.

3. As late as 1948 peasants still numbered 49.41 percent of CPY membership. K. Bulajić, "Ideološko-vaspitni rad komunista u selu," *Borba*, 13 January 1949.

4. *Službeni list DFJ*, br. 64/65. [Cited hereafter as S.L.]

5. The 1945 reform granted 316,415 peasant households an average of slightly over 2 hectares each. Velimir Vasić, *Ekonomska politika Jugoslavije*, 5th ed. (Belgrade: Savremena administracija, 1970), p. 275.

6. Mihajlo Vučković, *Istorija zadružnog pokreta u Jugoslaviji, 1918-1941* (Belgrade, 1966), p. 58.

7. Savo Čukanović, *Zemljoradničko zadrugarstvo u agrarnoj politici Jugoslavije* (Belgrade: Mala poljoprivredna biblioteka, 1971), p. 20, Table 1. I am leaving out of consideration the traditional extended family cooperative (*zadruga*), which had ceased to have any significant socio-economic influence in most regions by the end of the interwar period.

8. See, for example, E. Kardelj's warning in a radio address in December 1945. "Glavne karakteristike ustave Federativne narodne republike," *Borba*, 7 December 1945.

9. In the 1945-1948 period so-called General Farmer Cooperatives (OZZ) were set up in the countryside as a quasi-state goods distribution network to compete with private traders for peasant surpluses above the compulsory delivery quotas. They were given stocks of scarce goods to induce the peasants to sell their produce at below free market prices. Čukanović, *Zemljoradničko zadrugarstvo u agrarnoj politici Jugoslavije*, p. 40.

10. Vasić, *Ekonomska politika Jugoslavije*, p. 285.

11. Horvat, "Yugoslav Economic Policy in the Post-War Period: Problems, Ideas, Institutional Developments," p. 146.

12. See, for example, Tito's speech in Kraljevo in late 1951. Josip Broz Tito, *O poljoprivredi, izabrani govori i članci* (Belgrade, Zagreb, Sarajevo, Skopje: Zadružna Knjga, 1967), pp. 111-112.

13. Ljubo Božić, *Agrarna politika sa osnovima zemljoradničkog zadrugarstva* (Sarajevo, 1960), p. 461, Table.

14. *S.L. FNRJ*, br. 14/53.

15. See the comments by Tito in Tito, *O poljoprivredi, izabrani govori i članci*, pp. 162-163; and by Edvard Kardelj in his *Neki problemi socijalisticke politike na selu, izabrani govori i članci* (Belgrade, Zagreb, Sarajevo, Skopje: Zadružna Knjga, 1967), p. 74.

16. Tito, *O poljoprivredi, izabrini govori i članci*, pp. 166, 196.

17. *S.L. FNRJ*, br. 5/54.

18. Tito, *O poljoprivredi, izabrani govori i clanci*, p. 176.

19. The Resolutions of this session were printed as a special supplement in *Borba*, 28 April 1957.

20. Among these are the League of Communists (LCY) (as the party has been called since 1952), the trade unions, and local and national political authorities. Major decisions must be ratified by a "social compact" involving community political authorities; thus the workers are not entirely self-managing.

21. Noted in speech by Stevan Doronjski at the Tenth Party Congress, in *Politika*, 29 May 1974, p. 3.

22. Tito, *O poljoprivredi, izabrani govori i članci*, p. 176.

23. See, for example, the Resolution "On the Socialist, Self-Management-Based Organization and Development of Agriculture and on the Tasks of the L.C.Y." adopted at the Tenth Party Congress, *Socialist Thought and Practice*, no. 6-7 (June-July 1974), pp. 227-34.

24. At least the Yugoslav political and agricultural leaders profess to be convinced that the combines are most efficient. There is a strong element of ideology in their reckoning, a cult of bigness and vertical integration for its own sake. Similar proclivities may be found in the present Soviet leadership. On the other hand, peasants often point out that the combines swallow up an inordinate amount of capital to produce their higher average yields; in short that they are *less* efficient overall. See the opinion survey in Edhem Dilić, et al., eds., *Društvene promene u selu* (Belgrade: Biblioteka Sociologije sela, 1974), p. 178.

25. Ibid., p. 36; my rough calculations on machinery apply to 1969 (data from Čukanović, *Zemljoradničko zadrugarstvo u agrarnoj politici Jugoslavije*, p. 121).

26. For an extensive treatment of the legal aspects of the kooperacija contract see Radomir Djurović, *Ugovor o kooperaciji u poljoprivredi* (Belgrade: Izdanje Saveza udruženja pravnika Jugoslavije, 1964).

27. Tito, *O poljoprivredi, izabrani govori i članci*, p. 207.

28. Ibid., p. 199.

29. In 1955 and 1957 legislation was enacted on obligatory cultivation of private farmland. Land uncultivated for one year, or poorly cultivated in the opinion of special local commissions, was placed at the disposal of local authorities for assignment to cooperatives or estates on a temporary basis. One effect of these laws was to induce peasants to sell or lease their surplus land to the cooperatives. The author was informed that since the early 1960s this legislation has rarely been enforced. *S.L. FNRJ*, br. 19/55; *S.L. FNRJ*, br. 12/57.

30. Ibid., br. 5/54.

31. Čukanović, *Zemljoradničko zadrugarstvo u agrarnoj politici Jugoslavije*, p. 159.

32. This is the opinion of a group of Yugoslav specialists. See Mihailo Vucković, Veselin Rasović, et al., *Udruživanje poljoprivrednih proizvodjača* (Belgrade, 1970), p. 26.

33. Čukanović, *Zemljoradničko zadrugarstvo u agrarnoj politici Jugoslavije*, p. 159.

34. Savezna privredna komora, Glavni savez zemljoradničkih zadruga Jugoslavije, *Razvitak i problemi kooperacije zemljoradničkih zadruga i individualnih proizvodjaca* (Belgrade, June 1964), p. 24.

35. *S.L. SFRJ*, br. 7/67.

36. Dilić, et al., eds., *Društvene promene u selu*, p. 33, Table 15.

37. Ibid., p. 45 (emphasis in original). Had they possessed the data for 1974 and 1975, the authors would have been much less sanguine over the performance of kooperacija in animal husbandry. This may be a temporary phenomenon, however.

38. Milivoje Kovačević, *Komunalni sistem i komunalna politika* (Belgrade: Zavodza izdavanje udžbenika Socjalisticke republike Srbije, 1968), p. 236.

39. Ibid.; see, also, Slavoljub Popović, *Upravno pravo, opšti deo*, 5th ed. (Belgrade: Savremena administracija, 1970), pp. 247-50.

40. For an example of the positive activities of the SDK see Rajko Nenadic, "Rezultati sprovedenog kliringa u poljoprivredi," *Ekonomika poljoprivrede*, no. 12 (December 1968), pp. 577-86.

41. *S.L. FNRJ*, br. 22/62.

42. For a good description of the structure and functions of the komora see Zoran Antonijević, *Privredno pravo*, 4th ed. (Belgrade: Savremena administracija, 1972), pp. 111-15.

43. Čukanovič, *Zemljoradničko zadrugarstvo u agrarnoj politici Jugoslavije*, pp. 206-207.

44. See, for example, R. Gavranović, "Pet milijardi dinara za razvoj stočarstva," *Politika*, 5 June 1974.

45. For a composite model of the agricultural cooperative see Robert F. Miller, "Socialism and Agricultural Cooperation: The Soviet and Yugoslav Cases," Canberra, Australia: Australian National University Press, Occasional Paper no. 9, Department of Political Science, Research School of Social Sciences, Australian National University, Canberra, 1974, pp. 3-12.

46. Svetolik Popović, "Poljoprivredno zadrugarstvo i samoupravljanje," *Socijalizam*, no. 4 (April 1970), pp. 494-495.

47. See the interviews with Kustrak in *Politika*, 19 October 1974, p. 5; and ibid., 9 July 1972 (Biserka Matić, "Put do obilja").

48. Dragoljub Simonović, "Uticaj kooperacija na raspadanje klasične seljačke strukture," *Ekonomika poljoprivrede*, no. 9 (September 1969), p. 591.

49. The event was briefly covered in an interview program on Belgrade television. After one set of interviews very sympathetic to the peasants, coverage was discontinued, presumably on the orders of higher political authorities.

50. However, this was only the sixth most frequent response out of a total of ten. Edhem Dilić, et al., *Društveni položaj i orijentacija seoske omladine* (Belgrade, 1971), p. 115.

51. Ibid., p. 282.

52. Dilić, et al., eds., *Društvene promene u selu*, p. 174.

53. Ibid., p. 177.

54. Ibid., p. 179.

55. Ibid., p. 182.

56. There is a growing literature on the problems of the aging peasant by Yugoslav scholars. A good example of recent writings may be found in the special issue of the journal *Sociologija sela*, no. 37-38 (July-December 1972) entirely devoted to "Old People in the Village."

57. This fear is expressly stated in an article by V. Conić, "Livade na poklon opštini," *Politika*, 11 November 1974, p. 26.

58. Petar Marković, "Domaćinstva bez omladine u selima ravničarskog područja Jugoslavije," *Sociologija sela*, no. 37-38 (July-December 1972), pp. 88-89.

59. Ibid., p. 90.

60. Dilić, et al., eds., *Društvene promene u selu*, p. 61.

61. For a poignant discussion of this problem see Svetozar Livada, "Naše drustvo i stari seljaci," *Sociologija sela*, no. 37-38 (July-December 1972), pp. 5-14.

62. For example, Petar Marković, "Domaćinstva bez omladine u selima ravničarskog područja Jugoslavije," pp. 92-94.

63. Dilić, et al., eds., *Društvene promene u selu*, p. 29, Table 12.

64. Svetolik Popović, "Poljoprivredno zadrugarstvo i samoupravljanje," p. 508.

65. Stanko Mirić and Danilo Pejin, "Uslovi i rezultati proizvodnje poljoprivrednih preduzeća raznog nivoa integrisanosti," *Ekonomika poljoprivreda*, no. 11/12 (November-December 1971), p. 827.

66. Ibid., pp. 828-29.

67. Ibid., p. 835.

68. *S.L. FNRJ*, br. 24/63.

69. *Statistički godišnjak Jugoslavije, 1971*, p. 126, Table 107-1.

70. Computed from ibid., and Čukanović, *Zemljoradničko zadrugarstvo u agrarnoj politici Jugoslavije*, p. 102, Table 31.

71. Mirić and Pejin, "Uslovi i resultati proizvodnje poljoprivrednih preduzeca raznog nivoa integrisanosti," p. 828.

72. Čukanović, *Zemljoradničko zadrugarstov u agrarnoj politici Jugoslavije*, p. 87.

73. Miller, "Socialism and Agricultural Cooperation: The Soviet and Yugoslav Cases," pp. 58-59.

74. See the Draft Theses formulated by this Commission published in *Ekonomika poljoprivrede*, no. 4 (March 1970), pp. 233-318.

75. The Serbian branch of the Socialist Alliance held a conference on 30-31 March 1970. In the unpublished stenographic record of the conference peasant delegates catalogued the ills of the present system in much the same terms I

have described in this paper. See Miller, "Socialism and Agricultural Cooperation: The Soviet and Yugoslav Cases," pp. 70-71.

76. Milivoje Kovačević, *Ustavni amandmani XX do XLII, sa objašnjenjima tekstova i pojmova* (Belgrade: Zavod za Isdavanje Udzbenika Socjalističke republicke Srbije, 1971), pp. 49-51.

77. *The Constitution of the Socialist Federal Republic of Yugoslavia* (Belgrade: Dopisna delavaska Univerza, 1974), pp. 115-17.

78. See the speech by Živan Vasiljević reported by V. Dobrijević, "Neophodno je masovnije udruživanje poljoprivrednih proizvodjaca," *Politika*, 11 December 1974, p. 9.

79. See, for example, S. Nikolić, "Zašto zaostaje udruživanje zemljoradnika," ibid., 25 November 1974, p. 7; also, L. Ivković, "Čekaju jedni na druge," ibid., 10 December 1974, p. 11.

80. S. Nikolić, "Zašto zaostaje udruživanje zemljoradnika."

81. "Zeleni program stabilizacije," *Komunist*, 28 December 1972, p. 4; also, Biserka Matić, "Solidna osnova za solidan početak," *Politika*, 15 June 1973, p. 7.

8. Social and Economic Organization of the Chilean Reformed Sector During the Popular Unity Government (1971-September 1973)

Jacques Chonchol

When the Popular Unity government came into being in 1970, Chile had already experienced six years of land reform under the previous administration. This administration had greatly transformed existing rural social and labor legislation and had passed the National Agrarian Reform Law which permitted large-scale expropriation of private property. Thus the legal and institutional groundwork had been laid for a significant restructuring of the traditional land tenure system. During the Frei years, some 1,500 farms had been expropriated.

There can be no doubt, however, that expropriation constitutes only the first step in the agrarian reform process. Simultaneously, it was necessary to create a new social and economic organization for the reformed sector; an organization that would benefit the greatest possible number of *campesinos* and allow them to deal efficiently with an agricultural establishment that for 30 years had been unable to keep pace with the ever-increasing needs of Chile's population.

THE PERIOD OF TRANSITION AND THE FUNCTIONING OF THE ASENTAMIENTOS

Agrarian Reform Law 16.640 established a *period of transition* after expropriation, during which the Agrarian Reform Corporation (CORA) retained possession of expropriated lands before turning them over to the campesinos. The normal transition period would be three years, but it could be prolonged to five. This period of transition served several purposes. One was the avoidance of a temporary drop in production that might result from poorly thought out decisions on the future use of the expropriated farm. In fact, even if a farm had not been very productive before expropriation, it might afterwards find itself in a worse situation, at least initially, as the result of an inadequate redistribution of land, water, and labor.

The experience of the old Colonization Agency, which was the predecessor to today's Agrarian Reform Corporation, had shown that production dropped on some farms during their early years as land reform "colonies." Yields dropped off while plans were made as to how to divide the land and water, while the required complementary investments were made, and while additional beneficiaries were selected. Therefore, in carrying out a large-scale land reform, an attempt was made to avoid repetition of the mistakes of those earlier experiments and the resulting negative effects that a temporary decrease in production could have on the general economy.

On the other hand, it was duly recognized that it would be difficult for campesinos working on reformed farms who had lived their whole lives under a paternalistic, dependent regime to begin overnight to take charge, to take the full initiative in developing the new farm enterprises that would emerge as the reform proceeded. They needed an intermediate period to retrain themselves, to further develop their initiative, and also to learn, through their own experience, the practical advantages of cooperative effort which was considered vital to the future success of the reform.

Also, before granting permanent title to the campesinos, investments had to be made in irrigation and other infrastructural works, and it was necessary to provide enough operating capital (machinery, cattle, household goods, etc.) to work the land successfully.

For all these reasons, it was considered indispensable to have a period of transition during which the expropriated farms would be worked and their unity preserved under a contract between the campesinos living there (the future beneficiaries of the reform) and the State. According to this contract, the campesinos contributed their labor, and the State, through CORA, contributed the expropriated land, credit, technical assistance, and training. This transitory stage came to be called the "asentamiento phase" (literally "settlement period"), and it was hoped that at the end of it the campesinos would have enough technical and managerial skills and capital to enjoy full success in the future development of their farms.

However, experience with the asentamientos as they were established showed that they had many operational defects that negated the expected benefits and greatly affected the campesinos. Some of these defects originated from the behavior of CORA and its functionaries toward the campesinos, and others from the attitudes of the campesinos themselves. Among the foremost errors was excess paternalism on the part of CORA and too much dependency on the part of the campesinos during the Christian Democratic government. For example, the yearly plans for working the farms and keeping their accounts were all determined and carried out exclusively by CORA with very little or no campesino participation. All necessary operating capital came from CORA, which also covered all short-run costs, including daily living expenses. Remunerations were excessively egalitarian and did not serve to stimulate quality and effort in the farm work. To the contrary, they discouraged more responsible workers, who under the system of collective operation earned only as much (or as little) as the least diligent and least responsible workers. Hard work did yield a commensurate benefit on the individual family plots. As a result, work on these plots was often at the expense of the collective work and subsidized in part by resources intended for the collective enterprise. Thus, individualistic attitudes and interests were reinforced and cooperative attitudes were actually discouraged.

Theoretically, this should not have happened, since when the annual accounts were balanced any excess profits were to be distributed among the campesinos in accordance with their work contribution. But in fact, the accounting was done with great procrastination, sometimes falling years behind. And it was done in offices far away from the campesinos and without their participation. It was very difficult to know at the end of each year if there had been profits or losses, and when there were profits, how much they amounted to and how to distribute them.

None of this helped the campesinos in gaining a spirit of initiative or an interest in collective management of the farm. The faulty accounting system and a lack of information on the real costs of producing each crop left the campesinos without the necessary information and incentive for lowering costs and improving production.

Moreover, various conflictive issues emerged among the campesinos on the asentamientos because of differences in status. The oldest heads of families living on the farms at the time of expropriation were really the "settlers" and were to be the full beneficiaries of future land allocations. The rest of the campesinos--unmarried workers, future workers, workers not living on the farm, etc.--constituted a temporary or occasional "labor force" which received a salary but was not eligible for land and other future benefits. Frequently, children or relatives of the campesinos living on the asentamiento were an important part of the labor force, as were campesinos who had worked for years on the farm without living on it (salaried workers, sharecroppers, etc.).

These status differences generated conflicts between the campesinos who lived on the asentamientos and members of the outside labor force. This often led the nonresident workers to press economic demands which reestablished worker-employer relations among the campesinos.

In addition, the settlers were often hesitant to accept new campesinos on the asentamientos, even when there was land available which, with more effective management, would have offered new occupational opportunities to unemployed sectors of the campesino population. This hesitancy was due to a fear that the greater the number of permanent asentamiento workers, the smaller the plot of land that would be allotted each of them under the final assignment of land.

Another problem was that the size of the asentamientos was frequently inadequate for more rational planning, for efficient use of resources, and for new forms of operation that would overcome the campesino's traditional concept of his relationship to the farm.

One must remember that the expropriations were carried out at the level of individual farms and that the expropriated part often included an area smaller than the original farm since many owners were allowed to keep part of their land as a reserve. This practice had several negative repercussions. On the one hand, the number of campesinos who remained on the expropriated part was often excessive considering the traditional techniques employed in exploiting the land. On the other hand, since the fixed investments were geographically located near the owner's house (which was part of the reserve), there was an imbalance between the fixed and movable capital on the expropriated land, which was insufficient for its productive needs, and the fixed and operating capital[1] on the reserve, which was excessive for its productive possibilities.

Thirdly, limiting the expropriation to what had been the boundaries of the former farm, or part of it, tended to reinforce the "localist" mentality of the campesinos, in which the traditional farm was of great significance to their view of life and work.

All these drawbacks demonstrated the inadequacy of operating on a farm-by-farm basis and the need to build a different kind of campesino enterprise which would create opportunities for more workers--not only those living on particular farms, but also other campesinos of the district or region.

Another problem was the great irresponsibility in the use of funds provided by CORA. The campesinos considered the loans to be CORA's obligation and were therefore not sufficiently concerned with repayment. When President Allende assumed office, the loans to the asentamientos by CORA that had not been repaid over the contracted amortization period amounted to more than 260 million 1970 Escudos. Between 1967 and 1970 the asentamientos had received loans totaling E°1,726 million for production and investment costs; in those years CORA should have recouped E°758 million in short-term loans, but the total repayment was only E°497 million, leaving an unpaid balance of E°261 million. This situation created serious problems for CORA, which the new government deemed it indispensable to correct in order to avoid a serious financial crisis. It would not be possible to survive such a crisis once the agrarian reform was well underway if appropriate measures were not taken in time.

The new government first sought to correct this complex of problems by attempting to distinguish the function of financing the farming operation from the total operation of CORA. It encouraged the campesinos to more clearly perceive that the funds and resources received as farming capital were neither salaries nor gifts from CORA, but rather loans which were to be repaid. It also attempted, through this same mechanism, to diminish the paternalistic dependence of the asentamientos and to make the campesinos more conscientious about the repayment and efficient use of resources which were to be handled in the future through the State banking system rather than through CORA. Toward these ends it determined that from the second half of 1971 onward all the annual capital would be contributed to each reformed farm, new or old, through the State Bank (Banco del Estado). It was further established that this financing would be granted on a monthly schedule in accordance with yearly work plans formulated by the campesinos themselves with the assistance of the representatives of CORA and the Bank.

This job was not easy because of the resistance of CORA officials, who through the mechanism of financing had become accustomed to controlling the asentamientos. Objections were also raised by officers of the Bank since many of the work plans showed up on their balance sheets as financially unsound. It was a fact that many asentamientos showed higher projected costs than incomes in their first annual work plan. This was unavoidable because of inefficient use of resources or excessively large work forces which had to be retained in spite of low levels of production, or because of the need to replace scarce fixed capital and other investments (cattle, machinery, etc.). The Bank felt that it was not the place of a banking institution, national or not, to subsidize the difference. The State Bank, the principal organization for agricultural financing, had in the past worked mainly with the private sector of *latifundistas*, or medium-sized farm owners, and was at first neither socially nor psychologically prepared to work with campesinos, whom many officials considered "poor risks." All such resistance was finally broken down by agreement between the Ministry of Agriculture and the highest officials of CORA and the State Bank, but this effort required constant attention during much of 1971 and even part of 1972.

AGRARIAN REFORM CENTERS

Having achieved its first objective, the new government contemplated the necessity of creating a new form of transitory organization for the reformed sector which would avoid those defects observed in the asentamientos. It developed a different concept of the transitory period, and the newly expropriated farms in this stage were called "agrarian reform centers."

The definition of the agrarian reform center was the subject of innumerable discussions in the government parties, where it was proposed as a final organizational form by some and by others as a temporary form to replace the asentamiento. In either case it was agreed that all new reformed farms should be organized as agrarian reform centers and that the older ones (those expropriated by the Christian Democratic government or during the first months of the Popular Unity government), already constituted as asentamientos, should continue as such for as long as the asentamiento period might last. As for the subsequent permanent assignation of agrarian reform center land, some parties held that there should be none, that the property should be maintained under government control; but the great majority felt, in accordance with the Popular Unity program, that after a period of transition the land should be allocated to the campesinos as communal property unless they themselves objected, in which case the farm would become state property and be called a "production center."

It was finally established that the agrarian reform centers (CERA) would be temporary economic and social organizations to be instituted on newly expropriated lands and on which the defects of the asentamientos would be corrected.

The general organization outlined was one in which each CERA would be formed from the union of two or more neighboring expropriated farms. It was hoped that this would foment greater solidarity among the different groups of campesinos, who had previously been isolated on separate farms, while at the same time forming work units of sufficient economic scope to permit economies of scale and greater labor specialization. In uniting farms, different factors were taken into account: the contiguity or reasonable proximity of the farms, the functional relations between them or the complementary nature of their resources, and the social and cultural relations between the campesinos on the farms and those of the district or region. Under special circumstances, a single farm could become an agrarian reform center (if, for example, it was sufficiently large). As far as the total number of farms that a CERA could include, care would be taken not to make it too large or its population too numerous so that efficient campesino direction and management would be difficult.

There were many objectives established for the CERAs, all of them based on the experience of the asentamientos and intended to overcome their difficulties. One of the foremost was to guarantee maximum campesino participation in the direction, management, and control of the land, so that the CERAs would be fundamentally democratic organizations, without distinctions of sex or civil status, and in which all would receive maximum experience in production tasks and social development.

A second goal was to provide the greatest employment possible for campesino workers, not only those originally living on the farms but also other campesinos who could gradually be incorporated. To offer such employment, it was necessary to take into account the initial productive capacity of the farm, its potential capacity, the intensification of investments, the diversification of crops or the substitution of crops which would be more efficient from the point of view of profits and labor employed, the industrialization or processing of some of the production within the CERA, as well as the development of general services supporting production and the social development of the campesinos, etc. All this implies, of course, that the goals of the CERA would be part of a planning process for CERA activities linked, in turn, to broader regional and national planning.

A third objective of the CERA was to improve the campesinos' standards of living by closely tying the system of incentives and rewards to production and to the effort of each worker, with the aim of correcting the negative effects which equally remunerated collective work had had on the asentamientos. Improved standards of living would also be achieved by allocating part of the farms' profits for the public good--houses, schools, clinics, recreational and cultural activities, etc.

A final objective was bettering the conditions of other campesinos living in the community where the CERA was located. The government hoped to link the improvement of the CERA campesinos' living conditions to that of the other campesinos in the community and thus reinforce the bonds of solidarity between the groups. Toward this end it was decided that a part of the profits earned by the CERA would be contributed to a communal, regional, and national fund for compensation and capitalization. This fund would permit the raising to equal status of workers with the poorest social and work conditions and would endow each rural community with the infrastructure needed for its fullest development. Thus the different agrarian reform centers needed to be closely tied to the communal campesino councils in their respective communities.

The definition of the CERA provided that membership could be claimed by all campesinos over 18 years of age, men or women, who worked permanently in the area bounded by the CERA, whether or not they lived therein. Their immediate families were, of course, also included. The CERA was so defined in order to avoid excluding laborers working on farms which became part of the CERA but who for various reasons might not live on that land. In addition, all other campesinos who later became incorporated into the CERAs had membership rights.

The policy became one of effecting progressive incorporation of new permanent workers as the CERA became more efficient and intensively farmed. It would therefore be necessary to increase the demand for labor through increased cultivation, substitution or diversification of crops, or better utilization of land through development of previously nonexistent economic infrastructural works and establishment of industrial enterprises related to agricultural production. The campesinos to be incorporated could be workers who had had some relation to the farm--sharecroppers, children of small farmers, paid laborers, or other landless campesinos of the community or region. They would be accepted by proposal of the communal campesino council and approval of the campesinos living on the farm.

In the internal structure of each center, the basic organization was the campesino assembly, composed of all members of the CERA over 18 years of age, men and women, permanent workers on the CERA and their families. Seasonal laborers who worked on the CERA could also participate in this assembly.

By secret democratic election this assembly designated the members of the other directing committees: the production committee, charged with directing production activities; the social welfare committee, charged with all social welfare activities (education, health, recreation, etc.); and the control committee, charged with supervising the activities of the different organizations of the center.

For the purposes of the production process, the workers were to organize themselves into "production teams," according to the nature of their activities and/or the physical location within the center in which they carried out those activities.

All these organizations sought the broadest participation by the greatest number of campesinos, without distinction of sex, in the different activities of the center, such that members would grow accustomed to the process of democratic self-direction.

The system of remuneration, linked to the production process, functioned according to rules of days worked, type of work performed, etc. The salary advances campesinos required in order to live could be different depending on the task they executed, but of course no campesino could earn less than the minimum campesino salary. The CERAs also provided incentives or economic stimuli linked to the production process and traditional campesino rights such as the use of a private plot of land and the right to pasture some large animals on communal lands.

Upon calculating the balance at the end of the year, any profits were allocated in differing proportions to: a social welfare fund for the center's campesinos, a capitalization fund for new investments needed on the farm, the compensation and communal capitalization fund already mentioned, and a fund to be distributed among the center's workers according to the number, type, and quality of work-days contributed by each during the course of the recently completed agricultural year.

As can easily be observed, the entire structure of this new transitory organizational form, the agrarian reform center, was designed to overcome the faults and weaknesses of the asentamiento experience.

PRODUCTION CENTERS

In addition to constituting the CERAs as the new general form of transitory organization for farms expropriated under the agrarian reform, another mode of social and economic organization for the reformed sector--this one permanent--was the production center.

The agrarian reform law recognized the possibility that because of their very special nature some kinds of farms ought not be turned over to the campesinos but rather maintained as state enterprises. Such farms would include, for example, large forest farms whose exploitation and conservation required planning for a long cycle, farms which were the source of supply of certain inputs vital to agricultural development (e.g., farms dedicated to the commercial production of high-quality seeds), and farms serving as research or experimental centers or model centers for the demonstration of new production techniques.

Also falling into this special class were some regular farms; there were several cases in which the nature of the farming conducted required large investments and also often employed a small number of workers. Typical of this type of farm were the large cattle ranches bordering on the Straits of Magellan. It was considered illogical to permanently allocate such lands to a small number of campesinos who would have to pay dearly for that allocation, given the high investment capital existing there.

It was therefore decided that upon expropriation these farms would be organized as production centers, directed by the State's representatives, but with campesinos participating in the management and partaking of the profits. Production centers were also to be formed when, at the end of the agrarian reform center period, the campesinos overwhelmingly felt that to be the most suitable form of permanent organization.

IMPLEMENTING THE NEW POLICY

In accordance with the above-outlined criteria, since the end of 1971 and after months of seemingly interminable discussion by the Popular Unity's National Agrarian Commission,[2] newly expropriated farms began to be organized as either agrarian reform centers or production centers. In practice, however, this effort faced various obstacles.

In the first place, there was delay and confusion in clearly defining the characteristics distinguishing the agrarian reform and production centers. This was the result of the different criteria proposed by the different Popular Unity political parties, which led to endless theoretical discussions without definitive resolution. Meanwhile, each field agent conceived of the center as having those characteristics that his party believed it should have. Campesino leaders who had no Popular Unity affiliation and had no clear idea to follow, and those who opposed the government did everything possible to distort the objectives of the new organizational forms in the eyes of the campesinos.

As a result of this political strife and the consequent lack of convenient, precise definitions of these new organizational forms and their relationship to the State apparatus, to other campesino organizations, and to the rest of the productive system, many agrarian reform and production centers were officially conceived and set up by field agents, with the campesinos who formed them having no idea of their theoretical characteristics. There was no previous training for campesinos to provide adequate comprehension of the why and wherefore of these organizations. Sometimes the campesinos even received widely divergent opinions of different aspects of the organizations from different government officials. All this contributed in a crucial way to a general disorientation and lack of information on the part of most of the campesinos.

This disorientation greatly facilitated the political opposition to the new organizational forms, especially to the CERAs, carried out by opposition parties and their militants (officials, campesino activists and directors, parliament members, political directors, etc.). The opposition by members of the Christian Democratic party was particularly virulent. It was very difficult for them to morally oppose a policy of expropriation carried out in accordance with a law passed during the Frei presidency. Moreover, since the great majority of campesinos were pressing for accelerated expropriations, opposition to expropriation would have lost political support among the campesinos. On the other hand, they perceived that opposing the CERAs, distorting their true characteristics, and presenting them as organizations managed and controlled by the State, never to be campesino controlled, would give them an effective political banner with which to increase their campesino support.

All possible means were tried: radio and written propaganda, attempts by activists, politicians, and militant officials to show the country and the campesinos that the CERAs were nothing more than disguised state property which would in practice impede the campesinos from ever controlling the land. This campaign was, of course, backed by other opposition parties and by the right-wing press. It was further abetted by the lack of clarity with which the government's politicians, officials, and representatives faced the situation.

Along with these organizational and political difficulties, another problem facing the CERAs and production centers was the resistance of certain sectors of asentamiento members who were in practice being converted into privileged campesinos through the process by which the asentamientos had been formed.

Even when, in an attempt to avoid these problems, the government determined that CERAs would be formed only on newly expropriated farms and that already formed asentamientos would continue as such, many asentamiento members mounted an intense campaign against the new organizational forms. They were influenced by the Christian Democratic party and expressed themselves through the asentamiento federations. The reasons given by the groups mobilized against the CERAs were that the CERAs were not organizations constituted under the agrarian reform law (in spite of the fact that they were) and that the new organizational form had been established without first consulting the campesinos, with the sole intention of nationalizing the land. What really was behind this campaign, aside from politics, was the clear understanding that the CERA would put an end to the privileges enjoyed by those few heads of families who had been original residents of expropriated farms, which in practice had been becoming asentamientos. Also carrying some weight was the lingering in many campesino minds of certain cultural tenets of the traditional system which held that once the owner was displaced the possession and control of his land reverted to the old families living on the farm, and only to them. The incorporation of more campesinos would therefore mean that the power and its benefits enjoyed by the original campesinos and their families would be diminished.

After the end of 1971, there were asentamientos, CERAs, and production centers in the country, even when their characteristics and structure were unclear and confused, as well as "campesino committees," the form used to temporarily organize the campesinos on expropriated farms when they themselves could not agree on whether to form asentamientos or agrarian reform centers. Once the land was expropriated and occupied, the campesinos had to be organized somehow in order to receive credit from the Bank, design their production plan, and work the land. The campesino committee was therefore designed as a temporary solution when the confusion generated among the campesinos by officials and political activists was too great.

In the face of this disorientation, in mid-1972 the government agreed to adopt several measures. The principal one was that the reform sector campesinos could adopt the form of CERAs, production centers, or asentamientos in accordance with the will of the campesino majority. This decision could be made provided the internal structure of the asentamientos would be modified (as agreed to by the campesino organizations) so as to correct its defects and make it more closely approximate the internal structure of the CERAs. Simultaneously, there would be an acceleration in the assignation of lands to those asentamientos that had completed the time legally required before such assignation could be made, a problem uncorrected since the days of the Frei administration.[3]

This speed-up of land allocations was intended to counter the political campaign being waged among the campesinos maintaining that President Allende's government had no intention of assigning the land as outlined in the agrarian reform law.

Nevertheless, until mid-1973 it was not possible to bring about a widely recognized agreement between the State's agrarian organizations and the campesinos on the internal functioning of the asentamientos. Nor was there a clearer internal structure for the CERAs and production centers, all of which contributed to continued confusion in the social and economic organization of the reformed sector.

Clarification became increasingly more urgent as innumerable asentamientos, CERAs, production centers, and campesino committees were differentiated in name alone, and all showed the very same kinds of difficulties and defects that the new organizational forms were intended to correct.

It must also be pointed out that only 80,000 campesino families out of the 350,000 with little or no land lived on the reformed sector farms, which covered 40 percent of the cultivable agricultural land. An important disequilibrium existed, therefore, which was in great part the result of the *latifundios*' low population density. This mal-distribution made the 80,000 families living in the reformed sector seem privileged in the eyes of the campesino majority with little or no land. This tended to produce an important contradiction among campesinos, and one to be overcome as quickly as possible. However, it could not be overcome, as some intended, only through the expropriation and incorporation into the reform of more farms from the middle-sized sector. It also required an increase in the productive incorporation of campesino labor on farms already part of the reformed sector. Only as that integration was achieved could the benefits of the agrarian reform be extended to the campesinos most alienated from the land.

THE POLICY OF LAND ASSIGNMENTS

All these problems led the government and the Popular Unity parties to try to define with the greatest precision possible the policy of assigning land to campesinos. Given the differences among the government parties, the following criteria were finally adopted between November and December of 1971:

a) That productive land would mainly be assigned as communal land. There would also be family allocations--property for a house and a garden for personal cultivation by each family;

b) That in limited cases, and only with the full agreement of the campesinos involved, state property would be established, structured on the model of the production center;

c) That in no case would there be allocations to individual farmers of productive land, except for the house and family garden, which could not exceed one-half of one basic irrigated hectare;[4]

d) That for the purpose of guaranteeing land to the small landholder producers who needed it and showed a certain degree of organization, a certain amount of the expropriated land would be allotted to campesino cooperatives composed of small farmers who requested land and demonstrated the inadequacy of that which they already held. Of course, this measure implied resolving the problem of the campesinos who had been living on that land, allowing them to be the first to be incorporated into the cooperative.

In accordance with these criteria and after the early months of 1972, there was an effort to accelerate as much as possible the assigning of land, beginning, of course, with the oldest asentamientos, some of which had already completed the five years of transition or had existed for three to five years. One of the greatest difficulties in this process was, however, the demand frequently voiced by campesino groups for the individual home-garden allocation to be two basic hectares. This demand, along with the limited human resources, was one of the principal problems faced during 1972 in assigning land on the large scale outlined in the agrarian reform policy.

These were some of the major difficulties with the reformed sector's social and economic organization during the Popular Unity government.

NOTES

1. The agrarian reform law did not authorize expropriation of movable capital (cattle, machinery, farm tools, etc.). Only land and investments tied to the land (constructed works, fences, houses) were expropriable.

2. Formed by representatives of the government parties who, together with the administrative authorities of the Ministry of Agriculture, formulated the general orientation of the agrarian reform process.

3. In fact, by the end of the Frei administration, roughly only one-third of the asentamientos that had completed the three-year legal time period had been assigned land in any of the ways outlined in the agrarian reform law. The other two-thirds had neither been assigned the land nor had they had their existence legally extended the two additional years set forth in the law. This was due to the fact that each assignment project was very complicated, having to complete a minimum of economic and social investments in the farm, select the campesinos to be assigned land, appraise the land and the improvements that had been made, and go through a series of legal procedures. In the face of this large task, with CORA's human and economic resources being limited, the general process of assignments had become bogged down.

During the beginning of the Allende administration, this situation tended to go uncorrected. A large part of CORA's human resources and other State services had to be concentrated on the work of expropriation and subsequent reorganization of the farms. Only at the end of 1971 could the assignment process be accelerated, and with limitations, beginning with the oldest asentamientos, that is, those which had already completed the legal duration of their temporary status.

4. A BIH is an abstract unit which represents productive potential rather than raw size.

9. Land Reform and Group Farming in Peru

Douglas Horton

THE SETTING

Peru is the fourth largest country in Latin America and the third largest on the South American Continent. Of the total population of roughly 14 million, three-fifths is urban and two-fifths rural. The country is sharply divided into three geographical regions. Forty-three percent of the total population, and well over half of the urban population, are located on the Coast--a narrow, arid plain which skirts the Pacific Ocean; 47 percent of the population reside in the Andean Highlands; and the remaining 10 percent reside east of the Andes, in the Peruvian jungle or *Selva* (Table 1).

Until recently Peru has been a predominantly rural country with mining, agriculture, and livestock raising in the Highlands dominating the economy. In recent decades, however, most demographic and economic growth has taken place in and around the country's major coastal towns, creating a highly urbanized Coast which is now Peru's most dynamic and modern region; the Highlands remain largely rural and traditional; while the Selva is the country's immense and sparsely populated frontier area.

Land use, landholding patterns, and the internal organization of large estates (*haciendas*) differ markedly between the Coast, the Highlands, and the Selva, and the problems and prospects of Peru's land reform and collectivization programs are intimately related to the characteristics of these regions.

All coastal agriculture is located in fertile, irrigated valleys which are favorably located with respect to domestic markets and shipping terminals on the Pacific Ocean. The country's principal export crops--cotton and sugar--and most of the produce consumed in Peru's major cities are grown on the Coast. Lack of irrigation is the most limiting factor of production in most coastal

TABLE 1. LAND AREA AND POPULATION IN PERU'S PRINCIPAL REGIONS, 1972

Region	Total Land		Cropland		Population	
	Million Hectares	%	Million Hectares	%	Million	%
Coast	14.0	10.9	0.8	27.6	5.8	42.6
Highlands	34.0	26.6	1.7	58.6	6.4	47.1
Selva	80.0	62.5	0.4	13.8	1.4	10.3
Total	128.0	100.0	2.9	100.0	13.6	100.0

SOURCES: CIDA, *Tenencia de la tierra y desarrollo socio-económico del sector agrícola--Perú* (Washington, D.C., 1966), Chapt. 1; General de División, E.P., Enrique Valdez A., Ministro de Agricultura, "El sector agrario, exposición en el CAEM, 24 de mayo de 1973" (Lima: CENCIRA, 1973), p. 2.

valleys. Prior to the land reform, large estates controlled the best-watered lands. Areas of coastal *minifundia* (sub-family-sized farms) were generally located on the poorest and least adequately watered lands along the margins of the river valleys. The largest and most profitable coastal estates produced sugar and cotton, and were operated as centrally managed wage-labor production units. Rice and potatoes were grown on both large and small farms.

Before the reform, rice estates on flat coastal lands were operated as wage-labor operations, but in the upper valleys, where terrain was rugged, land was often parceled among sharecroppers and labor-tenants, known as *colonos*. Most of the corn, beans, and truck crops grown on the Coast were cultivated by small private farmers. In recent decades large coastal estates mechanized heavily and cut their employment drastically, and on most estates sharecropping and labor tenancy arrangements were eliminated.

In the Highlands, agriculture is less technologically sophisticated, less productive, and more subsistence-oriented than on the Coast. In much of the region natural conditions do not permit cropping, and nearly everywhere the absence of all-weather roads limits the viability of commercial crop production. Most crops are grown for home consumption, and livestock products (primarily wool and meat) are the main exports of the region. In the *Puna* (high tableland) regions of the Central and Southern Highlands, no crops can be grown and sheep ranching predominates.

Prior to the reform, on both the Coast and in the Highlands most land was held by large estates. However, estates were organized quite differently in these two regions. Most coastal estates functioned as large-scale, centrally managed production units employing wage laborers, but the majority of Peru's highland estates were organized around some variant of the colono system, with residents farming small plots of land and/or grazing livestock on the estates' pastures and paying rent in commodities and/or labor.

A significant part of the highland population resides in traditional peasant communities which derive from the pre-Columbian *ayllu* (extended family). Originally, all land within the jurisdiction of these communities was considered communal property. Local government was strong and autonomous, and various forms of labor pooling and sharing were employed in agriculture and in the provision of public services. Pastureland was generally grazed in common, but cropland was parceled among *comuneros* (community members) for individual farming. Integration of rural highland areas into the national economy and society has resulted in the gradual loss of community lands and the erosion of communal organizations and traditions. Since the Spanish conquest in the sixteenth century, peasant communities have fought a losing battle to protect land against the encroachment of haciendas. Consolidation of the nation-state has reduced the autonomy and authority of communal governments. Within the communities, many comuneros have acquired private title to individual plots of cropland, and positions of leadership have been taken over by enterprising comuneros who advocate the private enterprise system. Despite the gradual breakdown of communal structures, most peasant communities continue to own some farmland and each community retains a hierarchy of public officials for the management of communal affairs. Labor pooling and exchanges are still common, although the wage-labor system is gaining in importance.

Although since the conquest haciendas have grown at the expense of small farmers and peasant communities, during the last century expansion of commercial estates has been exceptionally rapid on both the Coast and in the Highlands. The most extreme cases of land concentration on the Coast were in sugar-producing areas. In the well-documented case of the Chicama Valley, some 60 estates and numerous peasant communities lost their lands to two foreign-owned corporations.[1] On all these lands traditional crops, technologies, and landholding patterns were replaced by large-scale, highly mechanized sugar production, and the social structure of the entire region was drastically modified. The expansion of foreign-owned plantations and the resultant social and economic dislocations prompted the formation of Peru's APRA party in the 1920s, and help explain its traditional, pro-land reform posture.

Within the Highlands, the most pronounced land concentration processes have taken place in the High Puna, where land is unsuitable for cropping but offers great possibilities for extensive sheep grazing. As in the case of sugar on the Coast, expansion and modernization of sheep ranching in the Highlands produced serious socio-economic dislocations. Acquisition of communal lands by large corporations, elimination of small-scale peasant production, and specialization in extensive grazing pushed the rural population increasingly onto the diminishing resource base of the remaining peasant communities and stimulated migration to the Coast. Over time, boundary conflicts between livestock estates and neighboring peasant communities proliferated, and in the early 1960s organized peasant groups with an estimated 300,000 members invaded the lands of several livestock ranches, initiating a process of de facto land reform.

PERU'S LAND REFORM PROGRAMS

Early Land Reform Measures

APRA, Peru's largest mass-based political party (Alianza Popular Revolucionaria Americana), advocated land reform as long ago as the 1920s, but no substantial reform measures were implemented by the government until four decades later. In 1956 a high level commission was set up to study the land reform issue and in 1959 an Agrarian Reform Settlement Institute was established. However, no hacienda lands were redistributed until 1963, when organized peasants took over estate lands in several parts of the country. Special land reform decrees legalized the peasant occupations of hacienda lands in two regions.

In 1964 President Belaúnde signed the country's first comprehensive land reform bill (No. 15037). However, the law's provisions were so complex, its exclusions and exceptions so numerous, and the compensation authorized for landowners so generous that land reform was extremely difficult and costly to implement.[2] Neither the Executive nor Congress actively supported land reform; the Reform Agency was inadequately financed and staffed; and the land reform bureaucracy was controlled by officials who were sympathetic to estate owners. Consequently, very little land redistribution took place under the 1964 law. Instead, the government attempted to diffuse the peasant movement by promising land redistribution and by threatening not to give titles to land invaders.

Despite the fact that very little land was redistributed during Belaúnde's administration, the reform had a significant impact on the expectations and actions of both landlords and peasantry. Nonenforcement of the 1964 law had a detrimental effect on agricultural production and investment, and increased rural tensions and peasant pressures for real land reform. Fearing expropriation, landowners "mined agriculture," extracting surplus for investment elsewhere in the economy, and failed to conserve or improve their estates' land, equipment, and installations. Landlords also attempted to expel their colonos in order to avoid future conflicts and possible expropriation under the 1964 law.

The political campaigns of the 1950s and early 1960s led Peruvian peasants to believe that once in office, Belaúnde would carry out a massive land reform. When this did not materialize their hopes gradually turned to frustration and cynicism. By 1968 government officials, realizing this fact, feared that if a substantive reform program was not implemented, peasant movements and land invasions would spring up again.[3]

The Velasco Government's Land Reform Program

By the time the military, headed by retired army general Juan Velasco Alvarado, deposed President Belaúnde in October of 1968, it was clear that neither land reform nor any of the other ambitious reforms promised by Belaúnde could in fact be implemented by his administration. APRA had become allied with conservative groups in opposing the President, and corruption, poor administration, and internal dissension had brought the government's major programs to a near halt.

The first indication that the new government intended to implement land reform on a broad scale came in its "Outline of the Socio-Economic Policy of the Revolutionary Government" (December 1968) in which land reform figured as the first of five structural reforms to be implemented.[4] In this document the government promised to accelerate land redistribution and to grant preferential treatment to land reform beneficiaries in the government's agricultural promotion, credit, and technical assistance programs. In January 1969 arrangements for expropriation of the landholdings of the Cerro de Pasco Corporation were finalized, and in the following months a new land reform law was drafted.

The announcement of the military government's new land reform law (No. 17716) in June 1969, signified the beginning of a radical land redistribution process in Peru. In many respects the law is more radical than Belaúnde's 1964 bill and its provisions are much clearer and much easier to implement. More important, execution of the reform has been given high priority by the government and is considered the strategic first step in creating a new agrarian structure which will both promote agricultural growth and spread its benefits more widely among the rural population.

Three fundamental innovations introduced by the 1969 law are the following: first, the law called for the creation of a specialized branch of the judiciary ("El Fuero Agrario") to deal with all legal questions related to land reform; second, the new law prohibited all forms of land rental and required that all landowners reside on and work their properties directly; third, it required that all agricultural workers participate in the management and profits of the farms on which they work. In addition to drawing up a long list of specific "causes for expropriation," the valuation procedure was streamlined and ceilings were placed on cash compensations made to affected landowners. The 1969 land reform law has been complemented with a new water law, a new set of by-laws for agricultural cooperatives, and numerous other pieces of agrarian legislation.

After announcement of the law, rapid and efficient implementation of the land reform became the main responsibility of the Ministry of Agriculture. The financing and control of all public agencies dealing with agriculture became centralized within this Ministry, and a comprehensive agricultural planning procedure was instituted.

Land Reform by Region--the PIAR

While in the past no comprehensive plans for land redistribution were prepared and expropriations were made on an ad hoc basis, since 1969 the National Land Reform Office has drawn up global plans and priorities which have been elaborated and implemented by Zonal Offices through regional planning units known as PIARs ("Integral Rural Settlement Projects").

Implementing land reform by regions (rather than on an estate-by-estate basis) allows the Reform Agency to erase old estate boundary lines and create more rationally dimensioned enterprises. Within each PIAR, not only estate lands but all agricultural lands are subject to land reform measures, which include parcel consolidation and integration as well as subdivision of estates. To date, actions have concentrated on the redistribution of large estate lands, but in the future the minifundia problem is also to be attacked at the PIAR level.

Elements of a New Cooperative Agrarian Structure

As reflected in the following statement by Peru's Minister of Agriculture, the ideology of the land reform program is decidedly collectivistic:

> The Central element in Peru's New Agrarian Structure will be the "associative enterprises" ["empresas asociativas"] characterized by joint ownership and self-management. These enterprises are of two forms: Agrarian Production Cooperatives [Cooperativas Agrarias de Producción--CAPs] and Agrarian Social Interest Societies [Sociedades Agrícolas de Interés Social--SAIS].[5]

As detailed on pp. 223-25 below, the CAP is a conventional type of production cooperative which is owned and managed by its workers. The SAIS is a novel quasi-cooperative which operates much like a production cooperative, except that its ownership, management, and profits are shared both by its workers and by members of neighboring peasant communities. CAPs have been set up throughout the country. SAISs have been set up only in the High Puna, where impoverished peasant communities surround livestock ranches which have been expropriated under the land reform.

In addition to the CAPs and SAISs, which receive top priority under the reform, land has also been adjudicated to peasant communities alone and, in exceptional cases, to individuals. Peasant communities receiving land under the reform program are required to organize Communal Cooperatives for the management of these assets. Smallholders are encouraged to organize Agrarian Service Cooperatives, or, preferably, to pool their lands in Parcel Integration Cooperatives for joint production.

Furthermore, two types of Central Cooperatives are being established to provide marketing, credit, processing, accounting, technical and managerial assistance, and other services to the new CAPs and SAISs: Regional Central Cooperatives at the PIAR level, and Specialized Central Cooperatives for producers of export commodities such as sugar, coffee, tea, and wool. Each Central Cooperative is owned and managed by its affiliated base-level cooperatives and SAISs.

Land Redistribution Goals and Achievements

By the end of 1975 the government plans to redistribute approximately 9 million hectares of land among 340,000 beneficiaries. Eighty percent of this land is to be allotted to Production Cooperatives and SAISs, 16 percent to Communal Cooperatives, and only 4 percent to individuals (Table 2). Although the redistribution program will reduce the total area held in private farms by 8.6 million hectares, private farms will continue to hold 43 percent of the country's farmland. After the reform peasant communities will control 32 percent and CAPs and SAISs will hold 13 and 12 percent, respectively (Table 3). In total, about half the country's cropland and pastures are to be redistributed.

TABLE 2. REDISTRIBUTION GOALS OF THE LAND REFORM PROGRAM, 1976

Type	Number	Area		Beneficiaries	
		Thousand Hectares	%	Thousand Families	%
Individual Title	61,376	348	3.9	61.4	18.1
Production Cooperatives	818	3,777	42.0	174.5	51.3
Communal Cooperatives	372	1,400	15.5	59.2	17.4
SAIS	68	3,475	38.6	44.9	13.2
	62,634	9,000	100.0	340.0	100.0

SOURCE: Convenio, Ministerio de Agricultura, Agencia Internacional para el Desarrollo, Fundación para el Desarrollo Nacional, *Estudio de Evaluación del Problema de Carnes en el Perú* (Lima, 1973), Tomo 6. (Based on official data of Peru's Land Reform Agency, the Dirección General de Reforma Agraria y Asentamiento Rural--DGRA y AR.)

Between June 1969 and August 1973 over half the estates to be expropriated under the reform program had been affected. By the latter part of 1973 six times as much land had been expropriated and eleven times as much had been redistributed under the 1969 land reform law as under all previous agrarian legislation. In addi-

TABLE 3. PRE- AND POST-REFORM LAND TENURE STRUCTURE

Type of Land Tenure	Pre-Reform Situation (1961)		Post-Reform Goals (1976)	
	Million Ha.	%	Million Ha.	%
Private Farms	21.3	73	12.6	43
Peasant Communities	7.8	27	9.2	32
Production Cooperatives	-	-	3.8	13
SAIS	-	-	3.5	12
Total	29.1	100	29.1	100

SOURCE: Convenio, Ministerio de Agricultura, Agencia Internacional para el Desarrollo, Fundación para el Desarrollo Nacional, *Estudio de Evaluación del Problema de Carnes en el Perú* (Lima, 1973), Tomo 6. (Based on official data of Peru's Land Reform Agency, the Dirección General de Reforma Agraria y Asentamiento Rural--DGRA y AR.)

tion, over 90 percent of all land reform beneficiaries had received land under the military government's reform program (Table 4).

TABLE 4. LAND ACQUIRED AND REDISTRIBUTED, AND NUMBER OF BENEFICIARIES AS OF 31 AUGUST 1973

	Previous Period (as of 6/24/69)		Velasco Reform (6/24/69-8/31/73)		Total	
	Number	%	Number	%	Number	%
A. Land Acquired (000 has.)						
1. State Lands	96	44	121	56	217	100
2. Private Lands	739	14	4,553	86	5,292	100
3. Total/Average	835	15	4,674	85	5,509	100
B. Land Redistributed						
Area (000 has.)	321	8	3,511	92	3,832	100
Percentage of A.3	39		75		70	
C. Number of Beneficiaries (thousands)	14	8	152	92	166	100
D. Area per Beneficiary (has.)	23		23		23	

SOURCE: Data supplied by the DGRA y AR.

Complementary Programs

Several new programs of the military government complement land redistribution by expanding the land base; by establishing, consolidating, and supervising new reform enterprises; by implementing the water law and other facets of Peru's agrarian legislation; and by providing credit, technical assistance, and other services to farmers.

As noted earlier, responsibility for implementing the government's agricultural programs rests in the Ministry of Agriculture. Within this Ministry, responsibility for specific programs is delegated to specialized branches, the most directly involved in the reform process being the Land Reform and Settlement Agency and the Land Reform Investigation and Training Center. The Agricultural Development Bank is the principal source of credit for farmers, and the only source of credit for the new cooperatives and SAISs. A ministry-level agency known as SINAMOS (Sistema Nacional de Apollo a la Movilización Social) collaborates with the Ministry of Agriculture in setting up cooperatives and SAISs and supervising their activities.[6]

Since 1969 all public agricultural programs have given preferential treatment to Production Cooperatives and SAISs. Therefore, in spite of expansion and restructuring of the Ministry of Agriculture, it has not been possible to carry out the land reform without cutting back other programs. One of the most substantial cutbacks has occurred in the area of technical assistance to smallholders.

Extent of Collectivization

Despite aggressive implementation of the government's land reform and cooperative programs, the degree of collectivization has been limited. Prior to the land reform about half of Peru's cropland and pastures was held by peasant communities and by small and medium-sized private farms which were not subsequently subject to expropriation. The government has encouraged comuneros and small farmers to organize groups for joint production, but as a result of peasant opposition practically no group farms have been set up outside the land reform sector.

Within the reform sector the only fully collectivized production units are the CAPs that have been set up on the Coast, where estates were centrally managed prior to the reform. These enterprises farm 12 percent of the country's cropland and employ 8 percent of the rural population (Table 5).

In the Highlands and the High Puna regions, where nearly all haciendas employed sharecroppers or labor tenants prior to the reform, land reform beneficiaries have refused to pool their usufruct plots or to turn over their livestock for joint management. Consequently, it has been necessary to "bend the law" and organize CAPs, SAISs, and Communal Cooperatives in which only part of the land is managed by the group and part remains under the control of individual workers. Where peasant opposition to group farming has been

TABLE 5. CHARACTERISTICS OF POST-REFORM AGRARIAN STRUCTURE

	No. of Farms	Families		Total Land		Cropland	
		Thousands	%	Thousand Hectares	%	Thousand Hectares	%
A. Land Reform Sector							
Fully Collectivized CAPs	550	90	7.5	550	1.9	450	12.2
Partially Collectivized CAPs, SAIS, and Communal Cooperatives	700	190[a]	15.8[a]	8,100	27.8	1,050	28.4
Recipients of Individual Titles	61,400	61	5.1	350	1.2	300	8.1
Sub-total	62,650	341	28.4	9,000	30.9	1,800	48.7
B. Non-reform Sector							
Private Farmers	500,000	500	41.6	12,300	42.3	1,200	32.4
Comuneros	350,000	350	29.2	7,800	26.8	700	18.9
Full-time Workers	-	10	0.8	-	-	-	-
Sub-total	850,000	860	71.6	20,100	69.1	1,900	51.3
TOTAL	912,650	1,201	100.0	29,100	100.0	3,700	100.0

[a]Of the 190,000 members of partially collectivized reform enterprises, 150,000 (12.5 percent of all rural families) are member-workers and 40,000 are comuneros in SAIS who do not work in the Production Units. See pp. 225-28 for a detailed discussion of the SAIS.

SOURCE: Estimates based on the CIDA-Peru Study, official data provided by the DGRA y AR and preliminary results of the 1972 censuses of population and agriculture.

very strong, individual titles have been distributed. The partially collectivized cooperatives and SAISs farm roughly 30 percent of the country's farmland and employ one-eighth of the rural population. Land reform beneficiaries who have received individual titles constitute 5 percent of the rural population and control 8 percent of the cropland.

Within the partially collectivized reform enterprises (the term "reform enterprise" refers to all cooperative and quasi-cooperative enterprises set up under the Peruvian land reform program) about one-half of the cropland and two-thirds of the pastures are managed collectively, and the remainder is under the control of individual workers. Hence, a total of about one-fourth of Peru's cropland and one-third of the pastures is managed jointly by new peasant groups set up under the land reform program. The membership of these reform enterprises constitutes roughly one-quarter of the rural population.

CHARACTERISTICS AND PERFORMANCE OF NEW REFORM ENTERPRISES

Organizational Features

Most land redistributed under the Peruvian reform has been allocated to groups of peasants organized in Agrarian Production Cooperatives (CAPs) and SAISs. Communal Cooperatives and Agrarian Service Cooperatives are also being promoted by the government, and a few Central Cooperatives have been established to provide specialized services for the new base-level cooperatives and SAISs. The following paragraphs describe the principal organizational features of these new reform enterprises.

<u>The Agrarian Production Cooperative (CAP)</u>: As defined in Peru's cooperative by-laws, a CAP is an indivisible production unit in which the ownership and usufruct of all productive assets--land, livestock, crops, equipment, and installations--is collective; no individual production is permissible. Management is democratic and profits are distributed equally among the members or in proportion to the number of days worked by each. The membership is the supreme authority in the management of cooperative affairs. In CAPs with less than 500 members all members meet periodically in General Assembly; in larger cooperatives an Assembly of Delegates is elected to represent the general membership. In both cases democratic elections are held to select an Administrative Council, a Vigilance or Supervisory Council, an Education Committee, and one or more Special Committees. One member of the Administrative Council is selected as the Cooperative's President. Table 6 outlines the composition and principal functions of these bodies.

In most cases a professional manager is hired, and in a few cases (most notably in the large coastal sugar cooperatives) several technicians are employed. In smaller cooperatives managerial functions are performed by the cooperative's President and other

TABLE 6. CHARACTERISTICS OF GOVERNING BODIES OF AGRARIAN PRODUCTION COOPERATIVES

Governing Body	Composition	Principal Functions
1. General Assembly (Asamblea General)	All members of the cooperative.	1. Elect members of Administrative and Supervisory Councils. 2. Elect members of Special Committees. 3. Consider and approve Annual Reports and Accounts. 4. Allocate surpluses (within legal limits). 5. Approve annual Development Plans and Budgets. 6. Set financial limits on powers delegated to Administrative Council.
2. Assembly of Delegates (Asamblea General de Delegados)	100-200 members elected by General Assembly when cooperative membership exceeds 500.	Functions 3-6 of General Assembly.
3. Administrative Council (Consejo de Administración)	Not less than 5 members elected by General Assembly.	1. Responsible for day-to-day operations. 2. Appoint managers and grant them the powers necessary to carry out their work. 3. Submit an Annual Report and Accounts to the Assembly.
4. Vigilance Council (Consejo de Vigilancia)	Not less than 3 members elected by General Assembly.	1. Supervise actions of Administrative Council and Special Committees to ensure that these are adequate and within the powers granted to these bodies. 2. Check the cooperative's accounts. 3. Examine claims by members against actions by Administrative Council and Special Committees. 4. Submit a report on its activities to the Assembly.
5. Education Committee and Special Committees[a] (Comites Especializados)	Not less than 3 members elected by the operational or geographical unit for which the committee has been formed.	1. Plan the policy for the corresponding functional or geographical unit. 2. Advise on its execution. 3. Such other duties as may be delegated by the Administrative Council.

[a]Establishment of Special Committees is optional.
SOURCE: Cooperative Bylaws: D.S. No. 240-69-AP, Nov. 4, 1969.

member-workers. Where a CAP's members are not adequately prepared to manage the group's affairs, and where they cannot afford the full-time services of a professional manager, officials of the Ministry of Agriculture and/or the Central Cooperative may provide necessary managerial assistance.

Although profits are distributed equally among the membership, or in proportion to days worked, most CAPs pay their members a daily wage or monthly salary, the level of which depends upon each member's functions and pre-reform pay rates.[7] Within small cooperatives pay differentials of two or three to one are common. In larger enterprises the salaries of administrative personnel may be from five to ten times as high as those of field workers.

In practice most CAPs function as mixed cooperatives rather than as fully collectivized production units, and in all cases there has been a high degree of continuity between pre-reform land tenure patterns and the extent of collectivization in new reform enterprises. On the Coast, where estates were centrally managed prior to the reform, cooperative operations are fully collectivized. But in the Highlands, where the colono system predominated, land reform beneficiaries have refused to give up their usufruct plots and individual herds of livestock. In these cases, although land reform officials often pressed for the collectivization of colono production, staunch opposition from the peasants has necessitated the organization of cooperatives in which the centrally managed lands of the ex-estates are worked collectively but members retain possession of their usufruct plots. As will be discussed below, in many of these mixed cooperatives sharp conflicts exist between the economic interests of members and the management.

The SAIS: Peru's cooperative by-laws define the SAIS as a pre-cooperative institution to be established when the government determines that a group of land reform beneficiaries does not fulfill one or more of the requirements demanded for legal recognition of a cooperative. According to the same document, when the special conditions which required the creation of the SAIS cease to exist, the SAIS should then become a Production Cooperative.

Most SAISs have been established in highland areas where livestock estates are surrounded by peasant communities threatening land invasion. In these cases the principal objectives of the SAIS are to preserve the administrative and physical unity of efficient livestock ranches expropriated under the reform, while at the same time spreading the benefits of land redistribution as far as possible among the neighboring population.

Legally, each SAIS is owned and managed by a Service Cooperative,[8] made up of the workers of expropriated estates, and by a number of neighboring peasant communities which have been designated by the Reform Agency as land reform beneficiaries.

Each SAIS has two principal spheres of activity: (1) agricultural and livestock production on the lands allocated to the SAIS--termed Production Units; and (2) technical and developmental assistance to affiliated peasant communities. The SAIS' operations in the former sphere fall under the jurisdiction of Agricultural, Livestock and Administrative Divisions; activities in the communities fall under the jurisdiction of a Development Division (Figure 1). All work in the Production Units is carried out by the workers of estates turned over to the SAIS. Comuneros residing in affiliated peasant communities do not work in the Production Units.

FIGURE 1. ORGANIZATIONAL DIAGRAM OF THE SAIS "CAHUIDE"

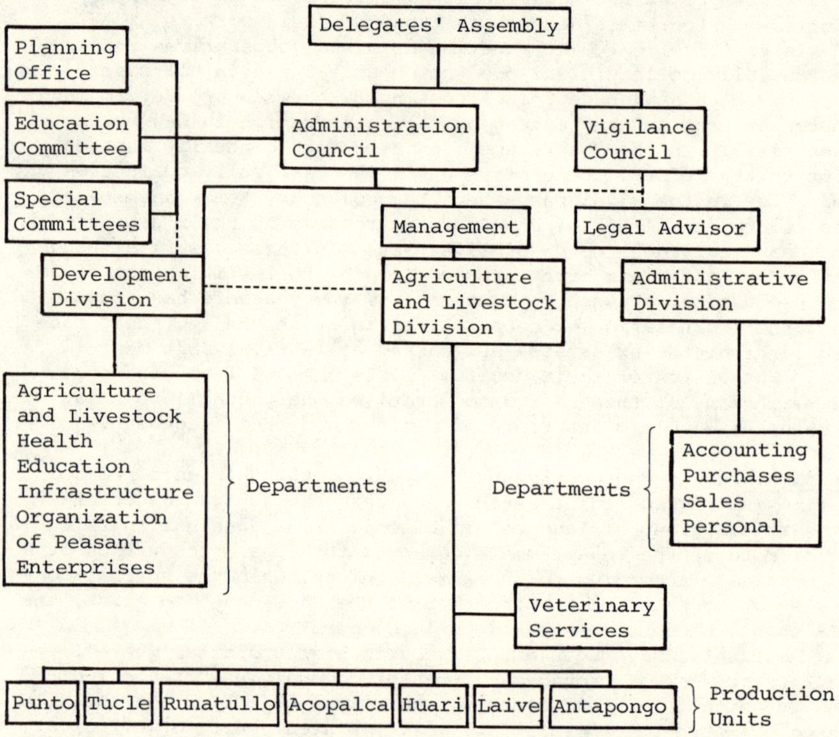

In determining which peasant communities should become members of a SAIS, the Ministry of Agriculture employs the following criteria:

(1) Communities must border on lands of the expropriated estates.
(2) Over 40 percent of the comuneros' incomes must come from agriculture and livestock raising.
(3) The productive capacity and per-capita family income of the communities must fall below the average for the area.

(4) Communities with border conflicts with expropriated estates are favored over those with no such conflicts.

Prior to land reform, the incomes of workers on highland livestock ranches were significantly higher than average income levels in neighboring peasant communities. The SAIS attempts to reduce the income gap between ex-estates and peasant communities through two principal mechanisms: (1) distribution of SAIS ownership and profits among the communities and the service cooperative in inverse proportion to their pre-reform income levels; (2) provision of technical assistance and credit to the communities via a "Development Division" which operates exclusively for the benefit of the affiliated communities. The poorer the community, the greater its share in profits and the more services it receives from the Development Division.

The governing bodies of the SAIS correspond to those of the CAP. The Service Cooperative and each affiliated peasant community have equal representation in the Assembly of Delegates. Since each SAIS has several community members and only one Service Cooperative, the communities not only receive most of the SAIS' profits, but also control management of the enterprise. Thus, in this complex reform-enterprise type the major shareholders (the communities) manage production and receive most of the profits but do not work on the SAIS, while the SAIS workers (ex-workers of the ex-estates) do not participate significantly in either management or profits.

Since most SAISs are very large enterprises (average membership 1,100, and land area 54,000 hectares) each employs a professional manager and highly paid accounting and technical personnel. Within the Service Cooperatives, pay rates approximate those on the estates prior to land reform. Most profits are distributed among the affiliated peasant communities, but for two reasons this profit distribution has little impact on the incomes of comuneros. First, the number of comuneros is very large and the average value of profits per comunero is very low.[9] Second, profits are not distributed in cash, but in funds for infrastructure investment, such as bridges, roads, and storage facilities.

As in the highland CAPs, most land reform beneficiaries organized in SAISs continue to farm small plots of land and/or graze their own livestock on the reform enterprises' land. Since most SAISs have been set up in areas of the High Puna, where livestock raising predominates, conflicts have arisen between management and membership over the use of pastureland. In many cases progressive charges are levied for member use of SAIS pastures in order to reduce the size of their herds. In the Central Highlands individual herds are small and are segregated from SAIS livestock. But in the South (Department of Puno), the number of livestock owned by SAIS members often exceeds the number owned by the reform enterprise itself. Since the reform, on most SAISs the size of members' herds has remained roughly constant, but SAIS managers hope to reduce these herds in the future.

Communal Cooperatives and Restructuring of Peasant Communities: Traditional peasant communities receiving land under the reform program are required to exploit this land collectively via Communal Cooperatives. The internal organization of these cooperatives is the same as that of Agrarian Production Cooperatives. The only difference between the two is that the members of the former must be comuneros, whereas in the latter case anyone may join the cooperative.

Title X of the 1969 land reform law calls for the formal restructuring and "cooperativization" of all peasant communities, and a 1970 Statute for Peasant Communities provides guidelines for restructuring. The new Statute requires three major reforms within the communities: (1) replacement of traditional communal authorities with administrative and supervisory councils; (2) purging community membership lists of individuals who do not reside full time in the community, who own farmland, and/or who derive most of their income from nonagricultural sources; and (3) collectivization of landownership and land use.

As noted earlier, most cropland within the communities is held in usufruct plots by individual comuneros. Many comuneros own farmland or run small shops, and most community leaders spend a great deal of time outside their community each year employed as workers, driving trucks, or engaged in commerce. Hence, many comuneros view restructuring as a threat to their economic interests and have opposed the government's attempts to implement the new statute.

By the end of 1973, very few peasant communities had been formally restructured. Communities benefited by the land reform have set up Communal Cooperatives to manage the new lands received, but in all cases individual comuneros have retained possession of their traditional usufruct plots.

Agrarian Service Cooperatives and Parcel Integration: Smallholders may form two types of agricultural cooperatives: Agrarian Service Cooperatives and Parcel Integration Cooperatives. In the former type, peasants who operate farms of less than three times the legally defined "family-sized unit" organize to provide themselves with input supplies, credit, marketing, or other services cooperatively. Agricultural production per se remains organized on an individual basis. In the second type, smallholders cede landownership or usufruct rights to a cooperative in order to farm these lands jointly.

Neither kind of smallholder cooperative has been actively promoted by the government since 1969. Most of Peru's landholding peasants are opposed to the pooling of land for joint farming, and no parcel integration cooperatives had been established to date. The previous administration promoted service cooperatives, however, and many of those set up continue to function. A few others have been formed in recent years, largely as a result of peasant initiative.

Central Cooperatives: Specialized Central Cooperatives: Prior to the land reform, large estates producing export commodities maintained offices in Lima to handle their financial and commercial affairs. In addition, producers' associations provided them with specialized services. After expropriating these estates and abolishing the producers' associations, the government set up Specialized Central Cooperatives to take over the functions of their original Lima offices. The organization and operation of the largest and most complex of these--the Central Sugar Cooperative (CECOAAP) --is outlined in the following paragraphs.

The CECOAAP's principal functions are the following: (1) management of credit for both itself and its member cooperatives; (2) purchase of inputs; (3) marketing; (4) research and scientific investigation; (5) technical assistance; (6) centralization of accounting, auditing, and control systems; (7) professional, technical, and cooperative training for members of affiliated cooperatives.

The Central Sugar Cooperative has twelve member cooperatives, each of which elects four representatives to the Assembly of Delegates. The Assembly, in turn, elects one representative from each cooperative to the Administrative Council. This Council, which is the principal decision-making unit within the CECOAAP's governing structure, is assisted in its tasks by several Special Committees, and its operations are supervised by a five-man Vigilance Council.

The Administrative Council hires a general manager, who is responsible for implementing its decisions. Implementation is carried out through a complex organizational structure which includes six specialized divisions. The manager is assisted in his functions by the CECOAAP's Administrative Division, the Legal Department, and a Comptrollership Unit (Figure 2).

FIGURE 2. ORGANIZATIONAL DIAGRAM OF THE CECOAAP

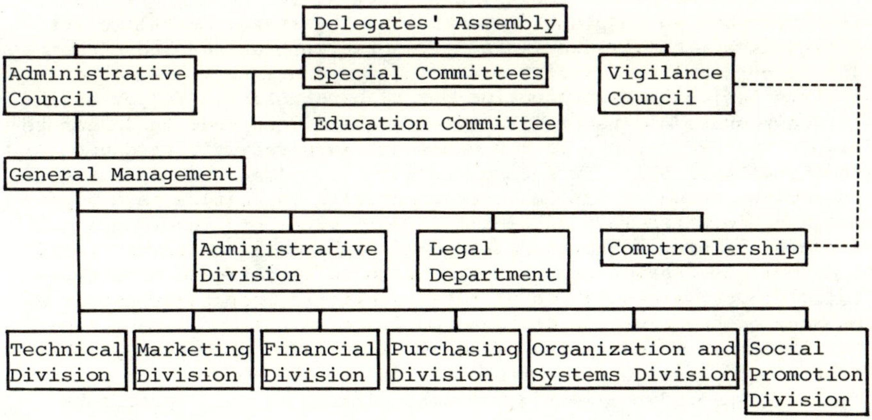

Among the six specialized divisions are the Technical Division, whose functions include scientific investigation, extension, and technical assistance to affiliated sugar cooperatives; the Financial Division, which administers the CECOAAP's finances, negotiates loans for members, and manages their insurance policies; the Organization and Systems Division, which operates a computerized data-processing center that processes accounting data for member cooperatives; and the Social Promotion Division, which provides advice, support, and coordination for member cooperatives in the areas of health, education, housing, and public relations.

The CECOAAP occupies 10 floors of a modern skyscraper in Lima, and employs a staff of nearly 400. In 1973 its operational budget was $4.4 million. Peru's twelve sugar cooperatives are the country's largest agricultural exporters; hence the impressive size and complexity of the CECOAAP. While the other Specialized Central Cooperatives have much smaller budgets and staffs and more limited functions, the CECOAAP has served as a model for their organization.

Regional Central Cooperatives: Until 1973 the Ministry of Agriculture gave top priority to the establishment and consolidation of base-level reform enterprises and a few Specialized Central Cooperatives. Since 1973 more attention has been paid to the creation of Central Cooperatives at the PIAR, or regional, level. According to government plans these second-level units are to facilitate the redistribution of income among base-level enterprises, and play a strategic role in implementation of the government's agricultural production plans and rural development policies.

The internal organization of Regional Central Cooperatives is similar to that of the CECOAAP, outlined above. The Regional Central Cooperatives are to plan production in accordance with national and regional plans drawn up by the Ministry of Agriculture. In addition, they are to be responsible for the purchase of inputs utilized by their members and the sale of commodities produced by them. Investments in processing facilities and economic infrastructure are to be planned and executed by these Central Cooperatives, as are socially oriented projects, such as housing and school construction.

By law, each base-level cooperative and SAIS is to pay a significant share of its profits to the Central Cooperative in its region. A part of these pooled funds is to be redistributed among the base-level units in accordance with social and economic needs in order to equalize incomes and productive capacities, and a part is to be invested in nonagricultural activities in accordance with the government's regional development plans. In the future, third- and possibly higher-level cooperatives are to integrate the base-level production units and PIAR-level central cooperatives into a broader cooperative sector.

At the time of my field work in late 1973, very few Regional Central Cooperatives had been established, and none of these functioned in accordance with government plans. Various factors have delayed the formation of these units and impaired their functioning. First, many cooperative and SAIS leaders oppose the establishment of Regional Central Cooperatives on the grounds that services can be acquired more economically elsewhere. Second, many cooperative and SAIS leaders fear that the Regional Central Cooperatives may become instruments of government control. Third, base-level reform enterprises adamantly oppose the pooling of profits for income redistribution or investments outside the agricultural sector. To date no Central Cooperatives have been able to collect profit shares from their members. Consequently, no significant investments have been made and attempts to diversify production and redistribute income have been frustrated. It seems likely that if the government's redistributive and diversification goals are to be achieved, other mechanisms, such as taxation and direct public investment, will have to be employed.

Performance of New Reform Enterprises

This final section presents an analysis of new reform enterprise performance based on a survey I carried out in late 1973 (Table 7).[10] Eighteen Production Cooperatives, three SAISs, and two Communal Cooperatives were visited. In each enterprise data were collected on 7 variables which can be employed as performance criteria:

(1) Intensity (mix) of crop and livestock production.
(2) Total value of output.
(3) Employment.
(4) Directly productive investments.
(5) Social investments (schools, housing, etc.).
(6) Profitability of the enterprise.
(7) Degree of member participation in management.

On the "average" reform enterprise the mix of crop and livestock production had intensified somewhat since the reform and the total value of production had increased slightly; employment had remained about the same and investments in social infrastructure and productive facilities had increased substantially. In most cases enterprise operations were profitable and the members participated in management, although not with great enthusiasm (Table 8). Thus, it seems that on the whole, in the short period since initiation of the land reform, cooperatives and SAISs have been able to maintain or modestly surpass pre-reform levels of production while investing heavily in both productive and social capital. However, they have not significantly increased employment. The poor employment performance of the reform sector is presently a major concern of both reform enterprise officials and agricultural policy-makers.

The overall or "average" performance of the 23 reform enterprises surveyed is of little significance due to the wide variance of performance scores. However, clear patterns of enterprise performance are discernible. The two Communal Cooperatives surveyed have lower economic performance scores than most CAPs and SAISs. The former enterprises were set up by indigenous groups without ac-

TABLE 7. SUMMARY DATA ON 23 REFORM ENTERPRISES

Case No.	Enterprise Type[a]	Enterprise Name	Location	Principal Crops and/or Livestock	Date of Govt. Intervention	Date of Adjudication	No. of Members	Total Area (has.)
1	CAP	Espinal	Highlands	Rice Production	6/69	1/73	163	8,212
2	CAP	Monteseco	Highlands	Coffee	6/69	1/72	272	1,960
3	CAP	Udima	Highlands	Cattle-Crops	6/69	12/71	512	39,500
4	CAP	Pomalca	Coast	Sugar	6/69	10/70	3,300	16,662
7	CAP	Cerro Alegre	Coast	Cotton-Potatoes	6/71	12/72	265	1,087
8	CAP	Santa Bárbara	Coast	Cotton-Potatoes	N.D.	6/72	137	443
9	CAP	Cerro Blanco	Coast	Cotton-Potatoes	8/71	12/72	150	532
10	CAP	Casa Blanca	Coast	Cotton-Potatoes	N.D.	6/72	72	668
11	CAP	Caldera	Coast	Cotton-Miscel.	N.D.	N.D.	150	900
12	CAP	Andahuasi	Coast	Sugar	6/69	N.D.	375	1,084
13	CAP	El Potao	Coast	Cotton-Corn	9/71	1/73	98	717
14	CAP	San Nicolás	Coast	Sugar	11/69	7/72	300	1,432
16	SAIS	Cahuide	Puna	Sheep	10/69	4/71	30[b]	268,182
17	CC	Quehuar	Highlands	Potatoes	N.A.	69/73[c]	280	N.D.
18	CAP	Túpac-Amaru II--Antapampa	Highlands	Cattle-Potatoes	1969	6/71	2,800	37,351
19	CAP	Tahuantinsuyo	Highlands	Cattle-Wheat	2/70	6/73	187	11,017
20	CAP	Mariscal R. Castilla	Highlands	Wheat-Potatoes	9/69	6/73	82	2,331
21	SAIS	Yocará	Puna	Sheep	1969	12/70	56	9,421
22	CAP	Santa Lucía	Puna	Sheep	N.D.	10/72	235	74,115
23	SAIS	Buena Vista	Puna	Sheep	N.D.	12/72	750	49,383
24	CAP	Micaela Bastides	Puna	Sheep	N.D.	8/72	586	60,250
25	CAP	Posoconi	Puna	Sheep-Cattle	11/69	1971	75	1,769
26	CC	La Unión de Soratira	Puna	Sheep	N.A.	9/66	87	6,539

[a] CAP: Agrarian Production Cooperative; CC: Communal Cooperative; SAIS: Agrarian Social Interest Society.
[b] One service cooperative and 29 peasant communities.
[c] A Communal Cooperative was set up in Quehuar in 1969, to which the Velasco Government redistributed more lands in 1973.
N.A.: Not applicable.
N.D.: No data.
SOURCE: Case Study Reports.

TABLE 8. AVERAGE PERFORMANCE SCORES BY TYPE OF REFORM ENTERPRISE

	(1)	(2)	(3)	(4)	(5)	(6)	(7)	(8)	(9)
		Changes from Pre-Reform Values							
	No. Surveyed	Intensity of Crop-Livestock Mix	Total Value of Production	Employment	Directly Productive Investments	Social Investments	Profitability of Enterprise	Total Economic Performance Score[a]	Member Participation in Management
A. Formal Classification of Reform Enterprises									
SAISs	3	33	100	67	167	200	200	128	0
CAPs	18	33	33	0	94	83	111	59	44
Communal Cooperatives	2	0	-100	50	-100	50	200	17	200
Total/Average	23	30	30	13	87	96	130	64	52
B. Classification of Reform Enterprises Based on Principal Activity and Natural Environment[b]									
Irrigated Crop Production (Coast)	9	67	122	44	111	78	200	104	122
Rainfed Crop Production[c] (Highlands)	6	-50	-117	-117	-17	67	-67	-44	-100
Extensive Livestock Production (Puna)	6	67	83	83	183	167	200	130	50

[a]The arithmetic average of scores for variables (2)-(7).
[b]CAPs and SAISs only.
[c]In 3 of the 6 cases both crops and livestock were produced, but in all cases crops predominated.
SOURCE: Case Study Reports.

Key: Variables (2)-(7) and (9):

	Variable Score				
Variable Number	100 to 200	0 to 100	0	-100 to 0	-100 to 200
(2)-(6)	Large Increase	Moderate Increase	No Change	Moderate Decrease	Large Decrease
(7)	Profits				Losses
(9)	Member Participation in Management		Ambivalence	Member Opposition to Adjudication Model	
	Active	Moderate		Moderate	Strong

TABLE 9. ECONOMIC PERFORMANCE SCORES FOR 21 CAPs AND SAISs

Case Number and Name			Economic Performance Score
Agricultural CAPs on the Coast			
8	CAP	Santa Bárbara	150
10	CAP	Casa Blanca	150
11	CAP	Caldera	150
9	CAP	Cerro Blanco	133
7	CAP	Cerro Alegre	83
12	CAP	Andahuasi	83
13	CAP	El Potao	83
4	CAP	Pomalca	67
14	CAP	San Nicolás	33
Average			104
Livestock CAPs and SAISs in the High Puna			
22	CAP	Santa Lucía	167
21	SAIS	Yocará	150
23	SAIS	Buena Vista	133
24	CAP	Micaela Bastides	117
25	CAP	Posoconi	117
16	SAIS	Cahuide	100
Average			130
Agricultural CAPs in the Highlands			
1	CAP	Espinal	50
3	CAP	Udima	0
18	CAP	Túpac Amaru II	-33
20	CAP	Ramón Castilla	-33
2	CAP	Monteseco	-83
19	CAP	Tahuantinsuyo	-167
Average			-44

SOURCE: Case Study Reports.

tive support from the government, and neither has had ready access to credit. Their production and investments have consequently been curtailed. Nevertheless, despite their rather weak economic performance, both these enterprises show profits, and member-participation in management is stronger than in any of the other reform enterprises surveyed. In both cases the membership's commitment to the group augurs well for the long-term viability of the Communal Cooperative.

The performance of SAISs is better than that of the CAPs surveyed on all six economic criteria. However, there is a conspicuous lack of member participation in SAIS management, and sharp conflicts among comuneros, ex-estate workers, and managerial personnel are frequent. Given the high level of internal conflict and rapid turnover of management personnel, it seems likely that over time the government will have to play an increasingly central role in SAIS management.

In the field there has been confusion over new property/management models, and many SAISs function essentially as production cooperatives, with little or no participation of the affiliated peasant communities in management and profits. This was the case in two of the three SAISs included in the survey. Thus, the superior performance of the SAISs derives not from their unique organizational features but from other variables, discussed below.

The survey results indicate that the two main correlates of reform enterprise performance are "principal activity" and "natural environment." Since the reform, most crop-producing enterprises on the Coast have increased production, employment, and investments, and operate profitably. The economic performance of livestock enterprises in the High Puna is similar to that of the coastal enterprises. However, in areas of highland crop production most reform enterprises have not been able to maintain pre-reform levels of production, employment, or investments, and they are losing money (Tables 8 and 9).

This finding reflects the fact that in the land reform sector, as in the pre-reform agrarian structure, the most efficient farm size and internal organization (degree of managerial centralization) depends largely upon natural and technological conditions. In rainfed crop production, specialization of personnel is limited by the sequential, seasonal nature of tasks. Use of large-scale, specialized equipment is limited both by the need for it to be mobile and by the short period each year during which it can be employed. Unpredictable weather and subtle variations in soils make it impossible for managers to schedule operations in advance; important adjustments must be made on a daily basis. Furthermore, the timeliness of decisions and operations at crucial seeding and harvesting times have a significant impact on the entire year's production. The biologically determined length of production cycles makes it difficult to evaluate the quality of labor inputs, and large land areas hamper the effective control and coordination of tasks. For all these reasons, managerial problems limit the size of labor force on crop-producing farms in rainfed areas. And for the same reasons many large farms had decentralized their operations by employing tenants and concentrating their managerial resources on activities which the tenants could not perform adequately on an individual basis (e.g., credit, purchasing, marketing, or processing services). Prior to land reform, this was the case of crop-producing colono-employing haciendas in the Andes.

Economies of Scale: Strong economies of scale are encountered in two areas of agriculture: (1) crop production in irrigated areas where neither weather nor soils are highly variable, and especially where the commodity produced is perishable, but can be profitably marketed or processed on a large scale; and (2) livestock production on natural pastures.

Sugar production on the Peruvian Coast is a classic example of the first case. Cultivation takes place in irrigated valleys where soils are quite uniform and the climate is hot and dry throughout the year. The absence of seasonality allows specialization of the labor force and performance of different tasks simultaneously in different fields. For optimal results the cane must be milled within a few hours after harvesting. Processing takes place in large, modern factories which operate year-round, 24 hours a day. The sugar mill and processing plant can be operated at a uniform rate near full capacity only if cane cultivation, harvesting, and transport are carefully scheduled in accordance with the mill's capacity. Thus, it is logical that on the Peruvian Coast sugar cultivation and processing takes place in large, vertically integrated, centrally managed units.

Economies of scale in sugar *processing* are greater than those in cane *cultivation*, and it is the need to coordinate fieldwork with processing which has led to the growth of sugar plantations of over 10,000 hectares. Most other coastal estates are centrally managed, but have less than 1,000 hectares of land.

Sheep ranching in Peru's High Puna exemplifies the second case. Here, fencing and centralized management of pasturelands (introduced in recent decades) have markedly increased the carrying capacity of pasturelands, while importation of new stock and initiation of crossbreeding programs have increased the productivity of herds. The rationalization of pasture use and stock improvements were both hampered by individual peasant livestock production on estate lands. Thus, since the outset of ranch modernization in the High Puna, managers have pressed hard for the reduction of colonos' herds.[11]

Support for Land Reform

The attitudes of Peruvian land reform beneficiaries vis à vis collectivization appear to be determined primarily by economic considerations. The strongest supporters of collectivization are coastal wage earners who realize that maintenance of existing large-scale production units offers strong advantages over subdivision of coastal estates into individual parcels. In the Puna most beneficiaries opt for partial collectivization with group management of pastures and group ownership of part of the livestock, but for continuing individual ownership of small herds, primarily as a security measure against failure of the new group enterprises. In highland crop-producing areas, where there are strong managerial diseconomies in large-scale crop production, most land reform beneficiaries are skeptical of the advantages of group farming, and would rather farm independently. In several cases land reform beneficiaries have subverted group activities in the hope that failure of the cooperative would prompt individual parcelization of the land. The most vocal opponents of collectivization have been ex-employees of highland haciendas, who have larger-than-average usufruct plots and herds of livestock, and whose economic interests are hence most threatened by the pooling of resources for group farming. Nevertheless, in the Highlands anti-cooperative sentiments seem to be nearly universal.

Lack of peasant support for group farming has been accompanied by severe managerial problems in highland reform enterprises. Most enterprises on the Coast and in the Puna are large enough to afford the full-time services of a professional manager, insuring reasonably adequate management. The smaller highland cooperatives must rely upon self-management, with occasional assistance from the Ministry of Agriculture. As noted above, the members of highland cooperatives possess neither the skills nor the motivation to manage their enterprises well. Fusion of several ex-estates has been attempted as a means of creating reform enterprises which are large enough to bear the fixed cost of a professional manager. But in these cases the managers have found the problems of planning, controlling, and coordinating field work to be intractable. Sporadic assistance from the Ministry of Agriculture has not been an adequate substitute for full-time enterprise managers. And it does not appear that Central Cooperatives can meet the reform sector's management requirements. The operations of large crop-producing units have been so poor that the Ministry of Agriculture is now considering subdivision of the largest ones.

Conflicts between members' economic interests and those of the group threaten the viability of all reform units, but they have been most serious in the Highlands where labor is scarce at crucial seeding and harvesting times. Wages are extremely low in highland reform enterprises, and to date no year-end profits have been paid. Thus, members are strongly motivated to increase their incomes by expanding production on their individual usufruct plots. Prior to the reform, landlords demanded that colonos work first in the haciendas' fields, leaving their usufruct plots for later. Reform enterprise managers cannot enforce such rules, and now at seeding and harvesting times members tend their own plots first, leaving the cooperative's fields for later. Since growing seasons are short and timeliness is an important determinant of yields, labor shortages have resulted in a contraction of both the seeded area and yields on the centrally managed lands of highland production cooperatives. Low productivity and resultant financial difficulties, in turn, help convince land reform beneficiaries of the futility of group farming.

Conclusion

For numerous reasons (related ultimately to natural conditions, production patterns, technology, and pre-reform estate organization, and more directly to peasant attitudes and managerial problems) fully and partially collectivized reform enterprises have the greatest potential for success in areas of irrigated crop production and in areas of extensive livestock production, respectively. The prospects for group farming in rainfed agricultural areas are not bright.

NOTES

1. CIDA, *Tenencia de la tierra y desarrollo socio-económico del sector agrícola--Perú* (Washington, D.C., 1966); and Peter Klaren, *Modernization, Dislocation and Aprismo: Origins of the Peruvian Aprista Party, 1870-1932* (Austin: University of Texas Press, 1973).

2. An excellent discussion of land reform implementation under Peru's 1964 law is presented in James Petras and Robert La Porte, *Cultivating Revolution, The United States and Agrarian Reform in Latin America* (New York: Random House, 1971).

3. Ibid., p. 97.

4. "Lineamientos de la política económico-social del gobierno revolucionario" (Lima, 1969). The other four major structural reforms were a new mining policy, a new industrialization policy, reorganization of the public administration, and fiscal reform.

5. Enrique Valdez A., "La Reforma Agraria Peruana, conferencia del Ministro de Agricultura en la primera reunión plenaria de Jefes de Misiones Diplomáticas Peruanas en el Exterior" (CENCIRA, 1973), p. 18.

6. In 1974 the Ministry of Agriculture was divided into two ministries; one was charged with land reform implementation, the other with production-related programs. In April 1975 SINAMOS was declared "in reorganization" for having failed to "carry the revolution to the people." *Latin America*, 9 May 1975.

7. A provision of the reform legislation aimed at protecting managerial personnel does not allow the reduction of beneficiaries' salaries below pre-reform levels.

8. The Service Cooperative differs from a Production Cooperative (CAP) in that its members share both ownership and management with the affiliated peasant communities. The Service Cooperative also differs from an Agrarian Service Cooperative (CAS) in that in the former case members are workers who exploit the resources of a SAIS collectively, whereas in the latter the members are independent smallholders who organize for the provision of services (e.g., delivery, marketing, and credit).

9. In the SAIS "Cahuide," which is a relatively profitable enterprise, profit shares were roughly 30 dollars (US) per family in 1972-73.

10. Based on my "Haciendas and Cooperatives: A Study of Land Reform and New Reform Enterprises in Peru" (Ph.D. diss., Cornell, 1976) and my "Land Reform and Reform Enterprises in Peru," Land Tenure Center and World Bank, 1974, 2 vols. References to case studies are from Appendix B of the latter.

11. Juan Martínez-Alier, *Los huacchilleros del Perú* (Lima: IEP, 1974), presents an insightful analysis of ranch modernization and the interplay of managers and colonos (*huacchilleros*, as they are known in the Puna).

10. Group Farming Experiences in Tanzania

Antony Ellman

INTRODUCTION: TANZANIA AT INDEPENDENCE

On December 9, 1961, Tanzania gained independence from Britain, which had administered the country under first a League of Nations and subsequently a United Nations Mandate since 1918.[1] The transition to independence was relatively smooth. A single party, the Tanganyika African National Union (TANU), dominated the political scene from the time of its foundation in 1954, and formed the first independent Government in 1961. TANU succeeded in uniting the 120-odd ethnic groups which make up the population of mainland Tanzania, no one of which was numerous or powerful enough to stand on its own.[2]

The population in 1961 was some 9.3 million, inhabiting an area of 360,000 sq. miles with an average density of 26 people per sq. mile. But this sparse density masked a variation from more than 1,000 per sq. mile in the mountainous areas of the north and south-west, the coastal strip, and the fertile surrounds of Lake Victoria, to densities of less than 10 per sq. mile over much of the dry sandy central plateau.[3] In the less heavily populated areas most farmers lived not in villages but on isolated holdings, partly due to the scarcity of water and partly as a result of moves to avoid taxation and labor conscription during the period of colonial rule.[4] This made the provision of agricultural education and necessary services very difficult, and these under-populated areas consequently remained the poorest parts of the country, with little cash cropping and relatively little social and economic differentiation.

The major agricultural products in the peasant sector were maize, sorghum, cassava, and plantains as food crops, and cotton, coffee, tea, pyrethrum, and cattle-rearing as cash enterprises. There was a small private estate sector, largely in the hands of foreign settlers, which produced sisal, tea, coffee, wheat, and cattle. However, the estate sector in Tanzania did not acquire the significant proportions it did in neighboring Kenya because, being a Trust Territory rather than a Colony, there were restrictions on alienation of land to foreigners, while most Tanzanians lacked the capital and experience to develop large estates, as well as being debarred for a long time by the colonial regime from establishing plantation crops.

In the peasant sector the average size of cultivated holdings was less than 3 acres. The tenure situation was rather confused. In some areas moves had been made toward granting freehold tenure, but most land was still held by the Native Authorities according to traditional custom, whereby land belonged to the community and inheritable usufructuary rights were granted to individual families.[5] There was little tenancy and, except in the over-populated highlands, there were few landless families. The average annual income per head including the value of subsistence production was some US$40.

ORIGINS OF THE GROUP FARMING STRATEGY IN TANZANIA

Tanzania is perhaps the country in Africa which has placed the greatest emphasis on group farming as the means of achieving both economic and socio-political objectives in rural development. It has certainly been one of the most persistent in striving toward this end. To trace the development of this strategy, it is necessary to return briefly to the period before Independence.

The objectives of the Colonial Government in rural Tanzania had been to promote increased production by individual farmers, and to prevent deterioration of soil, water, and forest resources. Financial incentives, encouragement of progressive farmers by agricultural extension officers, and imposition of numerous rules and regulations guarding use of the land and production methods were the means adopted to achieve these ends. For the most part the object of attention was the smallholder on his own land. But a number of "supervised settlement schemes" were established in the 1950s, partly with the objective of facilitating enforcement of the agricultural rules which were engendering much opposition, and partly in the belief that it was necessary to separate the peasant from his "traditional" environment before he could be transformed into a modern smallholder.[6] Although land on these schemes was allocated to individual tenants, and management was firmly in the hands of Government officials, services including credit, supplies, marketing, and machinery hire were cooperatively organized and considerable experience was gained in the imposition of cooperative farming by an outside authority.[7]

Somewhat similar thinking motivated the newly independent Government in its rural development strategy in the early 1960s. On the advice of a World Bank Mission in 1960, and of two Israeli advisers who came to Tanzania shortly after Independence, the traditional extension approach to agriculture was supplemented by what was called the "transformation approach," whereby Government resources were concentrated on groups of farmers settled in village settlement schemes in areas of high potential in an attempt to achieve rapid increases in production. Most of the new settlements were modeled on the Israeli *Moshav Ovdim*, with holdings allocated to individuals and only services cooperatively organized. A few followed the pattern of the *Moshav Shitufi*, with homestead plots allocated to individual members while most of the agricultural land was collectively farmed. It was felt that this villagization approach, by grouping together farmers who had previously been scattered and by introducing elements of cooperative production where appropriate, would overcome the economic disadvantages of small-scale peasant agriculture, would increase the efficiency of farmer education, would facilitate the provision of social and other services by the Government, and would reduce the resistance to innovations which was considered a major obstacle to economic advance. The socio-political objective of creating model socialist communities in the new settlements was not strongly emphasized at the start, but came to be regarded as more important after a few years. It was hoped that the technical and structural innovations adopted in the new settlements would spread by example to surrounding areas until the whole country was organized on a similar pattern.[8]

Twenty-three village settlement schemes were started between 1963 and 1965, with an investment of more than US$5 million. Government managers were put in charge, and heavy subsidies were given to the settlers. With few exceptions, progress was very poor. The settlers became dependent on Government patronage and their own work inputs and commitment were very low; the schemes were over-capitalized and poorly planned; most of the managers were reluctant to transfer responsibility to the Settlers' Committees. Even where the economic achievement was satisfactory, the level of investment in the settlements placed their example far beyond the reach of neighboring farmers and created jealousies and tension between the two communities.[9]

In early 1966, after economic evaluations had been made of all schemes, it was decided to start no new settlements but rather to transfer attention to "modernising existing traditional villages."[10] This would be done by injecting capital and technical services to enable the farmers to work together and obtain the benefits of modern farming techniques. The new approach took the form largely of Government-sponsored mechanical cultivation of blocks of land by groups of farmers, especially in the cotton-growing areas; and of expansion of smallholder cultivation of tobacco, for which there was an excellent market following the Rhodesian crisis. The cotton-growing experiment proved no more successful than the village settlement schemes, as the increased production failed to cover the high cultivation costs.[11] The tobacco-growing schemes, with a higher degree of self-reliance demanded from the farmers, were economically more successful. But the acreage allocated to each farmer was too large to be managed by family labor alone, and the layout of the fields did not encourage cooperation between neighbors, so that the degree of cooperation and equality which developed in the schemes was inevitably small.[12]

At the same time as these Government-sponsored attempts at group farming were taking place, with what we have seen to be a primarily economic motivation, there was a growing feeling that a more specifically Tanzanian approach was needed, which would not only achieve the economic objectives of the village settlement schemes but would also be more in line with the socialist philosophy which TANU had proclaimed. As early as 1960 a small group of farmers in an isolated part of southwest Tanzania had established a collective farming village without Government assistance, and this spontaneous initiative had spread until by 1967 some 20 such villages had been started in the area and formed themselves into an association, the Ruvuma Development Association.[13] Similar initiatives were taken in some other regions. Each of these villages included some 20 to 30 families who farmed their land collectively and distributed the proceeds equally. They attempted to build on the principles said to have characterized pre-colonial Tanzanian society--common ownership of productive assets within the "extended family," obligation on all to work, mutual responsibility for the sick and aged. The economic achievement in these villages was not dramatic, though their standard of living was undoubtedly higher than that of nearby farmers. But the level of services in the villages--made possible by specialization of labor, diversification of production, and reinvestment of surpluses--was impressive. The self-confidence and solidarity of the new communities illustrated in a practical way the potential for development of backward areas through self-initiated, self-managed cooperative farms.

The Ruvuma villages inspired and were in turn stimulated by President Nyerere's first policy statement on Tanzanian socialism: *"Ujamaa, the Basis of African Socialism."*[14] The achievement of these early villages, however, even with the President's backing, was not sufficiently dramatic to catch either the public or the administrative or the political eye. Though considerable practical experience had been built up by the end of the village settlement period, both in group farms set up by Government for primarily economic purposes and in those initiated by farmers with a strong ideological motivation, there was no widely accepted philosophy or direction in Tanzanian rural development. The gap was filled by publication in 1967 of two important policy documents, the "Arusha Declaration," which outlines Tanzania's socialist objectives, and "Socialism and Rural Development," which details the strategy for achieving these ends.[15]

UJAMAA VILLAGES: OBJECTIVES AND STRATEGY

The "Arusha Declaration" lays down the foundation of Tanzania's philosophy of socialism and self-reliance. Any type of development which leads to exploitation of man by man is rejected;

all who are physically able to work must do so; every worker must receive a just return for the labor he performs; and the income derived from work of different types must not be grossly divergent. The major means of production and exchange must be under the control of the people, through Government or cooperative institutions. Less emphasis must be placed on capital-intensive methods: "development is not brought about by money . . . but by people . . . . working with intelligent and committed leadership." A Leadership Code forms part of the Declaration: all TANU and Government leaders must be either peasants or workers, must have no association with capitalist enterprises, may not receive more than one income or own more than one house.

"Socialism and Rural Development" elaborates the policy and gives guidelines for its implementation in the rural sector. Tanzania should develop not as a nation of individual farmers, but as groups of families living together in villages and working together on land which is collectively held. These villages, called *Ujamaa Villages* (the literal translation of Ujamaa is "familyhood"), aim to reactivate and extend the principles on which the traditional extended family was based, and to use these as moral guidelines for future development. Not only marketing and services should be cooperatively performed, as in the old village settlements, but production and distribution also. Returns should be distributed in proportion to the work put in. Decision-making should be the responsibility of all the members. The Government's role should be to provide guidance, services, and finance, the latter given wherever possible in the form of loans rather than grants.

It is not supposed that the ultimate target of fully collective farming will be immediately achieved in every village. Each village or group of farmers should adopt that level of cooperation which the members understand and accept, moving to a higher level as and when the members agree. The path followed in each case will depend on the nature of the farming system, the extent of existing stratification, and the economic opportunities. Any step which increases the cooperative activities and understanding of the people is considered a step toward the target of Ujamaa.

This policy does not apply only to newly formed Ujamaa villages. While the Ujamaa system may be more easily adopted in new communities, concentration on these to the exclusion of established villages would result in the formation of a few pockets of socialism while the bulk of the rural population remained unaffected. The regional inequalities existing in Tanzania would remain untouched, and the objective of creating a society in which "all have a gradually increasing level of material welfare before any individual lives in luxury" would not be achieved. A "frontal approach" to Ujamaa development was therefore adopted in the Second Five-Year Plan (1969-1974), which requires that "the whole range of governmental and political institutions [be] mobilised behind the principles of Ujamaa, to ensure that some progress is made towards the long-term goal of socialist forms of production and distribution throughout the rural sector."[16]

In "Socialism and Rural Development," President Nyerere emphasizes that moves toward Ujamaa must be made voluntarily by the members, and through their own efforts. Here and in later policy documents, including the important *TANU Guidelines* published in 1971, the use of coercion for moving people into Ujamaa villages is strongly rejected.[17] The assistance and guidance of Government and Party are essential, with continuous education, persuasion, and encouragement, but "socialist communities cannot be established by compulsion."[18] "The need for vigorous leadership and official encouragement . . . has to be balanced against the need to avoid the dangers of bureaucratic control, coercion and over-capitalisation, which would negate the very principles of self-help and voluntary co-operation which the villages are intended to embody."[19] We shall see that in recent years actual practice has not always accorded with official policy, and there have been indications of a change in the policy. But for most of the period with which this review is concerned, the official commitment to voluntary transition was not in doubt.

PROGRESS TOWARD UJAMAA, 1967-1974

The objectives of the Ujamaa program being so far-reaching, and the paths along which progress can be made being so varied, achievements must be viewed in several different perspectives. First, progress in the newly formed Ujamaa villages must be considered; secondly, Ujamaa activities adopted by already established communities; thirdly, relations between Ujamaa villages or groups and the areas surrounding them; and fourthly, the institutional framework within which these developments take place.[20] This data base will enable us to analyze the achievements of the group farms in terms of their economic, social, and political objectives.

New Ujamaa Villages

Table 1 shows the number of Ujamaa villages and their population as it has grown from 1967-1974. These figures must be regarded with caution, for they include all villages graced with the name of Ujamaa by the local Government authorities, regardless of the progress they have made toward their various goals. We shall see that many of these villages were started by coercive means, or with a view to obtaining services and assistance from the Government which increasingly were becoming available only to Ujamaa villages, while little progress was made either toward collective cultivation or toward achieving self-reliance and other objectives.

The average number of members per Ujamaa village has increased progressively over the years. In villages existing at the time of the "Arusha Declaration" the average population was 104, while the average number of people in the new villages formed in 1973 was 657. This masks a wide variation in size of individual villages, from populations of 30 up to 1,800. The larger villages are all

TABLE 1. NUMBER OF UJAMAA VILLAGES

Year	No. of Villages	Total Population of Villages	Average Population per Village	Total Ujamaa Village Population as % of National Population
1967	48	5,000	104	0.04
1968	180	58,000	322	0.5
1969	809	300,000	371	2.5
1970	1,956	531,200	272	4.3
1971	4,484	1,545,240	345	12.3
1972	5,556	1,980,862	357	15.3
1973	5,628	2,028,164	360	15.3
1974	5,008	2,560,000	511	17.6

SOURCE: Rural Development Division; IBRD Agricultural Sector Survey 1974.

those started by Government officials for purposes of providing services and bureaucratic control.

Recently attempts have been made to categorize Ujamaa villages according to their level of economic and social development. This is primarily for purposes of determining their eligibility for loan funds, but it also gives a better indication of progress made than can be obtained from the figures in Table 1. Three stages are identified. Stage I is the formative stage, when villages are still dependent on Government grants for development finance and on Co-operative Society loans for recurrent expenditures. In Stage II the day-to-day running costs are covered by collectively earned revenues; these villages are registered as Agricultural Associations and are eligible for loans from the Rural Development Bank. Before they reach Stage III, the villages must be economically viable, and firmly established on Ujamaa lines with most of their activities collective. They are then registered as Multi-Purpose Co-operative Societies which can obtain commercial credit from any source.[21] Table 2 shows that in mid-1970, 90 percent of Ujamaa villages were in Stage I, 9 percent in Stage II, and less than 1 percent in Stage III.

The distribution of Ujamaa villages round the country is very uneven, as can be seen from Table 2. Most villages are located in the poorest regions, where there is plenty of uncultivated land on which collective cultivation can be started, and a low level of services which can be more easily improved after villagization. Where commercial farming is well developed and population density is high, few villages have been established. Thus, 70 percent of the Ujamaa villages in 1974 were in the five regions with the lowest per capita GDP, while less than 10 percent were in the six regions with highest per capita GDP.

TABLE 2. STAGES OF DEVELOPMENT IN UJAMAA VILLAGES, 1970-71

	No. of Villages				Approximate Population		
	Stage I	Stage II	Stage III	Total	Average per Village (a)	Regional Total (b)	(b) as a % of the Total Regional Population
Arusha	38	5	1	44	200	9,000	1.3
Coast	31	26	1	58	940	55,000	6.0
Dodoma	132	17	1	150	310	47,000	5.9
Iringa	349	-	1	350	240	84,000	10.8
Kigoma	93	15	-	108	300	32,000	6.3
Kilimanjaro	9	2	-	11	190	2,000	0.3
Mara	226	20	4	250	380	95,000	15.5
Mbeya	191	-	3	194	260	50,000	4.6
Morogoro	21	2	1	24	210	5,000	0.7
Mtwara	651	21	-	672	420	282,000	44.1
Linda	162	26	-	188	420	79,000	18.3
Mwanza	38	3	-	41	190	8,000	0.7
Ruvuma	105	15	-	120	100	12,000	2.7
Shinyanga	131	1	-	132	100	13,000	1.3
Singida	42	15	-	57	230	13,000	2.7
Tabora	35	43	4	82	240	20,000	3.2
Tanga	113	32	1	146	160	23,000	2.7
West Lake	43	-	-	43	250	11,000	1.5
Totals and Averages	2,410	243	17	2,668	315	840,000	6.3

SOURCE: United Republic of Tanzania, Economic Survey 1970-71.

The majority of the new villages started with groups of farmers coming together to clear new land, which they cultivate in addition to their individual holdings. There are special cases like the villages in Mtwara Region along the southern border with Mozambique, where farmers were brought together for security reasons, or the villages in the Rufiji valley, where farmers moved to higher ground to escape floods. In varying degrees these farmers were obliged to abandon their old land and commit themselves entirely to the new settlements.

In the new villages there is a nucleated settlement where each family has 1/2-2 acres for private cultivation. Outside the village centers are block farms which are collectively cleared and developed. In some villages all of this land continues to be collectively farmed, but in most cases part or all is allocated to individual members, though cooperation continues in weeding, crop protection, etc. The proportion of private to communally farmed land is very variable: in 15 villages in Handeni District it was 4 to 1 in 1971; in Tanga District it was 10 to 1. In some of the longer established villages, mainly the smaller ones, individual cultivation is very insignificant. One Ujamaa village in Ruvuma Region had 180 acres of collective cultivation in 1968 and only 70 acres individually farmed. But such cases are exceptional.

Production records in Ujamaa villages are very inadequately kept. Often not even the acreage of crops grown is known, let alone the production costs or the yields. In a few cases the acreage cultivated collectively has been measured: in the Handeni villages in 1971 it was said to average 0.6 acres per adult member;[22] in 7 villages in Ngara District in 1972 the average was only 0.33 acres apart from the individual cultivation, which ranged from 1.6 to 17 acres per household.[23] But expenditures, yields, and overall figures on which an assessment of the economic achievement of the program could be made, are regrettably not available.

Ujamaa Activities in Existing Communities

In land-hungry areas few new Ujamaa villages have been started, partly because most of the farmers already live in village clusters, partly because there is little uncultivated land on which new villages could be built. Voluntary pooling of individually held land has proved unworkable in view of disparities in holding size and quality, and the presence of permanent crops. There have been some instances of compulsory acquisition of large farms, mainly wheat, coffee, sisal estates, and cattle ranches taken from foreign settlers, but most of these lands have been used for establishment of State Farms, rather than Ujamaa villages. In one heavily populated area with extreme disparities in farm size--Ismani in Iringa Region, a pioneer maize-growing area settled by Tanzanian farmers, where 40 percent of the land is cultivated in units of more than 100 acres by 4 percent of the population--attempts were made in 1970-71 to take over part of the large farmers' land and redistribute it to landless Ujamaa groups.[24] This met with intense opposition from the large farmers, culminating in the assassination of the Regional Commissioner of the area by one of them.

In a few heavily populated areas landless farmers have moved away to settle in Ujamaa villages elsewhere. Thus the 11 villages in Kilimanjaro Region (Table 2) are all in the plains at the foot of the mountain. Nine hundred farmers from the mountain area also migrated in 1969 with Government assistance to Ujamaa villages in the ecologically similar Mwese Highlands of Mpanda District, 200 miles away. But for the most part progress toward Ujamaa in such areas has taken the form either of communal cultivation of a small piece of unused or unwanted land, or of labor-sharing on individual holdings, or of introduction of cooperative services or nonagricultural activities which do not require large extents of land.

Communal Fields: An example of communal farming in a high-density area has been described in Rungwe District, where from 1967 many villages cultivated small communal plots, often pieces of poor quality land temporarily loaned by the chairman of the Village Development Committee. These plots required the input of very little labor by the participants, perhaps half a man-day per week. They brought a limited advantage in the form of seed provided by the Government and occasionally a small distribution of profit. They constituted a marginal advance toward Ujamaa, and rarely disturbed the status quo to any degree.[25]

Labor-Sharing: Mutual assistance on private holdings has a long tradition in Tanzania, though in cash-cropping areas its significance has been greatly reduced with the growing use of hired labor. In some densely populated villages where labor-sharing has come close to dying out, attempts have been made to reactivate and extend the practice as a first step toward Ujamaa, but this has often met with strong resistance. In Rungwe in late 1967 all members of the village-level TANU cells--the "Ten-House Cells" (see below p. 251)--were persuaded by the local Government authorities to pool their labor in cultivating each other's plots, as a first step toward Ujamaa. Whereas traditionally the sponsors of such work-parties would invite only their friends and relatives and would reward them with food and beer, under the new system the participation of all members of the ten-house groups was obligatory and no food or beer was provided. The experiment was unpopular for several reasons: farmers feared that others would not give the same amount of assistance in return; there were difficulties in agreeing whose field should be tilled first; the system favored those with larger plots and less labor in their families. It was abandoned in 1968 after only one season.[26]

Cooperative Services: Service Co-operative Societies have been somewhat more successful in uniting groups of individual farmers in high-density areas. Cotton and coffee marketing cooperatives have been in existence in northern Tanzania since the 1930s, springing in these areas at least largely from grass-roots producer groups campaigning against exploitation by middlemen. The more successful of these cooperatives greatly expanded their activities after 1961, moving with Government encouragement and sometimes under Government pressure into the fields of processing, retailing, storage, credit, transport, and production services. But in most cases this expansion over-extended the cooperative's administrative resources and separated the leadership from the members, resulting in inefficient management and increased opportunities for corruption.[27]

The Government's response was to intervene further in management of the cooperative affairs--in 1966, 16 Co-operative Unions were taken over by the Government and their Committees dismissed--but this merely further alienated the leadership from the members. Increasingly, the cooperatives were becoming instruments of the wealthy groups, reflecting and reinforcing the growing socio-economic differentiation in the rural areas, which is greatest where the cooperatives are most active.[28]

Since 1967 steps have been taken by some Co-operative Societies to move from services toward production, either by establishing model farms (the Bukoba National Co-operative Union started several small banana and coffee plantations for demonstration and seed-multiplication purposes in 1968-69) or by providing services to small groups of members engaged in cooperative production. In Kilimanjaro and Arusha Regions a project was initiated in 1972 for introducing improved dairy cattle to smallholders, with cooperatively owned bulls, milking equipment, and trucks for transporting fodder from Ujamaa plots on the plains. Information is not available on the progress made in this project.[29]

Ranching Associations: Similar approaches have been taken to Ujamaa development in another type of existing community where there are particular obstacles to establishment of new villages: namely the arid or semi-arid zones inhabited by cattle-keeping nomads and semi-pastoralists. Here low rainfall and grazing shortages make permanent settlement difficult, and the wide disparities in size of individual herds form a barrier to collective ownership of livestock. In Masailand and Sukumaland Ranching Associations have been started, with collectively owned bulls, cooperatively supplied veterinary and other services, and regulations imposed on the members regarding limitation of stock numbers, rotational use of grazing, disease control, marketing, etc. There is evidence, however, that some of these Associations have been dominated by the larger cattle-owners, in the same way as large farmers came to control the cotton and coffee Marketing Co-operatives in the high-density agricultural areas.[30]

Regional Ujamaa Development

Although the "frontal approach" to Ujamaa development adopted in the Second Five-Year Plan required that whole regions be planned on Ujamaa lines, the tendency in the first years after the Arusha Declaration was to view each village or Ujamaa project independently. The hope was that if a large proportion of the population was settled in Ujamaa villages, their example would spread automatically to the intervening areas. Such integration has manifestly not taken place, even with active promotion by Government and Party cadres. The more advanced villages have tended to emerge as socialized islands in a nonsocialized sea.

The first twelve Ujamaa villages in Ruvuma Region foresaw the limitations imposed by such isolation and formed themselves into a Ruvuma Development Association through which they hoped to expand both their scale of operation and their influence on surrounding areas. The Association established a school, an adult education center, a workshop, grain-milling and timber-milling businesses, and a cadre of village activists known as the "Social and Economic Revolutionary Army," whose role was both to guide the Association members on the path to Ujamaa and to convince other farmers to follow the same course. This Association and its Army came to be regarded with great suspicion by the local Government and Party officials, and was disbanded on grounds of opposing Party policy in 1969.[31]

In 1970 the Government itself adopted a new approach to mobilization of farmers on a regional basis, preparing plans for resettling the whole population of Dodoma Region in Ujamaa villages. Dodoma Region is one of the driest and poorest parts of the country. It depends primarily on cattle-keeping, and is prone to recurrent famines. The plans for the new villages included cultivation of drought-resistant crops, establishment of cooperative grain stores, village water supplies, and cooperative creameries, proposals for upgrading local cattle, rotational use of grazing land, etc. By 1971 "Operation Dodoma" had resulted in some 50,000 people moving their houses to the new village sites, chosen for them by Government Planning Teams, where they started to clear new land for cultivation. The cost to the Government was considerable, and many of the demands for services in the new villages, particularly water, proved difficult to meet. The achievements in terms of production, and the effects on farmer attitudes to villagization and Ujamaa, have not yet been analyzed.

Since 1972 a number of other regions or districts have been chosen for similar "Operations," all in poorer parts of the country. Rufiji was chosen because of its need for flood protection, Kigoma and Chunya because of their historic backwardness. These regional developments seem to presage a change in the approach to mobilization, illustrated in a statement attributed to President Nyerere in late 1973 that all farmers *must* live in Ujamaa villages by 1976.[32] The implications of such massive upheavals, in terms both of short-run production effects, of the capability of the administrative machinery to cope, and of the impact on farmers' attitudes, cannot yet be known though some predictions will be made in the sections which follow.

INSTITUTIONAL FRAMEWORK OF UJAMAA DEVELOPMENT

Since Ujamaa is now seen as the foundation of an integrated program for rural development, it is important to understand the administrative and political framework within which it has been, and is to be, implemented. Two critical phases can be identified.

Pre-Decentralization, 1961-70

Up till 1971-72, responsibility for promoting, planning, and implementing the Ujamaa program had been widely dispersed and ill-defined. The initative for starting Ujamaa activities was supposed to come from the farmers, but guidance and support were required from the Party and the Administration. Each Government ministry or agency was expected to orient its programs in line with the Ujamaa objectives laid down by TANU, and in response to plans drawn up in the regions, districts, and villages. At regional and district levels there were Development Committees chaired by the Regional and Area (= District) Commissioners and comprising the representatives of the different technical ministries or agencies. There was also a local Government structure with District Councils responsible for provision of certain services. Below this there were elected Village Development Committees later merged into larger Ward Development Committees. TANU has a parallel organization, with committees at national, regional, district, and ward levels, and after 1964 a grass-roots organization was started: groups of ten neighboring households formed ten-house cells, each with its elected leader. The political and administrative hierarchies were horizontally integrated through the Regional and Area Commissioners: as heads of the Civil Service at their respective levels they chaired the Regional and District Development Committees, and as heads of the Party organization at the same levels they were also ex-officio Regional and District Secretaries of TANU. The Chairmen of the Ward Development Committees (WDCs) were also Branch Chairmen of TANU, and many of the members of the WDCs were chairmen of their ten-house cells.[33]

This structure had several shortcomings. First, the Regional and District Development Committees were unable to prepare integrated development programs, since the heads of each technical department were vertically linked to their parent ministries in Dar es Salaam, rather than to the Regional or Area Commissioners. Thus when a new Ujamaa village was being planned, its water supply had to be approved and funded by the Ministry of Water Development, its agricultural program by the Ministry of Agriculture, its clinic and school by the Ministries of Health and Education, etc. After 1969 a small Regional Development Fund was placed at the discretion of the Regional Development Committees, but it was insufficient to break the domination from the center. The system inevitably led to delays and lack of coordination at the local level. Secondly, the District Councils were largely controlled by local elites, and were more concerned with demanding services for their constituents and themselves than with initiating economically productive programs. Thirdly, the Party, though represented at all strategic points, felt that its programs were continually distorted by the bureaucracy. The technical officers at regional and district levels, with their greater expertise and access to financial resources from the center, could indeed circumvent or ignore the political directives of the Regional and District TANU Secretaries, and in many areas programs continued to be implemented which clearly encouraged rural capitalism or neglected poorer sectors of the community, long after the "Arusha Declaration" had set the path in the opposite direction.

Post-Decentralization, 1971-75

Decentralization measures in Tanzania aimed not just at bringing decision-making closer to the people and increasing the planning and implementing capacity of the regions and districts, but also at improving the capacity of the Party for guiding and overseeing the development process at all levels. The *TANU Guidelines* emphasized that "if development is to benefit the people, the people must participate in considering, planning and implementing their development plans." Some sacrifice of economic efficiency is considered tolerable if social and political goals are at stake: "any action that gives the people more control of their own affairs is an action for development, even if it does not offer them better health or more bread." But the need for Party guidance at all levels is paramount, to avoid the "problems that may arise when a Party does not guide its instruments. The time has now come for the Party to take the reins and lead all the people's activities."[34]

The structural changes consequent on these political guidelines were spelled out in a Presidential paper on "Decentralisation" in 1972.[35] The Regional and District Administrations became fully responsible for planning and implementing the internal development of their areas. The District Councils were abolished. The Regional Commissioners were given the status of Minister, with the Regional Directors of Development as their Principal Secretaries. The major part of the budget was decentralized. The heads of the technical departments in the regions became responsible not to their parent ministries in Dar es Salaam, but to the Regional Director of Development and the Regional Commissioner. Similar procedures were adopted at the district level.

The role of the parent ministries is reduced to preparing broad policies and priorities, and providing technical advice when requested. The Prime Minister's Office, however, to which the Regional and Area Commissioners are responsible, has enhanced powers for overall policy direction and coordination. It coordinates regional plans into a national rural development plan. Its newly established Ujamaa and Co-operative Development Division includes Planning, Research, and Training Sections, which are responsible for collecting data on Ujamaa and cooperative development, establishing standard reporting procedures, training village bookkeepers and other skilled personnel, preparing farm budgeting and planning guidelines, and promoting the merger of Ujamaa villages and Co-operative Societies.[36]

The Ward Development Committees remain unchanged, but additional Ujamaa Village Development Committees are formed, which in villages with more than 250 families become TANU branches in their own right. The ten-house cells are also unchanged, though the leaders are charged with a greater responsibility for bringing up the members' views to the Ward and for communicating information down again.

This decentralization brought the technical heads of the regions and districts under the control of the political leadership at each successive level. The caliber of technical and administrative staff posted to the regions has been upgraded. The demands on TANU cadres to supervise development and to mobilize the farmers have also increased. At TANU headquarters a Department of Political and Ujamaa Education has been established, and Regional Training Colleges are planned, to prepare the political leadership and village-level cadres to meet the demands which increasingly are made of them.

EXTERNAL INFLUENCES AND CONDITIONS AFFECTING THE GROUP FARMS

Analysis of the achievements of the Ujamaa program and of the obstacles facing it requires consideration first of the external influences affecting establishment of the group farms and second of the internal operating mechanisms of the cooperative enterprises. Though in several respects the latter will be seen to stem from the former, it will be clearer to examine the two separately.

International Influences

Those influences which are external to the group farms but internal to Tanzania will be our chief concern, but international influences are not without significance. Tanzania is as much affected by the pattern of international trade as any ex-colonial country in Africa.[37] Nor has the group farming strategy been developed in total isolation. British, Israeli, and Chinese advisers have been consulted at various stages; an American firm of management consultants was hired to prepare plans for the 1972 decentralization. At the same time, many foreign donors have regarded the Ujamaa concept with great suspicion, and have been unwilling to provide funds for the group farming program. But in all its dealings with foreign agencies or individuals the Tanzanian Government has made it clear that it follows its own line and stands by its own principles. The Government has been ready to accept foreign money and advice only on terms which do not conflict with Tanzanian ideals.[38]

Socio-Economic Influences

It has been argued that Tanzania's poverty and lack of socio-economic differentiation present ideal conditions for the peaceful socialist transformation of the whole rural sector which is the Government's objective. Most of the rural population being equally poor, and the major part of the agricultural land being officially public property,[39] it was felt that the pooling of resources would be a relatively simple process. Inequalities and privileged classes were known to exist: the need to prevent their further growth was recognized in the "Arusha Declaration." But they were not considered to pose a serious obstacle to Ujamaa development. In 1967 the President was able to say: "Fortunately this development [of two opposing rural classes] has not gone very far; we can arrest the trend without difficulty."[40]

Certainly the extent of differentiation in wealth and class in Tanzania cannot be compared with that in many more developed African countries, but the experience of implementing the Ujamaa policy over several years has shown three things. First, that social and economic inequalities in many rural areas have progressed very much further than was anticipated, such that there are powerful vested interests which see no advantage to themselves in Ujamaa; secondly, that those areas where Ujamaa is easiest to develop are generally the poorer areas with the lowest potential; and thirdly, that the sector of the rural population to which Ujamaa holds out most promise is in general the most under-privileged, which has the least experience and perhaps the least initiative for wielding power and adopting innovations.

Several studies have shown, for example, the existence of land markets, a landless labor class, and wide disparities in holding size in high-density areas of rural Tanzania. A study in Bukoba District in 1968 showed that 25 percent of the agricultural land had been purchased rather than acquired through inheritance. In one village in Tukuyu District, 10 percent of the households were completely landless in 1968, while 40 percent held more than 12 acres each and 4 percent held more than 50 acres. In areas producing labor-intensive cash crops, cotton and tobacco for example, up to 50 percent of the total labor input is supplied by hired hands, many of them landless migrants from other land-hungry districts.[41]

The obstacles posed to Ujamaa development by these conditions are very clear. We noted earlier that most of the new villages and Ujamaa activities have been started on unused lands, where existing inequalities need not be confronted. But even in such areas there are instances where the wealthier peasants and businessmen have turned Ujamaa to their own advantage, welcoming the new villages for the markets they offer for their shops or businesses and for the opportunity they bring to corner a larger share of the benefits. In one Ujamaa village in Handeni District the largest farmer-cum-businessman in the area managed to convince the members to cultivate one acre per member communally without touching a hoe himself, and dissuaded the others from starting a cooperative shop (which would compete with his own) on the grounds that even he had been unable to make a profit.[42] Such trends are even more glaring in some of the land-scarce areas, where moves to promote Ujamaa have been--if only temporarily--curtailed by the wealthy classes who have shown themselves ready when necessary, as in Ismani, to oppose the new reforms with force.

On the other hand, experience has shown that even the poor peasants will not always jump automatically at the opportunities offered by Ujamaa. Certainly villagization and the various services and hand-outs provided by the Government to the Ujamaa farmers are an attractive bait, and many farmers have been convinced by the improved economic prospects in the group farms, especially in areas such as Kilimanjaro and Ismani where they would otherwise remain landless. But in the poorest areas, such as Handeni and Kigoma, unfavorable ecological and socio-economic climates have over many years combined to induce a suspicion and reluctance to invest effort and resources in new and untried ventures, which will require strong leadership and convincing demonstrations to overcome.[43]

Socio-Cultural Influences

The cultural foundation of the Ujamaa policy--the extended family or clan-based cooperative community--has been shown by many writers to hold little practical validity as a basis for modern development. "The tangible individual benefits and the effective group sanctions that sustained the traditional co-operation of the Ujamaa family are unlikely to occur in larger-scale economic and social units."[44] Indeed the President had made it clear that it is not the actual practices and social institutions which he hopes to recreate in the new villages, but the attitude of mind which guided pre-colonial Tanzanian societies. Ujamaa is seen as a moral guideline, used for political ends to prepare the population to accept institutions and practices which are now seen as social and economic necessities.

Nevertheless, some elements of traditional culture remain as important influences on the Ujamaa program. In Ruvuma kin-based mutual aid groups are still active and formed the foundation of the new Ujamaa villages. In Bukoba members of one clan remained together in the new villages and the clan-heads emerged as the leaders.[45] But the failure of the labor-sharing experiment in the ten-house cells in Rungwe (which are not kin-based) illustrates the limitation of this approach.

The inferior place accorded to women in traditional Tanzanian society is a cultural influence regretted by Nyerere as a deficiency which persists and limits achievement of equality in the new communities. In one Ujamaa village in Ruvuma it is noted that while women do most of the hard work they play little part in decision-making.[46] In Iringa the reluctance of most wives to work with their husbands on the tobacco farms when other nonfamily members are present is an obstacle to cooperation.[47] Yet in other cases the cultural barriers have been overcome, with women becoming members of Ujamaa villages in their own right in some districts, undertaking jobs which are traditionally limited to males in others, and forming their own committees and running their own enterprises (such as poultry farms) in Ujamaa groups elsewhere.

Religious institutions have played little part either in promoting or in obstructing Ujamaa development in Tanzania. Both Muslim and Christian leaders have generally welcomed the policy (though in 1969 a Roman Catholic journal published an article equating Ujamaa with communism, and suggesting that common ownership in the new villages would soon be applied to wives as well as to land; this rumor was quickly quashed by the President, himself a practicing Catholic). In another part of the country a resurgence of traditional culture and religion has been noted in the new Ujamaa villages, but its significance has not been fully evaluated.[48]

Techno-Economic Influences

A major objective of group farming in Tanzania is to increase the scale of agricultural enterprise so that technical innovations requiring large-scale operation can be adopted. These apply mainly to the plantation crops--sisal, tea, and wheat--which are now grown principally on State Farms rather than group farms in view of the heavy investment in machinery and managerial expertise which they require. Sisal is grown in a few Ujamaa villages, which sell the crop to private estates for processing. Several of the State Farms buy crops and supply services to nearby Ujamaa villages; thus a Government-run sugar factory in Kilombero purchases most of its cane from cooperative producer groups in the neighborhood.

After the discouraging experience of the mechanized block farms and village settlements, tractor cultivation is not widely used or recommended in Ujamaa villages. Apart from the difficulty of maintaining expensive machinery, mechanization of one stage in the production process tends either to create labor shortages at other stages, or to supplant labor which may not find alternative employment opportunities elsewhere. Ox-cultivation is generally considered more appropriate, but in many parts of Tanzania it is as alien to the farmers as are tractors. Both ox-training centers and tractor-service centers were proposed on a large scale in the Second Five-Year Plan.

Farmers in the Ujamaa program stand to gain considerably from the readier access to credit and supplies, marketing facilities, water supply, and agricultural extension services, all of which are being concentrated in Ujamaa villages in preference to farmers elsewhere. The benefits to the farmers from extension services are not immediate, however, for several studies have shown that the opportunity which the Ujamaa villages offer for adoption of new agricultural techniques had rarely been taken. Both the crops grown and the methods used are generally similar to those found in ordinary villages, reflecting both the inappropriate training of the extension workers and the (frequently justified) reluctance of the peasants to adopt untried innovations.

Political Influences

Tanzania passed through no bitter struggle for Independence. It experienced no political revolution. No distributive land reform preceded the attempts at socializing agriculture. With some exceptions little awareness of exploitation is found among the poor peasants, and the level of political consciousness is rather low. Thus while the call for Ujamaa should have come from the farmers, except in a few instances it did not and the task of mobilizing and educating the rural population on the objectives and benefits of Ujamaa fell on TANU and the Government.

The Party's policy on rural development has been clearly spelled out in the various papers, declarations, and guidelines published since 1967. The structure of TANU described in the last section showed that it is well represented at all levels down to the village. However, being a mass Party rather than a vanguard Party, with no required qualifications for membership other than proof of citizenship and willingness to pay Party dues, the caliber and level of understanding of the TANU officials at village level, at least up to 1970, was rather low. Many of them were themselves from the wealthier class of the peasants, against whose interests the Ujamaa policy might be expected to act.[49] Since 1970 and especially since adoption of the *TANU Guidelines* in 1971, the quality of Party activists may be assumed to have improved, but still their capacity for mobilization of a reluctant peasantry must be somewhat limited.

Administrative Influences

With limited public understanding or popular demand for Ujamaa, and with limited political capacity for mobilizing the farmers, it was inevitable that the major influence in starting the new group farms should be that of the Government--the Regional and Area Commissioners and their Development Committees, the agricultural extension and rural development workers, the village-level executive officials.

The background and experience of most of these staffs made them, at least up till 1972, in several respects unsuited for the job in hand. With fixed salaries and living conditions much better than those of the peasants whom they were to lead, most of them--even those living in the villages--considered themselves of a separate class from the Ujamaa farmers and adopted a patronizing attitude toward them. Few had confidence in the peasants' ability to plan and run their own affairs. Many of the staff found it easier to keep a tight control of the Ujamaa developments than to promote participation and encourage indigenous leadership.[50]

It was thus almost inevitable that the staff should adopt authoritarian methods in getting the villages started and should view the new creations in bureaucratic terms. In many districts there has been a tendency to concentrate on starting the largest possible number of villages, and to measure progress in terms of the size of membership rather than in terms of less definable criteria such as the degree of communality, economic productivity, or democratic management.

Again, there has been a tendency where farmers have been reluctant to answer the call to Ujamaa to "bring back the stick," despite unhappy memories of the enforcement of agricultural rules in the years before Independence, and of the nationalist protest movement which rose up against it. In Handeni in 1968 the Regional and District Development Committees decided that 65 Ujamaa villages should be started. Village Development Committees were instructed to choose sites and recruit members. Farmers who refused to join were threatened with forced transportation to the new villages or with short prison sentences under the pretext of tax arrears or minimum acreage violation.[51] In Bukoba in the same year a similar policy was adopted, after a resolution passed by the Regional Development Committee that "all citizens aged 18 years and above without employment or without 2 acres of cultivated land should be encouraged and convinced to join the Ujamaa villages in the selected areas." The interpretation given to the words "encouraged and convinced" led to much debate: in a study of one of the villages 36 out of 42 settlers claimed that they had been forced to join, and 8 added that they had been intimidated, though an official inquiry into the matter did not find these allegations proven.[52]

In both Handeni and Bukoba the Central Government stepped in within a few months to stop the alleged use of force, and the officials responsible were transferred or dismissed. Use of coercive measures has not been continued after the initial period of mobilization. In many other districts the local administration has relied on the use of "carrots" rather than "sticks" to attract farmers to the Ujamaa villages, and even where sticks were used initially, they were always associated with carrots of some sort. Material incentives such as water supplies, dispensaries, seeds, and fertilizers have been offered, or at least promised, to the Ujamaa farmers. In one area which was hit by drought in 1971, a condition for obtaining famine relief was participation in some kind of Ujamaa activity. This encouraged large numbers of farmers to join the Ujamaa villages and start collective plots.

This is not to say that encouragement of voluntary initiatives through education and guidance of the Ujamaa farmers have been neglected. In most Districts, especially since 1972, the Administration and the Party have organized seminars in the Ujamaa villages; members have been trained as extension and health workers, mechanics and bookkeepers. Planning Teams have helped the farms to draw up their production plans, and agricultural extension and rural development workers have been posted to the Ujamaa villages to guide the farmers in implementation. But the mounting in 1970-71, and again in 1973-74, of "Operations" to move whole regions into Ujamaa villages over a short period suggests that periodic returns to administrative measures involving at least a degree of coercion may be expected.

INTERNAL OPERATING PROCEDURES AND RESPONSES

The internal procedures which have developed in the group farms must be related to the ways in which the farms were started and to the various internal and external conditions which confront them. This account refers largely to the newly established Ujamaa villages, though the procedures developed around Ujamaa activities in existing communities are in many respects similar.

Responses to Mobilization

In several of the "ideal" Ujamaa villages which started on the initiative of the members or as voluntary responses to political calls, a high degree of self-reliance and self-confidence is found, and productivity of the members' labor is very much higher than that of individual farmers around them. This increased productivity results from specialization of labor, systematic planning and work organization, and the tremendous release of enthusiasm that can be obtained in a cooperative system where the members know that they are working together for their own good. It illustrates the maxim that "the advantages of co-operative over individual farming are not automatically obtained by having the land pooled and manpower used in common. Pooling can create a new productive force and a new possibility for improving on individual farming. But it is only a possibility. Man's endeavours are necessary to turn the possibility to reality."[53] Ujamaa villages where the possibility has become reality are few in number, but in several of them the production targets which the members set themselves have been surpassed. These villages are characterized by great pride in their achievements and a determined spirit of independence; there is commonly a strong resistance to "interference" by outsiders in the villages' internal affairs.[54]

In the majority of Ujamaa villages, however, which were started as result of administrative measures or in response to material incentives, the reaction of the farmers has been at least initially very different. Though it is difficult to generalize, some common tendencies can be identified.

First, where coercion has been used to start new villages, resistance to it has been limited and generally ineffective. This is mainly because coercion has usually been directed against the weakest sections of the community who have the least power to resist. But even in the rare instances where large farmers have been involved, as in Ismani, the Government has shown itself ready and able to stand firm. Though the capacity of the Tanzanian government to enforce changes in its peasant communities has sometimes been questioned, at least on the scale used up to 1970, and with the limited objective of villagization, such doubts do not seem to be well founded.

This does not mean that the farmers as a whole welcomed the Government's initiatives or that the plans prepared for them were fully implemented. Many farmers joined the Ujamaa villages primarily to obtain the various benefits offered as incentives, and had little interest in communal production. Indeed, several studies made of the villages started in 1968 show immense shortfalls between the targets and the actual acreages cultivated in the first 2-3 years.[55] This reflects in part the unrealistic plans prepared by the Administration, often after minimal consultation with the farmers. In part it results from the farmers' indirect resistance to the Government's proposals, seen in such actions as planting manioc sticks upside down, or mixing salt with seeds to prevent germination and make it appear that the sites recommended by the Government were no good.

The farmers accepted the Administration's plans because they feared that the rations, water supplies, etc., might not be supplied if they questioned the officials' word. In Handeni it is reported that for the first two years most farmers regarded the communal fields as "something to please the Government and Party." Though they were told that productivity on these fields would be higher, they failed to see how, for they were growing the same crops with the same techniques as on their private farms. The only difference apparent to them was that "working with others brings more organisational difficulties on the one hand and more social enjoyment of labour on the other."[56]

The District and Divisional officers in Handeni feared that the Ujamaa villages might disappear or become indistinguishable from traditional villages when famine relief and other benefits were discontinued. They therefore tried to bind the farmers to the new villages by encouraging establishment of permanent crops, investment of surpluses in cooperative shops, etc. The farmers for their part sometimes used their new found strength to bargain with the Administration. One village in Pangani refused to send its women to work on the communal fields until the water supply promised by the Government was completed.

The studies made of Ujamaa development in Handeni District show that two years after their establishment there was a degree of apparent stability in the new villages. Most of the members felt that their lives were more interesting and more secure than before, that the social services in the villages were a definite gain, and that the communally grown crops might prove profitable, especially as they did not interfere to any extent with the private farms. Though the Ujamaa fields were in most cases still only a sideline to private production, they mobilized labor which was previously under-utilized and thus constituted a net gain. Indigenous leadership was emerging, often wealthy peasants who as we have noted saw advantages to themselves in joining the "Ujamaa bandwagon," but also hard-working farmers and those who had been active in political campaigns before--"protocadres," as they have been called--whose imagination was caught by the new program and who were ready to give it a try. Some members were urging their relatives to join them in the Ujamaa villages and were planning improvements to their houses, a sure sign that they intended to stay.

The situation in the villages started after 1970 in the regional "Operations" cannot yet be known. It is probable that production will have been disrupted in the short-run, especially in the dry areas where the ecology does not permit settled farming without heavy investment in water supplies and careful management of grazing and arable land. It also seems likely that many farmers will be alienated from the Ujamaa concept if their standard of living deteriorates through poor planning, lack of necessary services, and unsuitable leadership. The need to stabilize production quickly, supply services, and develop internal leadership will be very great.

Thus the effect of external mobilization of farmers must be viewed over a long period. Whether this approach to Ujamaa development can lead ultimately to as high a degree of productivity and commitment as is found in the best of the spontaneous Ujamaa villages depends on the attitudes of the farmers after the carrots and sticks are discontinued, on the suitability of the sites selected for the new villages, on the caliber of leadership which emerges, and on the procedures developed for overcoming the varied problems of cooperative life and work. These must now be considered.

Planning Procedures

In most Ujamaa villages Planning Committees are elected shortly after the members come in. Some of these Committees have adjusted the provisional plans given to them by the Administration, usually in consultation with the officers of the District Development Committees. Few of the plans have included significant technical innovations, and the crops, acreages, and budgets have been determined somewhat arbitrarily, for neither the farmers nor the advisory staff have much experience of farm planning and budgeting. Some villages have included processing and servicing enterprises to ensure that labor is fully utilized throughout the year. In the Ruvuma Development Association each village was allocated specialized production tasks and some inter-village trade developed. But this is an exceptional case: there is a strong need to apply the planning and budgeting guidelines prepared by the Ujamaa Development Division (page 252) and to train the village members and agricultural staff in farm planning.

Ownership Systems

Variability in the proportion of private to communally farmed land was mentioned earlier. Only self-initiated villages which have started from nothing have a high proportion of communal property. The land and other assets are held by the community, with equal and inheritable shares being allocated to each member. Private plots are given on usufructuary tenure, which is also inheritable. Machinery is usually held in common, but livestock are very rarely collectively owned. In the Ranching Associations attempts to encourage the members to pool part of their cattle in common herds have been resisted. Only where Ujamaa villages have obtained loans to buy cattle from outside (which are thus "depersonalized") has it been possible to establish collective ownership and management of stock.

If a member leaves the village or is expelled, he is theoretically entitled to compensation for his private property and share of the communal farm, but no cases are reported where such compensation has been paid. Most cases of drop-outs have been in newly formed villages where little property of value has been created. In some cases departing members have transferred their property and rights privately to the new members who replace them.

Family Participation

In most Ujamaa villages only the head of the household is the member, and he contributes the required amount of labor to the collective farm. Where most of the land is still individually held, this presents no problem: the other family members are occupied on the individual holdings, where food crops are usually grown. But where most of the land is collective, problems arise of providing for the different consumption requirements and different manpower availabilities of families of different size.[57]

These problems have been tackled in a variety of ways. At Kitete, one of the first village settlement schemes (which in 1967 was renamed an Ujamaa village), the commercial enterprises of wheat and cattle are farmed collectively by the men, while the women and children are mainly concerned with the 3-acre homestead plots. The women have also recently started their own collective poultry and vegetable enterprises.

At Liweta, one of the first Ujamaa villages in Ruvuma, all members of the family are members of the village but only one is obliged to work on the collective farm; for the others contributions are voluntary (though paid accordingly). Initially each family was permitted by the Village Committee to clear and cultivate a maximum of two acres of private land, in addition to its contribution to the collective activities. After two years this land was compulsorily incorporated in the collective farm, but the family was allowed to clear another two acres if it wished, similarly to be incorporated in the collective farm two years later. Thus the differing labor supplies and food requirements of the families were allowed for, and the collective farm was gradually built up. By 1968 as the members' commitment to Ujamaa had grown, most of the land was collectively farmed and most of the family members contributed to it.

At another long-established Ujamaa village, Mbambara in Tanga Region, all land was collectively farmed from the start, and all family members contributed to the collective work. This was possible for a number of reasons. First, the members had few dependents and there was little variation in family size: out of 60 members only 16 were married couples at the start. Secondly, the political consciousness of the members was very high; all belonged to the TANU Youth League and had a strong commitment to collective farming. Thirdly, sisal, the main crop which occupied most of the members' attention in the first years, benefits greatly from economies of scale and it seemed rational to grow the small acreage of food crops on the same basis. Both men and women are full members of the Co-operative Society at Mbambara: new members, including prospective wives or husbands of those who joined earlier, have to pass through a six-month probationary period before being finally accepted. The community organizes a day nursery so that mothers can work, and accepts responsibility for the aged. Cash distributions are made on the basis of labor inputs, while food is distributed according to family size.

Work Organization and Reward

The problem of maintaining incentive and high labor productivity in a cooperative work system has been tackled by the following procedures. First, most of the villages divide into small production teams, often based on the ten-house cells. It has been found that ten people work together more efficiently than a hundred: there is less chance of the level of collective effort being determined by the work input of the laziest or slowest member, and specialization of function is made easier. Secondly, daily or weekly tasks are subdivided among the individual members where the nature of the job allows. The work-group leaders are responsible for ensuring that the tasks are completed, and sometimes a spirit of competition develops between the groups. At Mbambara, where many of the members were previously estate laborers, the traditional plantation system of labor management is used, but the job of Work Superindendent is rotated weekly so that every member has the chance to bark orders at his fellows! Those who fail to complete their tasks are fined; those who finish early may either go off or take on an additional measure of work if it is available.

Thirdly, distribution of rewards is linked closely to the work contributed. Each person's daily work input is carefully recorded, and profits are distributed proportionately. In most villages every job is considered to be of equal value per unit of time. At Kitete, with mechanized wheat cultivation and several other enterprises, division of labor has proceeded very far. Tractor drivers, cattle herdsmen, and night watchmen who have to work longer hours than the average were initially compensated by extra leave allowances. After this system had been operating for six years (and after the Chairman of the Village Committee had visited China), a system of work-points was introduced with a small differential in payments for different jobs. The mechanic (the highest paid member) now receives a maximum of 13 points for a full day's work, the storekeeper and cattle herdsmen receive 11 points, the unskilled workers (the lowest paid) receive 9 points. As the range of rewards is small and mobility is maintained between jobs of different categories, there seems little danger of extreme stratification developing within the village.

Management Procedures

Once established, the Ujamaa villages have generally been left with a high degree of autonomy in running their internal affairs. The common pattern is that the General Assembly, comprising all members of the village, elects an Ujamaa Village Development Committee of 10-15 members, and a number of sub-committees for specific functions, as shown in Figure 1. Usually no member may serve on more than one Committee, so that as many people as possible are involved in village government. The committees are re-elected annually, to avoid domination by a few members.

FIGURE 1. UJAMAA VILLAGE MANAGEMENT SYSTEM

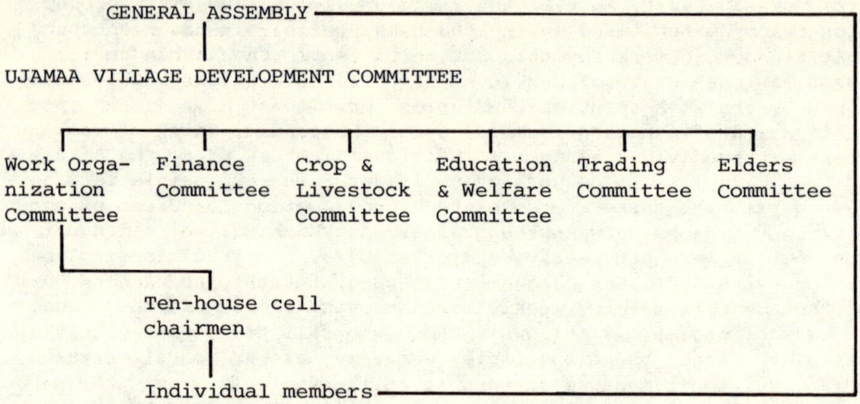

Each technical committee is advised by the local officer of the appropriate Government department; whether the views of the elected leaders or of the outside advisers predominate depends on the personalities involved and on the frequency of the adviser's visits to the farm. Since the end of the Village Settlement period the trend has been increasingly away from externally imposed authority in management decisions.

The Village Development Committee prepares a Constitution (many have based it on a model prepared by TANU) which then has to be accepted by the General Assembly. The Constitution covers the objectives of the organization, the maximum size of the community, the rights and obligations of the members, the functions of the management committees, the financial procedures, etc. The Committee of Elders hears cases of infringement of the rules made under the Constitution, settles disputes, and imposes punishments in the form of fines, extra tasks or, in the last event, expulsion from the village. This ultimate sanction has rarely been used; it must be confirmed by the General Assembly and usually by the Area Commissioner as well.

Financial Procedures

The various sources of finance at the three stages of Ujamaa village development were described above (see p. 245). After the experience of over-capitalization and dependence in the Village Settlement schemes, attempts were made to minimize subsidies in the Ujamaa villages through the provision of services as incentives to new members. Loans are normally available for cash crops only. The Village Development Committee, with the assistance of the Agricultural Department or Bank officials, prepares the loan application. Where the crops are individually grown the credit is passed on to the members, usually in kind, and debts are recovered by the Committee at the time of sale. Figures are not available on loan recoveries; they are always much easier to collect when the crops can be marketed only through a single channel.

Payments from gross revenue are normally made in the following order of priority:

(i) Repayment of annual crop finance loans.

(ii) Payment of installments on medium- and long-term loans (these are sometimes deferred if the balance for distribution to members is insufficient).

(iii) Deduction of levies for Co-operative Savings Account and Reinvestment Fund.

(iv) Payment of balance to the members, calculated on the basis of crop deliveries if tenure is individual, or of work inputs if tenure is collective. In the latter case this payment constitutes a dividend over and above the monthly subsistence allowance, which in most villages members receive as an advance during the cultivating season.

Accounts are kept within the Ujamaa villages by the Finance Committees, with training and guidance from the Co-operative Audit and Ujamaa Village Departments. Instances of misappropriation of funds, though not uncommon, do not compare with those in the Multipurpose Co-operative Societies, owing to the smaller scale of the Ujamaa villages, the lesser amount of money involved, easier supervision, direct answerability of the Finance Committees to the community, and the convergence of interest of those who produce and those who market the crops.

NATURE OF CHANGE OVER TIME

The progress of group farming in Tanzania has changed in several respects over the years through which it has been evolving. First, there has been a move away from private ownership of land toward public ownership. Second, the initial policy of villagization for primarily economic purposes has changed toward a policy of collectivizing land use (at least in the new Ujamaa villages) with broader ideological objectives. Third, the imposition of Government management of group farms has given way to self-management by the participants. Fourth, the policy of capital-intensive development with heavy subsidies has been modified to a policy of labor-intensive development and financial self-reliance. Fifth, the centralized Government structure has changed to a decentralized form, bringing development planning and implementation closer to the people it affects and more directly under political control. Sixth, there has been a trend, though uneven and by no means yet complete, away from sole emphasis on creation of new cooperative communities toward attention to transforming existing villages also. Lastly, the original strategy of a voluntary gradual transition to Ujamaa has been replaced by one requiring the whole rural population to move rapidly toward the targets of villagization and collective farming.

The last of these changes is perhaps the most significant, and the one raising the biggest dilemma for the Tanzanian Government. Spontaneous moves to Ujamaa have achieved a certain amount in limited areas, but it has become clear that--given Tanzania's historical background--voluntary advances will not on their own occur fast enough, nor on a sufficiently large scale, to stem the growing individualism and inequalities in the rural sector. More rapid progress necessitates the active involvement of the Government and Party in mobilizing the farmers. But departure from the policy of gradual and voluntary change introduces the risk--if the Government and Party mobilizers have insufficient experience and political awareness--of suppressing the farmers' initiative, disrupting production, and distorting the objectives of the Ujamaa program.

Pattern of Group Farm Establishment

It is possible to identify a series of waves in Tanzania's movement toward Ujamaa. A period of rapid expansion of cooperative activities is followed by a lull in which few new villages are started and existing ones are consolidated. Then comes another surge of activity and the cycle repeats itself. Each expansionist phase is associated with an increased involvement of the Government and Party in mobilization. This progression is illustrated in Figure 2. On the achievement of Independence in 1961 there was a proliferation of spontaneous cooperative activities which was formalized, after the initial enthusiasm had worn off, in the Village Settlement schemes from 1963. At the end of the Village Settlement period in 1965 came a two-year period of reappraisal and consolidation. In 1967, after the "Arusha Declaration," there was a further proliferation of Ujamaa villages, again many spontaneously initiated but a higher proportion being started by the Government. The expansionist phase continued until late in 1968 when there came a second period of retreat. In 1970 and 1971 expansion was renewed on a much larger scale, almost entirely on Government initiatives. Then followed a brief period of consolidation in 1972-73 before the current call from the President for the whole population to move into Ujamaa villages by 1976.

FIGURE 2. WAVES OF UJAMAA VILLAGE DEVELOPMENT

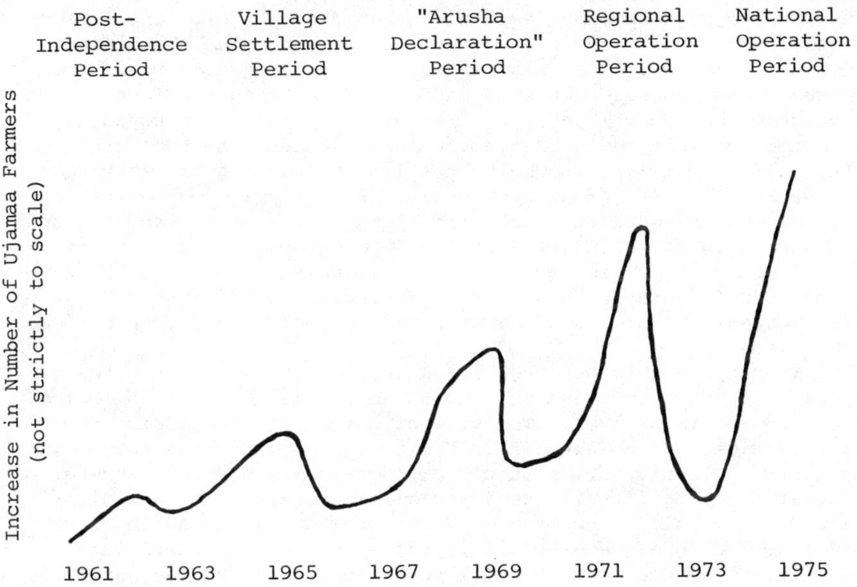

Official policy throughout this period has been that moves to Ujamaa must be made by the farmers voluntarily; the Government and Party must use persuasion but not force in getting the Ujamaa villages started. However, "persuasion" seems a euphemistic way to describe what actually happened in some districts during the expansionist phases. In the 1968 mobilization campaign local administrators in several Districts clearly exceeded their authority and used coercive means to start the Ujamaa villages; many of them were later criticized and dismissed by the Government for their action. Yet in the 1970-71 expansion and again in 1973-74, coercion was widely employed to move the farmers into villages, this time more openly. Throughout the period, however, once the farmers had been settled in the new villages the use of coercion was not continued.

I wish to suggest in this paper that the periodic departures from the official policy of gradual and voluntary change should not be seen as regrettable temporary aberrations by local officials, but rather should be viewed as a necessary element, and perhaps even a deliberate one, in the Government's strategy for moving the country to Ujamaa.

The argument I propose runs as follows: The Government realizes that a commitment to voluntary and gradual change is necessary to retain the peasants' support for the Ujamaa policy. It understands that the farmers will not commit themselves fully to the cooperative farms unless they see them as their own; also it realizes that it has neither the financial nor the administrative resources to run the Ujamaa program itself on a continuing basis. But to ensure that Ujamaa spreads fast enough to make a significant impact on the rural areas, the Government and Party have to step in and push the farmers into Ujamaa villages faster than they would move of their own accord. The expectation is that once the farmers are living in the new villages, even though they joined reluctantly, with the services and guidance that are available most of them will stay and they will soon come to realize the advantages of the Ujamaa way of life.[58] After each period of accelerated mobilization, the Government and Party therefore stand back and help the peasants who have been moved into the villages to establish themselves on a firm foundation. This period of consolidation also gives the Government administrators and Party cadres time to learn from their experience with the Ujamaa farmers before embarking on the next mobilization campaign.

To support this hypothesis, an analogy may be drawn between Tanzania's experience and that of China in the 10 years following the land reform of 1947. The view of the Chinese Communist party at that time, as in Tanzania in the 10 years after Independence, was that collectivization should pass gradually through a number of stages: mutual aid teams, semi-socialist cooperatives, and ultimately fully collective farms. Moves from one stage to the next should be voluntary and should depend on the principle of mutual benefit. "Chinese peasants will never accept collective ownership without a long period of socialist education to be provided by mutual aid teams and co-operatives."[59]

But in fact both the 1952 and 1954 campaigns to spread mutual aid cooperatives, and the 1955-56 collectivization campaign in which 63 percent of Chinese peasant households became members of collectives within a period of 12 months, disregarded the principles of voluntary participation and reform through stages. Each campaign had an adventurist quality with local cadres "blindly running after high level forms using political and economic sanctions to force the peasants to change." This stage was followed by a short period of retreat, Government criticism of local cadres, and a restatement of the need for stable advance toward socialism. The end of each round of consolidation came when the Government claimed that anti-socialist opportunists were using it to consolidate their position, and this was the starting point for a new campaign.[60]

The parallel between the two countries is, of course, far from exact. Tanzania has had no violent land reform; its cadres are only just beginning to be formed; and the gulf between landlords, rich and poor peasants is not comparable to that found in China before Liberation. But the same principles of voluntary and gradual development toward collectivization are stated by the two countries, while in fact both have used periodic enforcement campaigns to accelerate the process. Furthermore, the more cautious elements of Government and Party in both countries have tried to hold back the rate of collectivization, on the grounds that there are too few trained cadres to ensure success; and the argument used by Mao to counter this point of view--namely, that the cadres can only obtain their training through practical experience "on the job"[61]--is one which is also heard in Tanzania. Finally, it has been suggested for China, as this paper argues for Tanzania, that departure by the cadres from the official policy of voluntariness was inevitable and perhaps even intended. It was a more serious offense to be judged a rightist in China in the 1950s than to be accused of adventurism, and though this attitude was not found in Tanzania in the 1960s it is quite possible that it may appear in the 1970s.

CONCLUSION

In China the objectives of collectivization have already been achieved. In Tanzania progress in socializing agricultural production has still, with some exceptions, been largely limited to poorer areas; but there is evidence that the Government's expansionist strategy started in 1968 may now be having some success in broadening the collective base. As we have seen, farmers who were moved into some of the Government-initiated villages in 1968 were reported in 1972 to be settling down and to be realizing the advantages of Ujamaa. Government staff have been posted to Ujamaa villages, and are being forced to relate with the poor peasants. Some land has been taken over from large farmers; others are finding that they no longer have preferential access to their production requirements. "Protocadres" are beginning to emerge. The class conflicts in the villages are being brought more sharply into focus. It seems possible that some of the villages started by the Government campaign in 1968 will take the next step to Ujamaa of their own accord; or, if not, that they will at least accept more readily a further period of enforced advance should a follow-up campaign be mounted.

In regard to the 1970-71 and 1973-74 campaigns to bring whole regions into Ujamaa villages it is too early to speak. In view of the enormous scale of these "Operations," they must be making tremendous demands on the resources available to the Government and Party. Feasible village development plans have to be prepared; necessary services have to be provided; village leaders and technicians have to be trained; farmers have to be introduced to Ujamaa living and encouraged to higher productivity; incentive of the more progressive farmers has to be maintained. The risk of some process in the operation breaking down must be very great and the consequences could be calamitous.

The drought currently affecting many parts of Africa adds to the Government's difficulties. It has been argued forcefully that the current Ujamaa campaigns may have exacerbated the effects of the drought by uprooting households and further disrupting food production. But it is also possible that villagization was a necessary step for facilitating distribution of famine relief which would have been required anyway, and that in the long run it will prove to be the best way of countering adverse weather conditions in drought-prone zones.

Tanzania is fully determined to make its Ujamaa policy work. A great deal of practical experience has been gained of the internal problems of operating the group farms. There is a growing understanding of the nature of the external influences which act on the Ujamaa villages, and solutions are being sought to counter the effects of the adverse ones. Appropriate administrative and political structures have been developed and the education policy has been radically reformed to meet the needs of the rural development program.[62] It is not possible for an outsider such as myself, who has in any case been away from Tanzania for some years, to predict the long-term outcome of the Ujamaa policy. Clearly, it faces many obstacles. But the swings from acclamation to denunciation by some of Tanzania's observers abroad, as the waves of Ujamaa surge and fall inside the country, may be more extreme than the apparent changes in policy deserve.

Summary

1. Tanzania is one of the countries in Africa which attaches most importance to group farming as the key to rural development. The Government's ultimate objective, after grouping farmers into villages and introducing innovations through cooperative enterprise, is to transform the whole agricultural sector into one based on common ownership and exploitation of land by democratically governed village communities.
2. Several natural, demographic, socio-economic, and political features led to the emergence of this policy and made Tanzania at Independence a peculiarly favorable ground for its realization. First, the tradition of common ownership (though not common exploitation) of land and mutual assistance at village level; second, the poor economic base in most parts of the country, and the relative absence of social and economic stratification compared to many ex-colonial countries; third, the abundance of unutilized land in most parts of the country; and fourth, the single party political system based on a strongly socialist ideology.
3. In the first five years after Independence (1961-1966) a number of Government-financed and managed "village settlement schemes" were established. These proved expensive and unproductive. A privileged class of settlers dependent on Government assistance began to emerge in the new settlements and this approach to group farm development was abandoned.
4. Following the "Arusha Declaration" and subsequent policy papers in 1967-1968, the emphasis turned to voluntary formation of "Ujamaa (Socialist) Villages," with minimum investment by the Government and maximum dependence on the members' efforts both in production and in management.

5. Considerable progress in villagization was made in the poorer parts of the country, where population was previously sparsely scattered. Some extra land was brought into collective cultivation, and cooperative services were widely introduced. Progress toward socialization of production on already developed land was negligible. With few exceptions, collective farming was found only where it did not disturb existing property relations or power structures.
6. With a low level of political consciousness and a lack of trained cadres to mobilize the peasants into collectives, the conflict between voluntary initiation of Ujamaa villages and the need to proceed fast enough to stem the growing inequalities in the rural areas became increasingly apparent. Sporadic attempts were made in 1967-1968 by over-zealous political or administrative leaders to force the peasants into collective farms, and since 1970 endeavors have been made to move whole regions into Ujamaa villages. These attempts have had negative effects in the short-term, often disrupting production and prejudicing the farmers against the Ujamaa policy.
7. A series of waves can be identified in Tanzania's progression toward group farming. First comes a period where a policy of voluntary and gradual reform is pursued, and a slow advance is made toward socialism in limited areas. Frustrated by the apathy of the peasants and concerned at the increasing entrenchment of elites in the nonsocialized sector, cadres and Government leaders initiate an *adventurist* phase where the peasants are pressed into group farming through a combination of coercion and material incentives. This phase is generally insufficiently planned and many of its instigators have inadequate understanding of the objectives of the Ujamaa policy. Often the campaign disrupts production in the short term, engenders great opposition, and is quickly halted. A further period of *consolidation* ensues, with the commitment to voluntary and gradual change being recalled, and the cycle begins again.
8. It is suggested that the "adventurist phase," though contrary to the official policy and fraught with risks, is a necessary stage in the process of collectivization in Tanzania. However, more careful planning and preparation of cadres is needed if production is not to suffer unduly, if the peasants are not to be permanently alienated, and if the Ujamaa objectives are not to be distorted. In the "consolidation phase" more support and education of the farmers are needed to enable them to run their own affairs collectively and to permit both them and their guides to learn from the experience gained in the last "adventurist phase." If these two alternating phases are followed, more rapid and surer progress can be made toward Ujamaa than can be achieved by either phase alone.

NOTES

1. The name of the country at the time of Independence was Tanganyika. After the Union with Zanzibar in 1964 it took the name of (the United Republic of) Tanzania. This paper describes the experience of group farming in mainland Tanzania only. For simplicity's sake I refer to the country as Tanzania throughout.

2. In 1965 the absence of real opposition to TANU was recognized by the declaration of a One Party State. See Lionel Cliffe, ed., *One Party Democracy* (East African Publishing House, 1967).

3. Basic data on Tanzania at Independence are found in I.B.R.D., *The Economic Development of Tanganyika* (Johns Hopkins Press, 1961).

4. The scattered nature of settlement in much of Tanzania was an important determinant of later villagization policies. See N. Georgulas, "Settlement Patterns and Rural Development in Tanganyika," *Ekistics*, no. 24 (1967).

5. See I.B.R.D., *Economic Development of Tanganyika*.

6. A major factor behind the build-up of TANU was the widespread opposition to enforcement of the agricultural rules. See Lionel Cliffe, *Nationalism and the Reaction to Enforced Agricultural Change in Tanganyika During the Colonial Period*. Reprinted in Cliffe and J. S. Saul, eds., *Socialism in Tanzania*, vol. 2 (Dar es Salaam: Tanzania Publishing House, 1973).

7. Many of these schemes were established on land cleared for the ill-fated Groundnut Scheme. See W. P. Cocking and R. F. Lord, "The Tanganyika Agricultural Corporation's Farming Settlement Schemes," *Tropical Agriculture* 33 (1958).

8. The development of the village settlement policy is reviewed in L. Cliffe and G. L. Cunningham, *Ideology, Organisation and the Settlement Experience in Tanzania*; reprinted in Cliffe and Saul, eds., *Socialism in Tanzania*, vol. 2.

9. The Village Settlement experience in Tanzania has been well documented. See ibid.; H. Ruthenberg, *Small-Holder Farming and Development in Tanzania* (Munich: Weltforum Verlag, 1968); A. O. Ellman, "Introduction of Technical Innovation through Co-operative Farming," *East African Journal of Rural Development* (1970); and for a case-study of one of the more successful schemes, A. O. Ellman, "Kitete, A Land Settlement Scheme in Northern Tanzania," FAO, *Land Reform and Co-operatives*, no. 1 (1967).

10. The change in policy was announced in a speech by the 2nd Vice-President, Hon. R. M. Kawawa, reprinted under the title "New Approaches to Rural Development," *Mbioni* (Kivukoni College, Dar es Salaam), vol. 2, no. 11 (April 1966).

11. See J. D. Heijnen, *The Mechanised Block Cultivation Schemes in Mwanza Region*, Bureau of Resource Allocation, University of Dar es Salaam, December 1969.

12. See W. Scheffler, "Tobacco Schemes in the Central Region," in Ruthenberg, ed., *Small-Holder Farming and Development in Tanzania*.

13. Ralph Ibbott, "Ruvuma Development Association," in *Mbioni*, vol. 3, no. 11 (July 1966).

14. Reprinted in J. K. Nyerere, *Freedom and Unity* (Dar es Salaam: Oxford University Press, 1967).

15. Reprinted in J. K. Nyerere, *Freedom and Socialism* (Dar es Salaam: Oxford University Press, 1968).

16. These two quotations are taken from *Tanzania: Second Five Year Plan for Economic and Social Development, 1969-74* (Dar es Salaam: Government Printer, 1969).

17. *TANU Guidelines* (Dar es Salaam: Government Printer, 1971).

18. Nyerere, *Freedom and Unity*.

19. *Tanzania: Second Five Year Plan, p. 4.*

20. I have been out of direct touch with Tanzanian developments since the end of 1970. The picture in the last four years has been gleaned from various secondary sources, and may be inaccurate or incomplete in some respects.

21. Certain difficulties arise in this categorization of Ujamaa villages, both in establishing firm criteria for elevation to each successive stage, and in overcoming the legal problems of giving block credit to Stage I villages which are not members of a Co-operative Society. Necessary amendments to the Co-operative Societies Act to overcome the legal problems have been under consideration for some time.

22. M. von Freyhold, et al., "Rural Development through Ujamaa Vijijini; Some Considerations Based on Experience in Tanga," Department of Economics, University of Dar es Salaam, 1972.

23. J. Boesen, *Development and Class Structure in a Smallholder Society and the Potential for Ujamaa* (Copenhagen: Institute for Development Research, 1972).

24. The process of class formation in Ismani, and attempts to counteract the trend, are described in A. Awiti, *Economic Differentiation in Ismani, A Critical Assessment of Peasants' Response to the Ujamaa Vijijini Programme* (Dar es Salaam: Economic Research Bureau, 1972).

25. See H.U.E. Thoden van Velzen, *Some Social Obstacles to Ujamaa, A Case Study from Rungwe* (Leiden, Holland: Afrika Studiecentrum, 1970).

26. Ibid. It should be noted that in less densely populated areas labor-sharing is still commonly practiced.

27. Any number of instances of corrupt practices in Co-operative Societies can be given. In 1966 a Presidential Special Committee of Enquiry into the Co-operative Movement and Marketing Boards was set up and reported (Government Printer, Dar es Salaam, 1966). A major reason behind the failure of the cotton block farms referred to on p. 241 was that the tractors were operated by the local Co-operative Union, and "tractor-drivers and Co-operative Committee men divided the money received as hire charges, for which no receipt was issued . . . ." S. E. Mighot-Adholla, "The Politics of a Growers' Co-operative Society," in W. Luttrell, et al., eds., *Towards Rural Co-operation in Tanzania* (Dar es Salaam: Rural Development Research Committee, 1975).

28. For a sociological analysis of the marketing co-operatives see J. S. Saul, "Marketing Co-operatives in Tanzania," in Cliffe and Saul, eds., *Socialism in Tanzania*, vol. 2.

29. Other trends are reviewed in J. S. Saul, "From Marketing to Producer Co-operative," in Luttrell, et al., eds., *Towards Rural Co-operation in Tanzania*.

30. See R. Ole Kuney, "Ranching Associations in Masailand," Department of Geography, University of Dar es Salaam, 1970.

31. Ralph Ibbott, "The Disbanding of the Ruvuma Development Association," Mimeo., London, 1970.

32. Quoted in P. Raikes, "Ujamaa Vijijini and Rural Socialist Development," Economic Research Bureau, University of Dar es Salaam, December 1973.

33. For more comprehensive descriptions see Stanley Dryden, *Local Administration in Tanzania* (Dar es Salaam: East African Publishing House, 1968); Henry Bienen, *Tanzania: Party Transformation and Economic Development* (Princeton, N.J.: Princeton University Press, 1967).

34. These quotations are from *TANU Guidelines*.

35. J. K. Nyerere, *Decentralisation* (Dar es Salaam: Government Printer, 1972).

36. It had become increasingly clear that there was no logic in maintaining separate movements for cooperative and Ujamaa development. The Marketing Co-operatives have to move into production, the Ujamaa Villages when they become viable have to be registered as Multi-purpose Co-operative Societies, giving them the legal status to hold land, obtain Bank loans, etc. The logic of this trend was recognized by merging the Ujamaa and Co-operative Development Divisions in 1972 at national, regional, and district levels.

37. One of the objectives of the Ujamaa policy is to reorganize production in a way which allows farmers both to campaign for better prices for their products and through specialized production to develop internal trade and break the total dependence on export markets. See Cliffe, "The Policy of Ujamaa Vijijini and the Class Struggle in Tanzania," in Cliffe and Saul, eds., *Socialism in Tanzania*, vol. 2.

38. Several instances where international influences conflicted with Tanzanian principles, and were rebuffed, are included in the President's speech, "Principles and Development" (1966), reprinted in *Freedom and Socialism*.

39. The Government brought in measures from 1962 onwards to regulate land use, control land transfers, and ultimately to bring all lands into public ownership. These measures have not yet been fully implemented.

40. J. K. Nyerere, "After the Arusha Declaration" (1967), in *Freedom and Socialism*.

41. Several more illustrations of unequal development can be found in Cliffe, "Ujamaa Vijijini and the Class Struggle."

42. See M. von Freyhold, "The Potential for Ujamaa in Handeni," Department of Economics, University of Dar es Salaam, 1972. A series of papers by this author reveals clearly the ease with which poor peasants can continue to be exploited, even in Ujamaa villages.

43. The growth of this defeatist outlook in Handeni is traced ibid.

44. D. Feldman, "The Economics of Ideology: Some Problems of Achieving Rural Socialism in Tanzania," in Colin Leys, ed., *Politics and Change in Developing Countries* (Cambridge, Eng.: Cambridge University Press, 1969). See also I. Kopytoff, "Socialism and Traditional African Societies," in W. H. Friedland and C. G. Rosberg, eds., *African Socialism* (Stanford, Cal.: Stanford University Press, 1966); and R. P. Dore, "Modern Co-operatives in Traditional Communities," in Peter Worsley, ed., *Two Blades of Grass* (Manchester, Eng.: Manchester University Press, 1971).

45. B. B. Bakula, "The Effect of Traditionalism on Rural Development in Bukoba," in J. H. Proctor, ed., *Building Ujamaa Villages in Tanzania* (Tanzania Publishing House, 1971).

46. R. Lewin, "Matetereka," in Cliffe and Saul, eds., *Socialism in Tanzania*, vol. 2.

47. D. Feldman, "The Economics of Ideology."

48. M. von Freyhold, "Potential for Ujamaa in Handeni."

49. In one Rungwe village in 1968, the Branch Chairman of TANU and 9 of the 15 TANU cell Chairmen came from among the wealthiest 20 percent of peasant households in the village. H.U.E. Thoden van Velzen, "Staff, Kulaks and Peasants: A Study of a Political Field," in Cliffe and Saul, eds., *Socialism in Tanzania*, vol. 2.

50. Another factor is that in many areas village-level staff had traditionally formed alliances with the rich peasants or local political leaders, borrowing land from them in return for various favors; many, therefore, found the Ujamaa policy to be against their own interests. See for example H.U.E. Thoden van Velzen, "Staff, Kulaks and Peasants." In many of the new Ujamaa villages, however, this alliance showed a tendency to crack, as the wealthy villagers found it more to their advantage to ally themselves with the Ujamaa farmers, and therefore jettisoned their links with the staff who might expose their domination over the village which they were anxious to reinforce. See M. von Freyhold, "Potential for Ujamaa in Handeni."

51. Ibid.

52. I.K.S. Mussoke, "Building Socialism in Bukoba." In Proctor, ed., *Building Ujamaa Villages*.

53. Paraphrased from Le Duan, *On the Socialist Revolution in Vietnam* (Hanoi, 1965).

54. One of the best-documented such villages is Mbambara Ujamaa Village, a sisal farm started in 1963 by TANU Youth League members. See R. Lewin, "Ujamaa Living at Mbambara," *Mbioni*, vol. 4, no. 8 (1968); Binamu, et al., "Reports on Mbambara," Kivukoni College, Dar es Salaam, 1966; Kates, McKay, and Berry, "Twelve New Settlements in Tanzania," University of Dar es Salaam, 1968.

55. The best-documented examples are in studies of Ujamaa villages in Ngara District (J. Boesen, *Development and Class Structure*) and Karagwe District (A. Moody, *Study of Six Ujamaa Villages in Karagwe District* [Copenhagen: Institute for Development Research, 1972]).

56. Much of the information in this section is taken from the studies of Ujamaa villages in Tanga and Handeni by M. von Freyhold, et al., "Rural Development through Ujamaa Vijijini."

57. Some of the demographic obstacles to collective farming in three-generational communities are discussed in D. and R. Feldman, *Co-operation and the Production Environment*, Economic Research Bureau, University of Dar es Salaam, 1969. See also J. Goody, ed., *The Development Cycle in Domestic Groups* (Cambridge, Eng.: Cambridge University Press, 1969).

58. A similar argument has been expressed in relation to Indian peasants. "People can be pressured by carrots or sticks into doing things which they consider are evil or foolish; and a long enough experience may convince them that in fact these things are neither evil nor foolish . . . the conclusion must be that, given peasant resistance, a radical policy of political and economic modernisation can only be achieved by pressure and by continued success in material terms." F. G. Bailey, "The Peasant View of the Bad Life," in T. Shanin, ed., *Peasants and Peasant Societies* (Harmondsworth, Eng.: Penguin Books, 1971). (I

would add the requirement of success in social and political terms to Bailey's suggestion that success in material terms is enough.)

59. Sun Hsiao-ts'un and Hsiao Hsi'lin, *Some Problems in Promoting the Mutual Aid Co-operation Movement* (Peking, 1952).

60. The recurring sequence of adventurism and conservatism in China during this period has been finely described and analyzed by K. R. Walker, "Collectivisation in Retrospect: The Socialist High Tide of Autumn 1955-Spring 1956," *China Quarterly* (London), no. 26 (1966).

61. See Mao Tse-tung, *On the Question of Agricultural Cooperativisation* (Peking, 1955).

62. See J. K. Nyerere, *Education for Self-Reliance* (Dar es Salaam: Government Printer, 1967); and other articles in Resnick, ed., *Tanzania: Revolution by Education* (Longmans, 1968).

11. An Experience of Group Farming in Dahomey: The Rural Development Cooperatives

Moise C. Mensah

Group farming has been a very popular idea throughout independent Africa since the early 1960s. Conceptually, it is attractive to politicians and policy-makers in that it combines old African traditions with the ability to absorb the kind of progressive innovations that could be offered by modern cooperative movements. No wonder, therefore, that Dahomey, like many other African countries, has endeavored to use group farming as a spearhead of its rural development efforts.

THE REASONS FOR RURAL DEVELOPMENT COOPERATIVES

The Dahomean experience in modernized group farming which we shall be referring to in this paper has been based primarily on an attempt to solve land reform problems in southern Dahomey with a view to boosting production of the main export crop, oil-palm. It was therefore meant for an optimum use of land and labor available for maximum production of palm-oil. Hence, the denomination "Coopérative d'aménagement rural," which should be translated as "cooperative for rural area planning." It was, however, discovered very soon that the scope of the enterprise should be broadened to cover food-crop production, livestock development, cottage industries, health and education, so that the production units created became in fact rural development cooperatives, although the title "coopérative d'aménagement rural" was kept.

When we stated that the need for optimum use of land and labor was the basis for the creation of rural development cooperatives, we were referring to a situation where absentee landlords were owning land which they could by no means develop, while landless rural manpower was available in large numbers ready to accept any reasonable proposal in terms of partnership in the exploitation of the land. It must be added that small-scale, individual oil-palm groves had proved to be a total failure after more than a decade of experience, which led the governmental policy-makers to place a new emphasis on large-scale plantations. Hence, the need for large acreages to be developed. A few "latifundistas" apart, this called for a considerable amount of land consolidation, which was another aspect of land reform objectives. One should underline the fact that the area concerned is the most densely populated part of the country (from 50 to over 200 inhabitants per km^2), where, unlike elsewhere in Dahomey, land has been subject to private ownership for several generations.

WHAT IS A "COOPERATIVE D'AMENAGEMENT RURAL"? HOW DOES IT WORK?

Compulsory Cooperation for 50 Years

An obvious question is, how does one organize cooperation between landlords and landless labor? In the case of Dahomey, the policy-makers' choice was for a "compulsory marriage." Law 61-26 of 10 August 1961, in its articles 2 and 3 stipulates that "after the necessary surveys have been carried out on a suitable area, the President of the Republic may, by decree, based on appropriate report by the Minister of Agriculture and Cooperatives, decide on its full development, bearing in mind its soil potentialities and the market prospects." Such a decree goes into great details about the program to be implemented in the area which is declared to be of "public utility" (*utilité publique*). It is left to a ministerial decision to create one or several cooperatives covering the area. The important aspect of the ministerial decision is that all landowners in the area covered by a coopérative d'aménagement rural are obliged by law to adhere to the cooperative. They must bring their land as their contribution (shares) to the new venture for a minimum period of fifty years. There is no obligation to bring a labor force. The shares are transmissible to their descendants.

Land = Labor

By bringing his land to the cooperative, a landowner becomes a cooperative member known as shareholder of Category A. The number of his shares is proportional to the acreage of the land provided (e.g., one share per hectare). A landless agricultural worker who joins the cooperative may qualify as a shareholder and therefore become a full member if he works for a minimum required number of days per year. This minimum requirement is calculated to match the value of the unit of land equivalent to one share. For example, if one share = 1 ha. = $100, the landless would-be cooperative member will have to offer the value of labor equivalent to $100 per year. Assuming that his daily salary would be $2, he would have to accept $1 and give to the cooperative as savings $1 per day for one hundred days during the year. He could increase his shares as much as he is prepared to work hard the first five years of the life of the cooperative, which is the investment phase in the case of oil-palm plantations. The shares acquired by a landless cooperative member are called shares of Category B. Shares of Category A and Category B are equal in value and receive the same treatment as far as interest rates are concerned. Holders of Category A (landowners) may acquire shares of Category B by offering their labor force if they so desire.

One-Man-One-Vote

But for the differences in approach to membership, the rural development cooperatives are organized very much like ordinary cooperatives based on the so-called Rochdale principles. First of all, the most essential principle of one-man-one-vote is strictly adhered to. The assembly of cooperative members elects its own board of trustees. However, in view of the important contribution made by public authorities toward financing the required investments, and in order to ensure efficient management, it was found necessary that the Government should have representatives on the board (with voting rights, but no possibility of outweighing the united votes of cooperative members) and that it should also nominate the Director of the cooperative through the Ministry of Agriculture and Cooperatives.

As in the case of ordinary cooperatives, the interest borne by shares is limited (3 percent) in order to avoid speculation. Similarly, the rural development cooperatives are not allowed to keep profits. If these occur, they are distributed to cooperative members not according to their shares but proportionally to the number of days they worked for the cooperative during the relevant year.

Once fully established, a rural development cooperative takes total and effective control of its own destiny. While initial investment decisions which led to its creation have been made by the Government, further decisions to expand its activities are made by the cooperative members themselves. Their elected Chairman is duly entitled to represent them vis à vis third parties and may undertake any contractual arrangements or negotiations decided upon by the board.

Realistic Approach to Production Cooperatives

In their approach to collective farming, the rural development cooperatives have taken a pragmatic attitude of restricting collective farming to crops which lend themselves to such a system. The basic requirements are as follows:

a) Possibility of organizing effectively equal work participation by all cooperative members. This supposes standard jobs, easy quality control of performance, and guarantee of employment for a duration long enough to use the entire labor force available, either continuously or on a rotation basis.

b) Reasonable assurance that the crops will yield sustained production to cover the expenses involved. It was found that under the economic and agro-climatological conditions prevailing in southern Dahomey, these requirements were better met by a perennial crop, oil-palm. Therefore, the labor force available within a rural development cooperative will be used *collectively* only on the oil-palm plantations, which normally cover 50 percent of the land belonging to the cooperative. The other half will be organized into well-planned annual crop production areas where individual plots will be distributed to cooperative members. The cropping pattern on those plots is standardized but in a way which takes care of family needs: priority is given to food crops, mainly maize, but allowance is made for quick-return cash crops such as cotton and groundnuts.

The philosophy behind this approach is that one must not allow cooperation to kill the stimulus of incentive and thereby kill individual initiative. On individual plots, everyone gets the result of his own efforts. Inputs such as improved seeds, fertilizers, and pesticides are provided on a reimbursement basis, together with advice from extension workers, some of whom are paid by the cooperative itself.

On the plantations, a special system has also been developed to encourage high productivity per worker. Every activity from the clearing of land to production of seedlings, planting, weeding and harvesting, has been standardized and the labor requirements of each job quantified (practically all activities are manual to make maximum use of available manpower). Each day, cooperative members are organized into units under the control of a supervisor who has a job card defining the task to be performed. Each worker is then allocated a specific assignment (acreage of land to clear or weed, for example) for the seven-hour work day. Whenever he has completed his work, he calls for the supervisor who gives him a disc testifying that the job was done according to the standard. The cooperative member then goes to the cooperative office where a bookkeeper credits him with one day against delivery of the disc. It might happen that through hard work he was able to complete his assignment in four hours. Instead of going back home, he could take up another assignment and get eventually two days credit in the same day. On the contrary, a worker who has not been able to complete his task might get his disc (and credit) only the following day when the job is duly completed. This constant concern for efficiency through incentives for individual cooperative members to work hard is probably the most distinctive feature of the Dahomean rural development cooperative where, unlike the not uncommon experience of some cooperatives, one does not allow the lazy fellows to play the comfortable role of the parasite to the detriment and frustration of loyal and hardworking cooperative members.

It might be useful to mention that cooperative members may have land outside the cooperative area which would continue to be exploited along traditional lines (individually or by family groups).

THE ACHIEVEMENTS: A STEP TOWARD RURAL DEVELOPMENT

Facts and Figures

The first rural development cooperatives are now 12-13 years old. So it is possible to assess their achievements. By the end of 1973, total membership was in the neighborhood of 15,000 members, which means between 60,000 to 90,000 people located in the three Regions of southern Dahomey (Mono, Atlantique, Ouémé). The coopératives d'aménagement rural have developed 52,000 hectares of land into 28,000 ha. of oil-palm plantations and 24,000 ha. of annual crops, mainly food crops. They also own three palm-oil mills with a total capacity of 130,000 tons per year. They have been able to combine livestock raising with oil-palm production and have established very successful cottage industries mainly geared toward optimum use of the available female labor force (cassava flour industry).

Agrarian Reform Achieved

The initial purpose of the coopératives d'aménagement rural was land reform. This target was reached in those areas for which it was intended. To be sure, landowners have not been enthusiastic about the idea. Formal securities have to be given to them in the form of title deeds which guarantee that if the cooperative is dissolved after the 50-year term or loss of 3/4 of the capital assets, they could claim ownership over their land, which would be given back as consolidated plots. Firmness on the part of the Government and SONADER (the National Corporation for Rural Development, in charge of the operations) has helped to make it possible to bring landlords to accept the new venture. They were also encouraged by the payment of interest on the value of their land and the possibility of employment for their relatives.

Beyond land reform it was possible to induce agrarian reform. By agrarian reform, we mean a package of positive changes designed to bring improvements in the technical, economic, and social aspects of a rural community's life. The amount of investments involved in establishing rural development cooperatives was such that every possible measure was taken to ensure positive results. First of all, change in the land tenure system was by itself a major socio-political step with far-reaching implications. Then, improved technology had to be (and was) introduced to traditional farmers' communities. This included better seeds and seedlings, use of fertilizers and pesticides, and sophisticated oil-mills. Credit facilities were also made available, mainly by the Dahomean Development Bank, to support the various activities of the cooperatives. Traditional farmers have been introduced to new and more efficient management based on a combination of democracy and technical competence. Two social groups have been brought into the forefront--the youth and the women. Being more open to innovations, the youth learned faster than their elders the new techniques and were able to hold key posts in the cooperative structures. That by itself was a revolution. Another revolution resulted from the fact that a great number of women left traditional activities such as petty trading to join the cooperative labor force, and especially in activities such as seed sowing, weeding, etc., were able to earn as much, if not more, money than the men (sometimes their fathers or husbands). While presenting here the promotion aspects of the two main changes in social structure (youth and women) we shall in another section of this paper discuss the problems raised. We would like to mention also that the agrarian reform which took place made a definite impact in terms of slowing down rural exodus from the areas concerned by the creation of employment without distinction of age or sex. Finally, it should be noted that the rural development cooperatives are on the way to creating a much needed development factor in developing countries--a strong and constructive pressure group able to draw attention to all concerned with rural community problems and expectations.

MOVING FORCES BEHIND THE RURAL DEVELOPMENT COOPERATIVES: GOVERNMENT SUPPORT AND EXTERNAL AID

It would not have been possible to "sell" the idea of rural development cooperatives to traditional farmer landlords and landless alike without assurance that the program proposed would be thoroughly carried out. This implied strong financial support. A considerable amount of financial resources is necessary to go through the various steps of establishing a coopérative d'aménagement rural. These steps are:

a) Cadastral survey and land consolidation;
b) Soil survey to determine capability;
c) Allocation of plots for annual crops;
d) Oil-palm seed and seedling production (2 years);
e) Land clearing;
f) Planting of perennial crop (oil-palm) and maintenance of plantation for 5 years.

Parallel to these technical phases, tremendous efforts of information and education are made to persuade farmers to build up a viable cooperative enterprise. To give an idea of the financial resources needed, a standard type of cooperative handling 600 ha. of oil-palm plantation and 600 ha. of annual crops would require in investments $700,000 (May 1975 rate of conversion to CFA). This does not include the cost of the oil mill which in this case will belong to a group of several cooperatives.

In the past 12 years, the cost of establishing rural development cooperatives has been met partly by the Government, but mostly by foreign aid donors, prominent among which have been the French Aid and Cooperation Fund (FAC) and the European Development Fund, which offered grants, and the World Bank, which gave loans on "soft" terms. The government agency responsible for the implementation of the program is SONADER. It has always carefully planned every phase of its program implementation and made especially sure that efficient management was provided, not only during the creation phase of rural development cooperatives when it has specific technical responsibilities, but also throughout the life of the cooperatives for which SONADER remains the "god-father" and dutiful adviser. SONADER's approach in management is based on high technical competence as a means of reducing cost. The concern about low cost is essential since Dahomey, in spite of her historic reputation as an oil-palm growing country, offers marginal conditions for that crop as far as rainfall is concerned. Therefore, yields are relatively modest and profit has to come from low production cost. Nevertheless, for a standard 600 ha. cooperative, SONADER will assign during the plantation phase two senior officers, three technical officers, three assistant technical officers, one accountant, and two health officers. The core of the established cooperative's management, which is paid by the cooperative itself, includes one director, one deputy director, three extension workers, and one accountant. Additional assistance is provided free by the Ministry of Agriculture (agricultural research, extension services, community development, assistance from the cooperatives department, etc.).

THE PROBLEMS: DIFFICULTIES OF CREATING NEW MOTIVATIONS FOR DEVELOPMENT

Uncooperative Nature

Climatic conditions, which were already marginal for oil-palm production, have been worsened by several years of drought which resulted in yields far below expectations. The deficit in output represented between 50 and 75 percent per hectare. This has prompted SONADER to look into possibilities for irrigation. Experiments were carried out with the assistance of UNDP/FAO and preliminary results proved to be encouraging, indicating economic feasibility where water is available at low cost. New initiatives in oil-palm development by SONADER are based on those findings, but as far as existing plantations are concerned, it is likely that adverse climatic conditions, if they continue, will jeopardize their life and ruin years of sustained and costly efforts.

Farmers' Doubts about Production Economics

While production activities on the cooperative's plantations are quite easy to handle, it is a lot more difficult to manage the annual crop area where plots are farmed individually. One reason for this is a psychological constraint which farmers developed vis à vis extension workers' proposals for crop rotation. The proposals suggesting the abolition of mixed cropping (*cultures associées*) were resisted until a formula was found which provided for special plots where mixed cropping could be continued alongside plots where improved crop rotation techniques were being used. More serious was another difficulty raised by farmers' critical assessment of the value of some crops recommended to them. This refers in particular to cotton, which was recommended as a quick-return cash crop and above all as a much needed foreign exchange earner. Cooperative members found that the crop was too demanding in labor and inputs (expensive fertilizers and pesticides, great quantities of scarce water) for a rather insignificant profit. Here is an illustration of clashes between national (macro-economic) targets and individual producer (micro-economic) objectives. Needless to say, the producers have the last word. Good as this may be, such a situation can disturb the overall economics of the cooperative unit, for example, by overemphasizing crops for internal consumption to the detriment of commodities which could be sold in national and international markets and compensate for the eventual loss due to low productivity of the oil-palm plantations.

Anxiety about Reaching Economic Targets

There has been a great deal of uncertainty concerning economic goals largely as a result of the poor climatic conditions which led to deficit production of the palm trees. The concern in this respect is manifold. For obvious reasons, cooperative members want to see the remuneration of their work increase as time goes on, thereby improving their living conditions, which is the primary goal of the whole venture. Unfortunately, cooperative revenues have not matched the forecast figures. Therefore, members' income has remained stagnant within the context of a rising cost of living. Hope for improvement when agro-climatological factors become more favorable is marred by the need to repay part of the investments made by public authorities. As a matter of fact, whether initial investments in the cooperatives have been made from external financing in the form of either grants or loans, it is specified that the cost of establishment and maintenance of oil-palm plantations up to the first production year should be reimbursed and paid into a special fund called Fund for Expansion (Fond d'Entretien et d'Extension de la Palmeraie) which is meant for extending the benefit of the rural development cooperative scheme to other areas of southern Dahomey suitable for oil-palm plantations. If the rules were to be strictly adhered to, the chances for the cooperatives to improve the level of income of their members would be remote. Public authorities have fortunately realized this and accepted softer terms for repayment of investments so as to ensure that a decent level of remuneration is always guaranteed to the cooperative workers. A vital element is the price of palm-oil which has shown, in recent years, some favorable trends but on which national authorities have virtually no influence. Therefore, a real risk exists that frustrations will rise if cooperatives do not develop new resources to compensate for uncertainties in the oil-palm production activities. A reasonable answer might be to boost production of annual crops through a new imaginative program.

Convergent Attacks from Diverging Socio-Political Interests

The rural development cooperatives, being an instrument of agrarian revolution, have inevitably met with all sorts of criticism and opposition. The compulsory nature of participation by landowners was for them a source of deep psychological shock. Even when it was clearly established through delivery of property title deeds that they still owned their land which was only "rented" to the cooperatives on a long-term contract, the fact that they lost their right of exploitation was a bitter pill to swallow. Hence, their strong resistance to the scheme at its inception and persistent attacks at the early stages of the newly formed cooperatives. Political persuasion, thanks to the one-party system, made it possible to overcome initial difficulties and in most cases time has created a feeling of belonging which now helps in maintaining and strengthening the cooperatives.

Opposition came also from politicians who feared a loss of influence with their traditional constituencies. Experience proved their fears were justified as well-organized cooperatives rapidly become better informed and less vulnerable to political maneuvers. Some radical political groups fought against the basic concept of coopérative d'aménagement rural which associates landownership with a landless labor force. They were opposed to the two categories of members and mobilized landless cooperative members to serve their own political cause. It may be of interest to note that there have also been some attacks from certain government circles, mostly due to impatience over the multifarious new issues with which the cooperatives confronted them. Finally, the social changes brought about by the rural development cooperatives were also the subject of criticism. The practice of promoting youth and women was a special target. Promotion of youth disorganized to some extent the traditional scale of hierarchy based on age. Allowing women to compete successfully with men on the labor market destroyed the myth usually associated with division of tasks based on sex. Moreover, female wage-earners were no longer fully available for handling household affairs such as cooking, childcare, etc. The situation thus created was resented since it disturbed severely the traditional social values. Although that might be the price for progress, youth and women were found quite ready to pay the price!

The Frustration of Self-Employment and Temptations of Wage-Earners' Status

Detailed information and explanations are given to cooperative members to convince them that they are their own employers. In order to develop this feeling, they are closely associated with the management of the cooperative. As "employers," cooperative members have shown great interest in management especially with regard to financial matters. For example, they have strongly criticized and made representation to SONADER and the Ministry of Agriculture about having to bear the cost of extension workers attached to the cooperative while other farmers enjoy free assistance from government extension services. However, cooperative members are more inclined to behave as employees. The reason for this kind of mental attitude is simple. The status of employee is advantageous in that one can claim wage increases, increase in the interest paid on shares (particularly land-shares A), and improved working conditions. In one instance, cooperative members requested a change from 6 to 5 working days a week while keeping the same pay. Their demand was prompted by a government decree giving such benefits to the private sector.

Time and continued educational efforts together with positive economic results will be needed to bring about the changes in mental attitudes required to make a full-fledged cooperator.

CONCLUSION

Through her coopérative d'aménagement rural scheme, Dahomey has been able to carry out a quiet but effective land reform program in the land-hungry southern part of the country. Beyond land reform, agrarian revolution took place and rural development was induced. Technical, economic, and socio-political difficulties were encountered, but 12-13 years' experience have shown that adequate resource allocation, good management, and persistent educational efforts will succeed in making the scheme the spearhead of rural development in Dahomey.

12. Group Farming in Sri Lanka

Nimal Sanderatne

INTRODUCTION

Group farming is essentially a new endeavor in Sri Lanka. The short period of experience and the limited scale of group farming provides us with an opportunity for only tentative ideas for further experimentation. Although several forms of group farming have been attempted, all of them have been on a small scale; some have been very recent; others have been abortive or short-lived exercises. This paper focuses attention mainly on the Youth Settlement Schemes which were commenced in 1965 and *Samupakara Janawasas*, or Cooperative Settlements, the first of which was established in 1973. The former were an important initial experiment in group farming but have been mostly turned into family farms or individual plots. The latter, though recent, is the most systematically organized form of group farming. Since systematic group farming in Sri Lanka is still in its infancy, it is premature to attempt to gauge the performance and directions of group farming at this stage. The paper will, therefore, discuss the background for the establishment of group farming, and focus attention on the nature and problems of the Youth Settlement Schemes and Cooperative Settlements. There are a few other experiments in group farming under different auspices, but this paper will not draw on these experiences.

BACKGROUND, EXTERNAL INFLUENCES, AND CONDITIONS BEARING ON GROUP FARMING

Although Sri Lanka is a relatively small country of 16 million acres, with an estimated cultivated area of only about 4 million acres, it has a highly differentiated tenure structure, owing to both climatic differences in several regions and historical developments. However, most of the country's agricultural land is privately owned, either individually or by companies.

The country is conventionally divided into two broad agro-climatic regions known as the Dry and Wet Zones.[1] The Dry Zone is defined as the region receiving less than 75 inches of annual rainfall. The concentration of this rainfall within a short period of less than three months, coupled with a high rate of evaporation, renders this area uncultivable without irrigation, especially for the staple crop, rice. The Dry Zone constitutes about two-thirds of the total land area, or 10.7 million acres, while the remaining one-third, or 5.3 million acres in the southwest quadrant, is the Wet Zone.

A brief description of the evolution of the country's tenure structure is useful for an understanding of the factors bearing on the establishment and management of group farming.

The development of the Dry Zone dates as far back as the third century B.C. when the difficulties of cultivating this region were overcome by constructing a large number of huge reservoirs--great artificial lakes known today as tanks--and skillfully connecting them by a network of canals to form a vast irrigation system. This civilization based on an irrigated rice and food-crop culture continued until the thirteenth century when perhaps invasions from South India and internal dissension led to the breakdown of the administration required for maintaining this irrigation complex. Consequent malarial infestation made the Dry Zone unhealthy and nearly uninhabitable and the population migrated to the Wet Zone, where plentiful rainfall enabled easier cultivation of rice and other crops. A sparsely populated Dry Zone and a relatively densely populated Wet Zone resulted from these changes.

Foreign occupation of the country from the sixteenth century onwards led to the concentration of development in the Wet Zone. The cash crops which interested foreigners either grew in the Wet Zone or were cultivable there. The Portuguese (1505-1658) and Dutch (1658-1796), who occupied only the maritime provinces, were especially interested in the export of cinnamon, pepper, cardamom, and other spices which grew in the Wet Zone. The Dutch began systematic cultivation of cinnamon in the coastal region, but since the cultivation of cardamom, coffee, and pepper were difficult in this region, they encouraged, through trade, the cultivation of these crops in the hinterland, which was not under their rule.

The British gained the maritime provinces from the Dutch in 1796 and conquered the central highlands in 1815. At first they were content to exploit existing agricultural resources, but from about 1823 onwards they opened the central highlands for coffee cultivation. Later these lands were converted to tea cultivation. Rubber cultivation was also commenced about this time.

The presumption of crown ownership of all land not occupied or under settled cultivation made possible the sale of about 1.5 million acres of land at very cheap prices (5 shillings per acre). This enabled the development of large estates specializing in plantation cash crops. At the same time it deprived traditional villages of their grazing rights, rights to *chena* (slash and burn) cultivation, and the opportunity to expand their cultivated area to meet increases in their population.

The development of plantations in the nineteenth century created a dichotomous tenure structure. On the one hand, there are the large holdings primarily devoted to tea and rubber cultivation and secondarily to coconut, cocoa, and cinnamon cultivation, and on the other hand, there are a large number of small holdings devoted to rice and other food crops. Plantation crops are also grown in small units by peasants in the plantation regions, but they represent a significant proportion of cultivated land only in the case of coconut.

In the tea and rubber cultivation region the country's larger holdings co-exist with some of the smallest holdings and Sri Lanka's version of the Latin American *Latifundia-Minifundia* problem exists. Especially in the rubber-growing area, but in some of the tea-growing areas as well, where even prior to the advent of plantation agriculture there was a dense population pressure on the land, the incursion of plantation agriculture has created particularly acute pressures on land. Outside the plantations, paddy (rice) is grown in small holdings in valleys or terraced hills. A high degree of tenancy, high rentals, and very small holdings characterize this cultivation.

A large proportion of tea land is in large holdings; about one-fourth is owned by Sterling Companies and another fourth by Ceylonese companies. Approximately 40 percent of rubber holdings are below 50 acres. About one-fourth of the rubber area is company owned while roughly 10 percent is foreign owned.

Coconut is cultivated in both large and small holdings. Over 70 percent of the coconut area is in small holdings. Foreign ownership in coconut land is negligible. The larger holdings are concentrated in a small area known as the coconut triangle and are generally owned by individuals and families rather than companies. A significant proportion of these holdings has been expropriated under the land reform of 1972, and group farming as cooperative settlements is being attempted on them.

Since the 1920s the government has attempted to resettle in the Dry Zone people from the densely populated regions of the Wet Zone. This resettlement is based on the restoration of the ancient irrigation works mentioned earlier. Resettlement gained momentum after 1947 when malaria was eradicated. Until 1953, each family was given 5 acres of irrigated land for paddy cultivation and 3 acres of dry land for a homestead and cultivation of other crops. The allotment was reduced to 3 acres of irrigated and 2 acres of dry land in 1953, and further reduced to 2 acres of irrigated and 1 acre of dry land in 1958. Land was granted under the Land Development Ordinance "in perpetuity on a restricted tenure." The restrictions are the inability to mortgage the land, lease the land to others, fragment it through inheritance, or sell without the prior permission of the government. However, since 1971, this law has been amended and settlers can now purchase the land.

This discussion of land settlement is of significance for group farming for several reasons. A system of family-farm resettlement has been developed in the country since the 1920s; this is in contrast to the group farming systems that are now being developed. The family-farm method of resettlement continues at the same time as the establishment of group farms. Therefore, the country is sponsoring both private ownership of family farms as well as collectively owned farm units. This creates contradictions in a social structure in which the ownership of a plot of land, however small, is highly valued and confers a significant social status.

One of the first references to group farming was in the Paddy Lands Act of 1958. The objective of this legislation was to control and regulate tenancy conditions on paddy lands. Among its provisions was the establishment of Cultivation Committees for the organization of paddy cultivation at the village level. One of the provisions of the Act stated that these committees may organize paddy cultivation on a collective basis. This provision was seized by the opponents of tenancy reform to develop a campaign that the intent of the Paddy Lands Act was to nationalize paddy lands. The provision itself had no effect in practice and no paddy tracts were converted to collectives.

In 1960 the government commenced a system of electoral farms where young farmers were given farms not exceeding 50 acres to be cultivated on a collective basis. These were an unqualified failure. On eight schemes (six of which have survived) unmarried youths were settled. When the electoral farms were closed in 1963/64, it was also decided to open a number of such schemes to settle unemployed youths, but in fact only four were founded. All these schemes operated in much the same manner as other settlement or colonization schemes.

In 1965 the problem of unemployment among educated youth created a renewed interest in settling such youth in agricultural enterprises. The first group of Youth Settlement Schemes was begun in that year. The reasons for establishing these group farms were neither a political nor an ideological commitment to collective farming, but rather as a means of containing the problem of increased unemployment among educated youth. Although these schemes were another means of land settlement involving single men below 25 years of age, a greater emphasis was placed on growing crops other than paddy, and they were designed to achieve somewhat higher income levels than on family farms. After the initial years of operation as group farms, the youth were given an option to continue operation on a group basis or on individual plots. In all these schemes the settlers have opted to divide the farm into individual units.

The change of government in May 1970 was of significance for agricultural policy. The new government--The United Left Front--which had a socialist policy stance, advocated the nationalization of private sector industries, and greater state control of the economy. The new constitution of Sri Lanka, adopted in May 1972, does not specifically recognize private property rights but does specifically mention the right to own property collectively.

Within a year of the new government's assumption of office, a serious youth revolt or insurgency occurred in April 1971. Subsequent to this, and partly no doubt as a response to it, a land reform was enacted in August 1972. The Land Reform Law of 1972 specifically mentions collective exploitation of land as one of several ways in which expropriated lands may be utilized.

The Land Reform Law of 1972 placed a ceiling on the individual private ownership of agricultural land. The ceiling was 50 acres, except in the case of paddy land where the ceiling was 25 acres. Lands owned by companies were exempted from the reform. Under this law about 560,000 acres of land were expropriated and transferred to different forms of management and to individuals. Although several organizations that have been given land under the reform ultimately intend to exploit it in group farms, these have not yet been systematically established. As yet only lands given to the Samupakara Janawasas or Cooperative Settlements are systematically being organized as group farms. At present about 45,000 acres, or about 8 percent of the expropriated land and 1 percent of the country's agricultural land, have been allotted to about 150 cooperative settlements absorbing about 20,000 settlers. Indications are that lands given to some of the other organizations will in due course be converted into cooperative settlements. It is estimated that ultimately 85,000 to 90,000 acres, or about 2 percent of the country's agricultural land, will be organized as cooperative settlements.

Several changes wrought in the tenure structure by the land reform of 1972 provided a basis for the development of group farming. About 136,000 acres, or 23 percent of the country's tea land, and 115,000 acres, or 10 percent of the country's coconut land (about 20 percent of the larger coconut holdings), hitherto cultivated on an estate or plantation basis were expropriated. There was a fear that the break-up of these large units into private plots would adversely affect productivity. Some of these lands were therefore retained as estates under state-owned organizations, like the Sri Lanka State Plantations Corporation or the Upcountry Estate Development Board. But such a transference of land did not constitute a basic change in the tenure structure of the country. The estate system, as we have seen, had its origins in the country's colonization under the British and was based on a system of low-paid labor imported from India. The political and social developments of the country since independence in 1948, and especially in more recent years, suggested a change in this mode of production. Group farming seemed one means of retaining those lands in relatively large units and maintaining high productivity by a participatory work-force. However, as no creative thinking on how a plantation agriculture could be transformed into an efficient and yet participatory system of agriculture has been evolved, and group farming is a relatively untried system of farm management in the country, the more productive estate lands have been given to the State Plantations Corporation.

Reference has already been made to the problem of unemployment among the educated youth population. Since the problem of finding employment for these youth has been an important objective of recent government policy, the development of a more labor-absorptive mode of production had a high priority. Group farming was envisaged as a means of absorbing some of these educated youth.[2]

To understand the labor-absorptive potential of group farming, some reference must be made to the labor-utilization rates on estates. Tea, rubber, and coconut lands generally absorb 1.1, 0.5, and 0.1 units of labor per acre, respectively. In the case of tea lands, this coefficient applies to fairly well developed lands, while less developed lands absorb only about 0.85 units of labor per acre and very well developed lands could absorb 1.5 units per acre. Since a significant proportion of tea lands that have been expropriated are deemed to be either not very well maintained or agro-climatically marginal lands, two strategies are suggested. The first strategy is to cultivate the less well-maintained lands more systematically with regular weeding, pruning, and fertilization, and the filling of spaces among the tea bushes with high-yielding vegetatively propagated planting material. The second strategy is to diversify crops on marginal lands by uprooting tea and cultivating other crops. Both these strategies could increase employment and production, if systematically done. Such systematic cultivation could not be expected to be undertaken in small individual plots. Group farming is therefore envisaged as an alternate to the estate system.

Coconut lands have traditionally been the least labor absorptive. The strategy for increasing production and employment on these lands is to intercrop coconut with other crops.[3] Such intercropping has not generally been practiced on coconut estates and group farming is looked upon as a possible means of developing an intensive cropping pattern on these coconut lands.

Group farming is being organized on a limited scale in Sri Lanka. This raises several problems and contradictions--an island of collectivist agriculture in a sea of a private-property-oriented agriculture. The possession of land is a highly esteemed and entrenched value. Consequently, participation in an agricultural enterprise without ownership rights to a physically identified piece of land has generally been considered an inferior participatory involvement. This is particularly so as government policies of land settlement have hitherto transferred land (together with several other grants and subsidies) to settlers. Group farming therefore has the disadvantage of its participants not having a sense of complete involvement arising from the ownership of a salable or otherwise transferable plot of land. This is the underlying reason why in most Youth Settlement Schemes the youth opted for the break-up of the land into individual units. In such a context those farming collectively tend to be less certain of the continuity of the group farming system; they also harbor a desire to own a part of the farm individually. In other words, the group farmer plays a schizophrenic role: a collectivist attempting to make his group farm a success in a context of hostility to group farming and at the same time harboring an interest in the private ownership of property himself.

The hostility to group farming takes several forms. There is propaganda that these farms are not successful; that they are composed of insurgent and revolutionary groups; and insinuations of corruption and sexual misbehavior. The stealing of produce and instigation of thefts are other forms of opposition.

To summarize the external influences bearing on group farming, we may say that until very recently historical developments have been in the direction of establishing privately owned agricultural operations, whether as company estates, large or small individually owned lands, or government-alienated family farms. A large number of educated youth without employment opportunities gave impetus to the establishment of the first form of group farming. Yet the lack of an ideological commitment to group farming on the part of the government and the private-property-oriented value structure of the youth militated against the continuation of these farms as group farms. Since 1971 an ideological commitment to group farming and the development of collective forms of property ownership is evident among certain sections of the government. The availability of land expropriated under the Land Reform Law of 1972 has provided the physical base for the development of group farms. The fact that a significant proportion of expropriated land is already operated in large units under plantation crops along with the need to intensify land use and diversify agriculture as a means of providing employment and increasing agricultural production, have provided the economic rationale for group farming as an organizational means of utilizing the abundant labor resource by an intensive use of the scarce land resource. However, since group farming is recent, on a very small national scale, and is surrounded by an economic and social system largely based on private property ownership, the exterior opposition to these enterprises and the conflicts generated within can be considerable. Given this background to the development of group farming, we turn to observe some of the internal dynamics of two forms of group farms.

YOUTH SETTLEMENT SCHEMES

The need to find employment for an increasing number of educated youth was the main motivating factor for commencing the Youth Settlement Schemes in 1966. At the end of 1974, there were about 4,500 youths on 51 schemes. The total land area under the schemes was about 15,000 acres, of which about 10,000 were cultivated. The intent of the project was not to develop a group farming system, although the initial activities of the project were of a cooperative nature. Although Youth Settlement Schemes are not a group farming system, they illustrate some of the difficulties and problems that arise in attempting group farming in Sri Lanka.

Internal Operating Mechanisms and Procedures and Changes

The youths selected for the schemes were restricted to men between the ages of 18 and 25 having education at least through grade 6. Preference was given to those having education through grade 10 or the Farm School Certificate. Selection has generally been on the basis of an interview at the *Kachcheri* (the district administrative office). A few experiments for the selection of settlers have been tried. These include a ten-day work-camp performance and the selection by school principals from among their former pupils.[4] Politicians had increasing influence in the selection of settlers on some schemes.[5] Vacancies created by dropouts have been filled by the resident youth inviting their friends and relations.[6]

A division of the Land Commissioner's Department is responsible for planning, establishment, supervisory control, and obtaining financial assistance for the schemes. Prior to August 1969, these functions were performed by the Ministry of Lands "as it was felt that this subject needed special attention and also with the idea of preserving the distinctive character of these schemes. Another objective was to obtain better co-operation from the Departments involved resulting in the speedy execution of development programmes."[7]

At the district level government agents (the chief district administrative officer) are in charge of implementing the plans for each scheme and exercising immediate supervision. Each scheme has a Project Officer in charge, who is expected to submit a monthly progress report.

Each scheme is organized around a cooperative society of which all settlers are members. The functions of the cooperative are comprehensive and include the organization of production, transport, marketing, sale of inputs, and the provision of credit. The land is initially leased to the cooperative and subsequently, when the land is privately allotted, transferred to members.

The project received foreign assistance from a few organizations. The World Food Program gave food aid to supplement the nutritional requirements of the settlers during their initial years. The Organization of Netherlands Volunteers (Novib) and The Asia Foundation have given assistance in the form of machinery. A few foreign advisors have also been attached to the project at times.

Size per scheme has ranged from as little as 8 acres to as much as 2,100 acres. However, most schemes are within the 100- to 500-acre size group. The individual allotment ranges from 1 1/2 acres to 7 acres, depending on the type of crop and soil conditions. Most schemes have fewer than 100 settlers. Over two-thirds of the schemes have fewer than 50 members each.

Since all schemes have been on hitherto uncultivated lands, the first step in their establishment was a feasibility study which considered access to the land, water resources, a soil analysis, and an assessment of cultivable acreage. The reaction of the traditional villages and marketing facilities were also considered.

The construction of irrigation facilities and the building of a community dormitory were the first steps in the development of a scheme. The community dormitory provides living accommodation for 25 to 35 youth in the first two years or so of the scheme. Once settlers shift to individual homes, this is converted into a community services center and recreation hall.

The first two or three years (sometimes the period has extended beyond this) are devoted to the construction of irrigation facilities, the clearing of jungle, land development, and slash-and-burn agriculture. Since little or no income is derived from farming at this time, the settlers are paid a very minimal subsistence allowance which is supplemented by some food rations. They are also paid certain allowances for particular aspects of work such as jungle clearing, stumping, etc. They are encouraged to save a part of these allowances for the purchase of shares in the cooperative as well as to build a savings account. It is during these initial years that the schemes function as group or cooperative farms. The settlers both live and work as a group.

After the initial period of development the youths were given the option of working it as a group farm or as individual plots. In nearly all schemes they opted for individual holdings. During this transition the dormitory type of living is given up, at first for group houses of 4 to 10 persons who volunteer to live together. Sometimes these "living groups" also function as "work groups," cultivating a block of individually owned plots. Finally, group living is given up for individual houses. Both types of houses are constructed by the settlers themselves.

The experience of inadequate work output during the initial years as well as the general experience of most youth opting for individual plots resulted in a government decision that all Youth Settlement Schemes will operate as individual holdings, except for the initial years when land clearing and construction of houses will be done collectively. There was also a change in the method of giving assistance to the settlers. Instead of the subsistence allowance, settlers were given Rs 1,600 per acre up to a maximum of Rs 4,000 per allotment for capital expenditures for such items as land preparation, fencing, farm buildings, equipment, and a temporary house. In addition, the government met the costs of infrastructure such as road building and irrigation. This system has been applied in ten of the eleven schemes commenced since 1972. (In the eleventh the maximum amount of assistance for capital expenditure was also restricted to Rs 4,000 per allotment.)

Measures of Socio-Economic Performance

It is useful to analyze the experience of Youth Settlement Schemes in terms of the two phases--the communal phase and the subsequent phase of individual plots. The first few months of group farming are both a period of difficult work conditions and one of orientation to a new way of life. Susil Siriwardane, who was the officer in charge of the Youth Settlement Scheme Project from its inception in 1965 till 1970, summarizes both the advantages and problems of the experience:

> Especially during the first six months this enforced grouping produced satisfactory outputs of work and also brought one another into close inter-personal contact. The shocks of new life were more easily absorbed amidst this pooled camaradarie. But once the initial contacts were completed, output began to dwindle because the system of subsistence payment was not geared to *individual* performance. Since the slacker and the hard worker were paid the same daily allowance, it was the hard worker who decided to reduce his output as he saw no valid reason for working without a direct reward.[8]

Others have also noted the unsatisfactory work inputs during the period of group farming. The Government agent in charge of two schemes in south Sri Lanka (Galle district) commented that the "speed of the Co-operative [tends] to be that of its slowest member."[9] Ellman and Ratnaweera, commenting on the experience of five schemes, note that "when the collective system was operating without a clear commitment to divide up in the future, progress was in every case highly unsatisfactory"; once the land was divided up the work output was strikingly satisfactory.[10]

Siriwardane, writing in 1970, was of the view that the initial cooperative experience was of subsequent value even though the farming was on individual plots. He wrote:

> strands of the original pattern persist in two ways. First, even after individual blocking out, there is a communal work-day once a week, where all have to contribute their labour for communal work like the maintenance of roads and fences. Second, the youth voluntarily utilize the methods of large or small group work according to the particular task to be done. For example, harvesting of paddy and chillies or excavating a well would demand a group rather than a purely individual effort. Hence the past experience of communal living and working persists as a positive experience.[11]

Ellman and Ratnaweera found the five schemes they investigated to be highly individualistic societies. In their assessment:

> After the land was divided into individual holdings, many of the members forged ahead and quickly became self supporting. But a highly individualist society began to develop, with heavy dependence on hired labour, little use of co-operative work or services and appearance of (as yet limited) social and economic inequalities.[12]

Even the service cooperative has not functioned well on most schemes. Settlers have not repaid their loans, they market their produce outside the cooperative, and the managerial capability and financial standing of the societies have not enabled the government to hand over the maintenance of tractors to them. An official report concludes: "The functioning of Co-operative Societies in Youth Schemes has not been very different from the working of similar Societies in the country as a whole."[13]

The factors bearing on the disintegration of group farming on the Youth Settlement Schemes are instructive in assessing the influences and conditions bearing on similar experiments currently undertaken. Fundamentally, the problem arises in attempting group farming without an ideological commitment and in a socio-economic context of entrenched private property values, especially in land. These factors influence group farming in various ways.

First, the youth themselves displayed it in their work outputs, which were inadequate when there was uncertainty as to whether they would obtain the land. The lack of motivation to contribute to a collective effort was accentuated by the system of payment. Equal payment for unequal work tended to affect the work output of the more efficient settlers. This experience underscores the need to develop a system of payment based on labor inputs.

Second, the youth given a choice between a group farm or individual plot opted for the latter. This arises basically from the esteemed value of having a plot of land for oneself. However, the youths' natural disposition was strengthened by social factors. Parents and wives (once the settlers were married) were an important influence. The Project Managers, far from instilling the need to farm collectively, appear to have themselves been convinced that group farming cannot work and that group farming encourages laziness.[14] This indicates the need for a new cadre of officers committed to group farming to manage such enterprises. Finding such officers in a society with very little in the way of collectivist enterprises would, of course, pose a serious problem.

Third, from the commencement of the schemes there was no commitment to continue them as group farms--only an option by members to do so. It is conjectural whether the group farming in the initial years would have been different in performance and whether a compulsorily continued group farming scheme would have performed better. However, the knowledge that this was a possibility might have been a reinforcement. Subsequent government decision to make all schemes individual plots removed both the uncertainty and the possibility of the schemes becoming group farms.

Productivity on most farms has been fairly satisfactory. An official evaluation of the 41 farms operating at the end of 1971 considered 12 to be problematic.[15] The report found that in 13 of the 41 schemes the average income had exceeded the target income of Rs 150 per month and some youth in these schemes obtained incomes of above Rs 500 per month. In about 3 schemes the average incomes exceeded Rs 1,000 per month. After 1971 the performance was even better, owing to a substantial increase in price of commodities cultivated.[16] Ellman and Ratnaweera have observed that "Yields of the major crops grown on most of our five Youth Schemes have been exceptionally high."[17]

Since cultivation is done on an individual basis there are differences in methods of cultivation and the incomes derived. But generally a satisfactory level of production efficiency is considered to have been achieved.

The impact of the Youth Settlement Schemes on the problem of unemployment is minimal. Absorbing, as it has, fewer than 5,000 youth in nine years in a country where about 793,000 are estimated to be unemployed, it has had very little impact on the problem.[18] However, it has shown the possibility of agricultural employment of educated youth in a new settlement pattern.

SAMUPAKARA JANAWASAS

The first Samupakara Janawasa, or Cooperative Settlement, was established in January 1973. While 19 Cooperative Settlements were established in 1973, 124 were established in 1974. About 150 Cooperative Settlements were functioning by April 1975 on about 45,000 acres with about 20,000 settlers. Unlike the traditional settlements or the Youth Settlement Schemes, these were established on already cultivated lands expropriated under the Land Reform Law of 1972. Most lands were coconut estates; some were tea, rubber, cocoa, and mixed plantations. Prior to take-over these lands were operated as estates or were parts of estates managed by a superintendent or a proprietor planter. The important point here is that these estates were operated as relatively large units with hired labor.

Several background factors are important to understand the present situation on, and management of, these Cooperative Settlements. There were lapses of time between the announcement of the land reform, the vesting of the land in the Land Reform Commission, and the ultimate handing over to the Cooperative Settlements. During this period such cultivation practices as fertilization, weeding, and pruning (in the case of tea) were neglected. There has also been slaughter tapping of rubber trees, plucking of immature coconuts, and, in some instances, felling of trees and wanton destruction of plants and property. Not all these have been the work of the previous owners; persons from the villages surrounding these lands have often been responsible for these acts. This is partly owing to the frustration of their expectation that the land reform would break up the plantations and distribute small plots to them. They therefore consider the produce on these lands to be rightfully theirs. Other landowners, politically opposed persons, and traders opposed to the land reform also appear to be responsible for the destruction and stealing of crops with an intent to discredit the reform. Stealing of crops still continues to some extent.

The foregoing factors are important for several reasons. The conditions of several properties were such that they required additional investments to bring them to a state where normal production could be achieved. But drought conditions, the shortage of fertilizer, and the high prices of fertilizer have also been responsible for the inability to bring these lands to their full production potential. Current production on many lands is below potential. Since under the cooperative settlement program these lands absorbed a larger number of settlers than under the previous system, the current low production income divided by the larger number of persons implies a lower per capita income. At the end of 1974, on 70 of 99 settlements for which accounts were finalized, current expenditure exceeded current income. On 29 others in which income exceeded expenditure, the total surplus was only Rs 406,706. This indicates a very modest "profit" on each individual settlement, especially as wage costs were low. The settlements operate on loans from the Land Reform Commission, banks, and government schemes. This is particularly so since the strategy to intensify land use by inter-cultivating with new crops, the development of poultry and animal husbandry, and the development of agro-based industries requires additional investments which have a considerable gestation period. All these factors have an important bearing on the organization, management, and performance of these settlements; for instance, it is difficult to pay an adequate living allowance out of current income during the first years of each scheme's operation and this dampens the enthusiasm for collective enterprise.

Internal Operating Mechanisms and Procedures

The Cooperative Settlements Division of the Land Reform Commission has the overall responsibility for policy formulation and supervision of the Cooperative Settlements. Apart from its own staff headed by a Directress, and including several administrators, it has recourse to technical agricultural personnel from other governmental departments and organizations. The functions of the Cooperative Settlements Division include the preparation of plans for the development of each settlement and the supervision of implementation of these plans. It is also responsible for conducting training classes on the settlements and obtaining resources from other agencies, when its own resources are inadequate. This Division manages the funds obtained from both the government and the banks for the development of the settlements. A very significant amount of financial control of the settlements is maintained by this Division. Although self-management of the settlements is the ultimate intention, much of the policy formulation and implementation is still the responsibility of this Division.

The District Land Reform Authority performs an advisory and supervisory role at the district level. This body consists of a Director appointed by the Land Reform Commission and the government officials responsible for agricultural policy at the district level. The District Land Reform office controls the accounts of the settlements, obtains supplies and markets some of the produce, supervises the field staff and the implementation of the settlement's plan. The District Land Reform Authority acts as an agent of the Land Reform Commission's Cooperative Settlements Division and reports the progress of the settlements periodically to it.

The management of the settlement is in the hands of a main committee consisting of six members: president, secretary, treasurer, and three other members. Each of these six members is in turn chairman of a functional committee. These six committees deal with finance, development, marketing and production, discipline, cultural matters, and political matters. The names of these committees are indicative of their functions. The political committee's function is to organize the dissemination of ideas regarding the collectives and socialism as well as to politicize the neighboring villagers. This committee has not been appointed in some cases. In practice, the Land Reform Commission's Settlements Division exercises a significant amount of control and offers guidance. (There is a greater degree of self-management on the coconut schemes than in the rubber and tea projects.)

The day-to-day work in the settlement is organized on the basis of work teams; each work team consists of a few persons (about 6 to 8) and is responsible for completing a particular task. The group leader is responsible for the discipline of the group and this leadership is rotated.

The training given to settlers includes such areas as the importance of land reform to national development; the social, economic, and political significance of cooperative farming; methods of self-management and work-team organization. About 15,000 of the 20,000 settlers have been given a general training in these aspects while another 2,000 settlers have been trained at a more intensive level. Field officers have given settlers training in agriculture and, where appropriate, in animal husbandry at a very rudimentary level. There is also a limited transfer of settlers from their own farm to others for a period to acquire new knowledge and skills.[19]

The lack of adequate agricultural training is a significant deficiency. Since these settlements are involved in the cultivation of a number of crops, as well as poultry and animal husbandry enterprises, such training is particularly important. As it is, each settlement is highly dependent on advice from the national agricultural extension service which is itself inadequate. Dependence on outside assistance is unrealistic for day-to-day needs. The best strategy would be for each settlement to send one or two persons for an intensive training in the growing of particular crops and animal husbandry. Such development of skills within the settlement will be an important determinant of the capacity to successfully manage the farm.

The hostility from several external sources induces those in charge of group farms to place considerable emphasis on ideological commitment. The political or ideological commitment to group farming and socialism is considered a must for membership and a considerable amount of time and organizational energies are spent on instilling this ideological commitment.

The emphasis on ideological commitment has several advantages. In a context of hostility and antipathy to reform from the outside and an underlying skepticism about the viability of cooperative farming, this ideological commitment or "indoctrination" is a useful cement binding the membership and a means of enlisting their enthusiasm. Such commitment also helps mitigate the disadvantages arising from the low allowances paid, the failure of certain agricultural enterprises, and the gestation period through which most of the schemes are passing. But this emphasis tends to lead to a certain neglect in developing the technical management and accounting systems of the settlement. It also places a certain limitation on the persons who can be drawn into assisting the settlements as well as reinforcing the opposition to group farming from some political quarters.

Until recently, settlers were paid a daily allowance of Rs 3.50, which was raised in some cases to Rs 4.50. This was lower than the going wage rate for labor, which was Rs 5 or Rs 5.50. In addition to this allowance members were given produce on the settlement at cheaper rates or in the case of a few commodities, like bread fruit or mangoes, the settlers were allowed to take a share of the produce free. Despite the low remuneration the settlements have, however, retained their membership and the drop-out rate is low owing to the unavailability of alternative jobs, the hope and promise of better remuneration, and the ideological stance or commitment.

The payment of this inadequate allowance and its unrelatedness to work inputs created both dissatisfaction and conflict among the membership of several schemes. The former workers who were now members were particularly dissatisfied as they were earning a lower allowance than the wage they would have obtained had they continued on the estates and had the estates not been taken over.

Some of these deficiencies have now been mitigated to some extent. A new method of payment has very recently been introduced.[20] Each worker is paid a basic allowance of Rs 3.50 for which a minimum work norm has been prescribed. For each additional unit of work, an extra payment is made up to a maximum number of work units. The maximum that can be given in each scheme is fixed. These amounts are expected to serve as advances. Once the return on the scheme is calculated, the amounts paid to each settler would be related to this return and his work input.

The work norms have been developed by the Land Reform Commission's Cooperative Settlements Division, in consultation with agronomists and settlers. Yet these are general work norms and will have to be adapted to each scheme. It is expected that each settlement will discuss the norms among its membership and decide upon its own coefficients. The general scheme gives the minimum number of units of work for obtaining the basic allowance, the additional payment per unit of additional work, and the maximum payment. These have been worked out for all the major crops and major work operations of each crop. Sometimes the work norm has to be applied to a work team as a whole, and the responsibility of ensuring a satisfactory compliance within the team rests with the team leader.

At present most settlers live outside the farm in nearby villages. Travel to the settlement is sometimes costly, tedious, and time-consuming. It is the intent of the program to settle the members on the farms on a housing lot of 20 perches (or one-eighth of an acre), but financial constraints are binding at present.

Where limited housing facilities have been provided or where such facilities are to be provided soon, the question has arisen whether each household should be permitted a small private plot. This is a difficult question to resolve on some settlements as previous workers on some estates were given a small plot surrounding their houses. A small amount of garden crops were grown and the workers often raised some poultry. Loss of this privilege was difficult, owing not only to their having enjoyed this opportunity in the past, but also because of the payment of a low allowance. The supplementary produce would provide an important income supplement. On the other hand, the continuation of a private plot leads to the likelihood of cooperative inputs being utilized on the private plot. The members on some settlements have decided that settlers will be allowed only the barest minimum of land surrounding the house and that no crops shall be grown. Where settlers already had poultry, they were required to get rid of it or sell it to the cooperative. In one settlement the prohibition was to be made effective from the date of increasing the allowance.

One of the major deficiencies of these settlements is the lack of a proper system of accounting. Some of the produce is sold indirectly to agencies and the returns credited to the accounts of the settlement. Members have only a general idea of the price for their produce where sales are made in this manner. Similarly, inputs are ordered through and paid for by the Land Reform Commission. These factors, as well as the limited amount of discretionary authority in financial matters and the system of accounting, make it difficult for the membership to determine whether particular enterprises are economically viable, the relative profitability of the different enterprises, and the nature of returns from their inputs. A better system of cost accounting for the individual enterprises which enables easier identification of costs and returns is very much needed. A simple yet meaningful system of accounts should be developed for all the settlements and one or two members in each settlement should be trained in these accounting methods. This should not be an insurmountable problem as in many schemes there are youth who have had some basic or elementary training in accounting.

Socio-Economic Evaluation

A socio-economic evaluation of the Cooperative Settlements is not possible owing to the brief experience of these farms and the inadequacy of data.[21] Yet a few comments indicating the directions of these farms can be made here.

The 45,000 acres of land presently under Cooperative Settlements include a membership of about 20,000. There are no data on the number of persons previously absorbed on these lands which would enable us to assess the increment in employment. However, the absorption of 2.25 persons per acre of land is relatively high given that a substantial proportion of these lands were under coconut. Many of these settlements have commenced inter-cropping and the rearing of dairy cattle and poultry. There has not been a high drop-out rate even though there is discontent about the allowance received. For reasons discussed earlier, only a small proportion of farms have shown a profit on the current account and these too have been modest. This, however, is not a reason for discouragement considering the state of the lands at take-over, the investments required for new activities, the gestation period of several new crops that have been grown, and the larger work force employed.

CONCLUDING REMARKS

While the experience of the Youth Settlement Schemes offers little encouragement for the prospects of group farming in Sri Lanka, several differences in the present context should be noted. Primary among these factors is the government's commitment to continuing these farms on a group-farming basis and not allowing their break-up into individual plots. This is enhanced by a strong ideological commitment to collective farming among those in the Land Settlements Division of the Land Reform Commission.

The highly esteemed value of private ownership of land has been identified as an important factor bearing on the difficulties of continued group farming. This perspective derived from past experience may be less valid today. The limitation of private ownership of agricultural land to 50 acres, an expectation that this ceiling may be reduced further, and an increasing encouragement of collective forms of property ownership may influence this. There are indications that the new generation of settlers are ideologically more committed to collective forms of property ownership than the earlier settlers; that they are more reconciled to the lack of employment opportunities elsewhere; and that a reasonable regular income may be an adequate incentive to continuation and involvement in group farms.

The efficient management of the group farms enabling a reasonably high income for each settler will therefore be an important determinant of success in group farming. In turn, much of the success of these schemes will depend on the training of the youths in agriculture, the implementation of incentive-oriented work-sharing arrangements, and an accounting system which enables decision-making on more economically rational bases.

The rationale for group farming in Sri Lanka lies not merely in substituting this for other forms of tenure with production and productivity at least as high as achieved under the earlier forms; nor is its justification confined to doing better what was done before; rather, it must be justified in terms of doing well those things that were not done before: the intensive utilization of land by inter-cropping with new crops, the development of dairy and poultry enterprises, and the establishment of agro-based industries, among others.

Not least among the contributions which group farming could make is the development of a new psychology toward technology which is more consistent with the resource endowments of the country. Other large farming units tend to have a preference for capital-intensive technology and this preference is supported by their financial resources. Unlike these, group farms must develop a sense of self-reliance on their own resources and develop technologies and techniques which they themselves can collectively handle with resources readily available to them. In other words, these farms must be in the forefront of developing a "barefoot doctor" technology or, as it is sometimes known, an intermediate technology. Such developments are not merely useful for the group farms but serve also as an example for surrounding village agriculture.

Group farming is gaining ground as a socially and politically more acceptable form of managing land. Group farming has been conducted under several other auspices, such as cooperative farms on an electoral basis, farms under the National Youth Council, and cooperative farms under the Upcountry Estate Development Board. The total extent of land under all these forms of group farming is still relatively small compared to that under private farming.

However, group farming on a limited scale as in Sri Lanka raises important broader issues. Are these group farms to exist as efficient farms in isolation from the surrounding villages? Should they develop an agriculture different from that of village agriculture? Is their contrasting mode of production an object lesson for village agriculture and an inducement for family farms to collectivize? Are these pilot schemes to prove that plantations could be transformed into economically viable group farms with considerably greater participatory involvement? Are they to demonstrate the capacity for a more intensive agriculture absorbing a higher labor input and increasing the total output per unit of land?

We may perhaps raise these issues as one basic question: Are group farms expected to remain as one form of tenure among others, or are they expected to develop into the major form of land tenure in the country, absorbing both family-farm village agriculture and the wage-labor plantation agriculture? The answer to this question is not clear and, in any case, may be decided by movements in the body politic without reference to the performance of group farms.

NOTES

1. An Intermediate Zone of lesser rainfall than the Wet Zone is also demarcated, but is not particularly significant for this discussion.

2. The rate of unemployment was observed to be higher among those who had obtained a secondary school or university education. Department of Census and Statistics, *Socio Economic Survey 1969-70*.

3. Among the crops found suited for such inter-cropping are ginger, bananas, pineapple, turmeric, groundnuts, yams, fodder, and gingerly. Adequate fertilization could ensure high yields on coconut and the inter-cultivated crops.

4. Susil Siriwardane, "The Youth Settlement Scheme Project--Some Descriptive Notes," Mimeo. Paper presented at a Meeting of the Ceylon Agricultural Economics Society, 1970, p. 4.

5. A. O. Ellman and D. de S. Ratnaweera, *New Settlement Schemes in Sri Lanka*, Agrarian Research and Training Institute, Research Study Series no. 5, August 1974, p. 35.

6. Siriwardane, "The Youth Settlement Scheme Project."

7. E.M.D. Wickremasinghe, et al., "Report on the Evaluation of the Youth Scheme Project," Mimeo. Colombo, 1972, p. 4.

8. Siriwardane, "The Youth Settlement Scheme Project," p. 7.

9. P.A.T. Gunasinghe, "Youth Schemes in Galle District," quoted in Ellman and Ratnaweera, *New Settlement Schemes in Sri Lanka*, p. 15.

10. Ibid., pp. 41-42, 46, et passim.

11. Siriwardane, "The Youth Settlement Scheme Project."

12. Ellman and Ratnaweera, *New Settlement Schemes in Sri Lanka*, p. 60.

13. Wickremasinghe, et al., "Report on the Evaluation of the Youth Scheme Project," p. 19.

14. Ellman and Ratnaweera, *New Settlement Schemes in Sri Lanka*.

15. Wickremasinghe, et al., "Report on the Evaluation of the Youth Scheme Project," p. 10.

16. Land Commissioner's Department, "Review of the Position of the Youth Scheme Project as at End of November 1974," Mimeo., p. 1.

17. Ellman and Ratnaweera, *New Settlement Schemes in Sri Lanka*, p. 48.

18. Economic Research Department, Central Bank of Ceylon, "The Determinants of Labour Force Participation Rates in Sri Lanka 1973" (1974), p. 142.

19. "Annual Progress Report of the Samupakara Janawasas for 1974," Sinhala typescript.

20. This method was proposed in February 1975. Some schemes commenced operation under the new method in April 1975.

21. The study by Ellman and Ratnaweera is the only one available. Since this study is of only 5 settlements after a few months of operation, it is not valid to draw generalizations based on it. The main observations made by them are: there has been a modest increase in employment; a significant increase in activities on the farms; the settlements have been well organized; settlers have accepted the group farming system; and the technology used is labor intensive. The shortage of capital, the continuation of an estate type of organization, the lack of an easily identifiable accounting system are among the unsatisfactory factors. Ellman and Ratnaweera, *New Settlement Schemes in Sri Lanka*, pp. 161-217.

ADDENDUM

In October 1975, an amendment to the Land Reform Law of 1972 expropriated the large estate holdings owned by companies. This resulted in the expropriation of about 418,000 acres, which included 237,600 acres of tea and 95,000 acres of rubber land. Most of these lands continue to be operated as large estates under state-owned corporations. Only 835 acres of these lands have been given in Janawasas. In mid-1976 the number of Janawasas had increased to 175.

Part V: Group Farming in Industrial Market Economies

Editor's Introduction

In the following two chapters, experiences with group farming in two of the more industrialized nations having systems of private property and relatively competitive market economies are discussed. These concern group farming in Japan and in France. These experiences, included under Galeski's Type Four, are, of course, quite different from any discussed thus far. In Japan, a major reason for farmers joining together in various forms of group activities is the shortage of labor as well as, at least in some cases, the more efficient use of water and machines in rice farming. With rapid growth in the industrial sector increasing the demand for workers, part-time farming with attendant labor shortages, especially in some seasons, has become a phenomenon of overwhelming importance. Likewise, labor shortages have led to more extensive land-use patterns (e.g., eliminating winter cropping). The Japanese forms of group farming (from joint use of machinery and transplanting, coordination of farm operations with formal agreements, through cooperative management in which production, marketing, and profit-sharing are all included) are essentially a spontaneous development undertaken at the initiative of the farmers themselves.

The French GAEC (Groupements Agricoles d'Exploitation en Comun), by contrast, have been authorized by specific legislation adopted by the government in the early 1960s. Here the motive was essentially to take advantage of economies of scale and the more efficient use of land and capital. The GAEC are considered corporate entities under law, and they are intended for the partial (and perhaps temporary) merger of small- and medium-sized peasant farms.

The situation in both Japan and France is, of course, quite different from that in the less developed countries. The countries are industrialized; the farm population is rapidly declining; population growth rates are low; farmers are generally more advanced in the use of modern technology; etc. Nevertheless, these experiences are of interest and could provide significant lessons in the future. But only time and more research can provide an objective evaluation of their prospects and the possible lessons to be learned that might be of use elsewhere.

In his discussion of Kanazawa's paper, Professor King[1] points out that the case of Japan suggests that, even where landownership and land tenure provisions are established to provide for small family owner-operated farms, group farming arrangements may emerge if these facilitate increased productivity or if the changing structure of opportunities warrants such adjustments. The Japanese land reform and subsequent legislation included strong provisions to discourage the re-emergence of tenancy and absentee landlordism. Thus recombination of farming units through formal leasing arrangements was not usually a feasible alternative. The earlier assumption that recombinations would occur through land sales also proved incorrect. The latter was the result of a combination of factors--preferential assessment of agricultural land for tax purposes, and very large prospective profits if permission could be obtained to convert land to residential, industrial, or other urban uses. The capitalized value of land within commuting distance of urban areas is so high as to make its purchase for agricultural use prohibitive. All these factors led to the expansion of part-time farming and the associated group farming activities rather than the expected exodus of large numbers of households out of farming and into urban areas.

NOTES

1. Adjunct Assistant Professor of Agricultural Economics and the Land Tenu Center, University of Wisconsin-Madison, now Agrarian Reform Research Advisor, Agrarian Reform Institute, University of the Philippines-Los Baños.

13. Problems and Direction of Agricultural Group Activities in Japan

Natsuki Kanazawa

EARLY ORIGINS OF AGRICULTURAL GROUP ACTIVITIES

A simple definition of group activities in Japanese farming is quite impossible because of the great organizational diversity of these activities. There are many reasons for the appearance of such a variety of group activities in recent times; for example, the rapid outflow of labor from rural to urban areas, a remarkable increase in the number of part-time farmers, the introduction of large machinery and facilities accompanying land improvement, and the provision of governmental subsidies. In particular, Government policy for "agricultural structure improvement" has played a very important role in promoting group activities by encouraging large-scale, highly productive farming.

To analyze the character of group activities in Japanese agriculture, however, we have to examine the long history of their development. Group activities of an earlier age were of course quite different from those of today, but it seems important to start with accurate knowledge of the traditional labor organization in village communities.

Until recent times, group activities had a strong traditional character, particularly in rice farming. The reason is that for many centuries rice was the main crop within a structure of subsistence farming. Even after the development of commercialized farming, rice continued as the most important commodity. But its production helped shape, and was shaped by, an element of the Japanese village community called *buraku*; the latter helps to explain the character of traditional group activities.

What are these buraku and how were they developed? A detailed explanation would be difficult and complicated. Essentially these communities were originally a kind of group production unit, particularly for water utilization, but also a group living unit.

Roughly speaking, a traditional Japanese buraku consisted of 20-30 farm households (in some cases 30-50 households), a very suitable size for group water utilization and management. Generally, the size of each buraku depended on the stage of technical development of the irrigation system, and they became larger, or several small burakus combined, in response to the development of irrigation technology.

Historically, Japanese irrigation started from the utilization of water from swamps. Later irrigation from small rivers and reservoirs was developed. The final technical development was the use of water from large rivers in plain areas. We can think of a buraku as the terminal unit of water utilization and management for rice production in a single area or district. To attain the aim of fair water distribution, the buraku itself had to be a self-governing body capable of developing and enforcing rules for water utilization.

In very early days, irrigation facilities were constructed by Emperors and later (in the ninth to fifteenth centuries) by landlords who had large private lands called *shoen*. The shoens were somewhat like manors in England, *seigneuries* in France, and *Grundherrschaften* in Germany. Shoen landlords, who were usually men of the peerage or temples or shrines, held many farmers under their control and governance in their territories. They made considerable effort to reclaim paddy lands by means of irrigation, and buraku were established following the expansion of rice lands through new irrigation developments. So the buraku was the smallest unit of organized farmers and also the terminal unit of water management.

Within the buraku were leaders, usually the larger farmers, who received directions from the landlord. Nevertheless, the buraku was a self-governing body within a limited meaning. For example, a buraku had definite rights as to the amount and time of water intake for irrigation and these rights were supported by customary law. Moreover, a buraku frequently held forest and grassland in common property. These properties were managed and maintained by all members of the buraku, and their utilization by members of another buraku was not permitted. Thus the buraku was the smallest community of farmers who shared the same base for production and living.

The significance of transplanting rice cannot be overlooked in discussing the character of the Japanese buraku. Ancient records show that by the third century rice farming had appeared and the basic transplanting technology was already understood. But it was not until the tenth century that the practice of transplanting prevailed over almost all of Japan; until that time direct seeding was general. The Ise Shrine (Shrine of the Emperor's ancentors) is the oldest shrine in Japan. Usually these old Japanese shrines have some ceremonies related to agricultural production. But the Ise Shrine has no ceremony related to transplanting practices. This shows that transplanting had not been introduced at the time of its origin. It took about 700 to 800 years for transplanting to become popular throughout Japanese rice farming, and its spread depended largely on the expansion of irrigation. Transplanting is possible only where irrigation conditions are reasonably good; thus the history of transplanting runs parallel to the history of irrigation.

The productivity of rice per hectare increased rapidly with the introduction of transplanting. However, this practice required considerably more labor than the direct seeding method. The fact that transplanting required so much labor input strengthened the social powers of the man who could procure and control such labor forces. In other words, the most important qualification for being the head of a farmers' society (or buraku) was the ability or power to procure and organize labor.

Gradually, transplanting itself became a kind of ceremony which displayed the social powers of the leaders of farmers' societies. The old traditional music called *dengaku* was used to encourage transplanting laborers. So we can infer that transplanting was a ceremony performed by means of the labor of all of the members of the buraku. From early days, all members of the buraku offered their labor for transplanting on the land of their head (dominator) without remuneration. This kind of labor for their dominator by all members was called "common work" and was considered the same as the repair and maintenance of roads and water ditches. This could be considered the early forerunner of agricultural group activities.

Following the weakening of the powers of dominators, as a consequence of economic and social development, this system gradually changed and labor exchange without payment became popular among individual farmers. This labor exchange, called *ui*, is in principle an exchange of equal amounts of labor, and at present we see ui everywhere in Japan as the most simple form of group activity. The meaning of ui in the Japanese language is "combination" or "binding." But it should be emphasized that this most popular and simple form of group activity, ui, came originally from the dominator's transplanting ceremony performed by the contribution of labor from all members.

A large share of the many kinds of group activities in Japan is built on the base of buraku. Understanding the relationships between buraku formation and water utilization and management, and between rice transplanting and the development of irrigation, seems basic to understanding agricultural group activities in Japan today. It is important to emphasize this fact. Using Japanese experience as a model for group activites in Southeast Asia or elsewhere requires recognition of and adaptation to the differences in the nature of village communities, irrigation development stages, rice production techniqeus, and landownership institutions.

TYPES OF GROUP ACTIVITIES IN JAPAN

The following classification of group activities is that of the Japanese Ministry of Agriculture:

(I) Farmers' Production Organizations
1. Joint operation of farming practices
2. Joint use of farm machinery and facilities
3. Group farming organization
4. Contract farming

(II) Cooperative Management
1. Cooperative management of a single enterprise
2. Cooperative management of all enterprises

Farmers' production organizations are those in which more than two farmers work together in part or all of the production processes. These organizations are sometimes juridical persons, sometimes not. Contract farming is included in Japanese statistics within the broad meaning of group activities.

Cooperative management covers the organization of more than two farmers who operate cooperatively their production, sale of products, settlement of accounts, and profit-sharing by cooperative investments in land and capital. We exclude father-and-son farming cooperatives from this category in Japanese statistics.

In Japanese statistics we give some special meaning to each organization of group activities. Joint Operation of Farming Practices [(I)1] refers to the sharing of responsibility for hand operations. Transplanting of rice is still the most important hand-labor operation. About 35 percent of Japanese farmers transplant rice jointly, and 75 percent of this joint operation of transplanting is done by hand labor. Ui--organization of labor exchange between farmers as mentioned above--is the most popular type of joint labor operation.

Joint Use of Machinery [(I)2] refers to organization for joint use of large machinery and facilities. The most popular machines for joint use are large tractors for plowing and combines for harvesting. (Transplanting machines are generally not used jointly, but individually.) This organization does not mean only joint ownership; if two farmers own a large piece of machinery jointly, but use it individually, that case is not included in this form of organization.

Group Farming Organization [(I)3] requires that farmers carry out their farming practices according to certain mutual agreements. Each member of the group operates his own farm, but all members agree to use the same varieties, the same applications of fertilizer in both amount and in time of application, the same treatment of agricultural chemicals, etc. Thus, Group Farming Organization depends on technological agreement among members, but with farms operated individually.

We can consequently divide Group Farming into three sub-types: (a) technological agreement only; (b) technological agreement plus Joint Operation of Farming Practices; and (c) technological agreement plus Joint Use of Machinery. The third is currently the most popular form of so-called group farming.

Contract Farming [(I)4] is a form of organization in which others operate one's farm, partly or wholly, under contract. This is a type of spontaneous and informal tenancy without formal legal recognition. After the enactment of the Land Law (Land Reform), the right of cultivation became very strong and it was difficult for the landowner to reclaim his land once he had let someone else cultivate it. Moreover, the rapid increase in land prices has made it very difficult for the farmer to buy or otherwise obtain land for enlarging his farming operations. These circumstances have promoted the growth of contract farming in the past ten years; they have also made it difficult to get exact figures as to the number of farmers and land area involved.

Of course, this form of organization cannot assure legal cultivation rights. The actual period of the contract, amount and conditions of payment, etc., are decided by negotiation between the two parties. Usually these contracts are verbal, so that a third person's presence might be desirable.

All of the group activities mentioned above are concerned mainly with crop production. Group activities in livestock production can be classified according to the statistics of the Ministry of Agriculture as follows:

1. Joint use of land (land for feed crops, pasture, etc.)
2. Joint use of facilities and machinery (livestock, buildings, equipment, etc.)

It is clear that there are in Japan today many types of group activities. Most of them are combinations of two or three of the types mentioned above. It is very difficult to specify the exact number for each type of group activity. According to the statistics of the Ministry of Agriculture, the total number of all types was 28,064 in 1973 (excluding shared effort in hand operations). The approximate proportion of each group activity is shown in Figure 1.

FIGURE 1

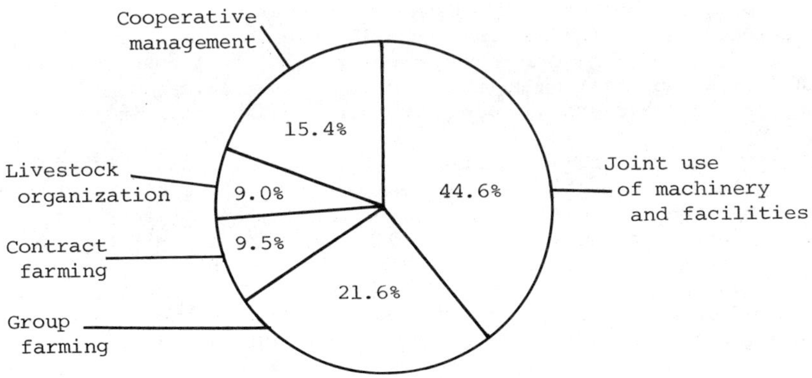

THE FEATURES OF GROUP FARMING

Group farming organizations in Japan have been developed and promoted spontaneously by farmers themselves rather than by Government. As a result, these organizations are very flexible. They can adapt easily to changes in the economic environment, on the one hand, and can readily disappear when no longer appropriate, on the other.

What is Group Farming?

It is difficult precisely to define group farming because its ways and means are so multifarious. Perhaps the largest change in Japan's agriculture in recent years, brought about mainly by developments in the nonagricultural sectors, is the outflow of farm labor to nonagricultural occupations. Farmers have been induced to take countermeasures such as the adoption of joint water management and the introduction of large machinery for joint use to replace individually owned small machines. To assure a high yield of paddy rice, farmers spend many hours carrying out the crucial operations of timely transplanting and irrigation. These high yields require carefully adjusted depth of irrigation water, high rates of fertilization, and disease and insect control. There are certain scale economies to be realized if these activities are performed jointly and on larger areas of land than that under control of an individual family. Since it is a prerequisite for rice-growing conditions on participating farms to be uniform, all participating farmers have to adopt the same variety and transplant at the same time. Without this timing, the adjustment of irrigation water cannot be carried out cooperatively because water depth must differ by growth stage and condition. The same is true for disease and insect control. Thus the simplest form of group farming operations is agreement to adopt the same variety, to transplant and apply fertilizer at agreed-upon times, and to manage water use and disease and insect control cooperatively. Usually the operations under the agreement are carried out individually on each farm except for cooperative water management and insect control. Of course, some operations do not fall under the agreement nor require cooperation; for instance, harvesting operations are usually on an individual basis.

The central characteristic of this kind of group farming is the agreement of participating farmers; it is the one kind of group activity based on such agreement. Group farming is sometimes called *joint farming* (translated as *group cultivation*); however, agreements among members are its basic foundation.

When group farming develops to a more advanced stage, it becomes advantageous to carry out plowing and harrowing jointly as a preceding operation to simultaneous transplanting. A large tractor and tillage equipment is jointly owned and used, with one member who can operate it being commissioned to do all the tractor work. To ensure the maximum result from the use of such a large machine, field plots must be consolidated and irrigation and drainage ditches must be re-aligned and improved.

At a stage even further advanced, the volume of operations under simple agreement gradually decreases and the volume of joint work increases. For instance, transplanting is done jointly and the same is true for harvesting. When practically all farming operations are done jointly, it resembles cooperative farming. Yet, this group farming is not cooperative farming.

At present, a major part of all group farming is at the second stage. However, the more advanced types as well as cooperative management in rice farming are increasing, though still relatively small in absolute numbers. It should be noted that group farming operations do not inevitably or naturally develop into cooperative management; in general, there is a distinct difference in the thinking of farmers engaged in group farming operations and those participating in cooperative cultivation. There is some opinion that group farming operations are an initial step toward large-scale individual management. Since group farming operations have emerged naturally as a countermeasure of farmers against external pressures, they are very flexible. Accordingly, it is not incorrect to speak of group farming operations as temporary. Yet it is wrong to assume that group farming will develop soon either into cooperative management or into large-scale individual management. The fact that group farming operations are flexible can be interpreted to mean that Japanese farmers are capable of meeting ever-changing situations. Rather than drawing hasty conclusions about the direction of its development, I believe it will be useful to ascertain fully the characteristics and existing conditions of group farming operations.

The Current Situation of Group Farming

According to statistics of the Ministry of Agriculture, Japan has 6,275 group farming enterprises, and 85 percent of them are engaged in rice farming. As far as rice group farming is concerned, most are of the "technical agreement plus joint use of machinery and facilities" type.

The number of households in one group farming organization varies considerably, with 21-50 households being the most usual size (45 percent); another 20 percent have from 11-20 households. Acreage of paddy in one group farming organization is usually between 10 and 50 ha.; 70 percent of all rice group farming organizations fall within that range.

Fifty-five percent of all rice group farming organizations have their own machinery operators within the group, and 45 percent hire operators from outside. Forty-five percent of all group farming organizations are managed on the initiative of a buraku, and in these cases all households in the buraku participate. Forty percent of all organizations operate on the initiative of voluntary groups of small farmer members. The remainder are cooperative associations, etc.

Evaluation of Group Farming Operations

Productivity and Rice Farming Technique: In terms of productivity, the 1967 investigation by the Ministry of Agriculture and Forestry shows that 35.1 percent of participating groups reported a good yield increase and 43.7 percent some yield increase, which means that nearly 80 percent of all groups realized some improvement in yields. Group farming also contributes to the stabilization of rice cultivation since 47.9 percent reported that stabilization has greatly increased and 38.6 reported some stabilization. Yield increases and stabilization effects differ to some extent by region. For instance, the regions which experienced a large yield increase were Kyushu, Chugoku, and Shikoku, followed by Tokai and Kinki regions. A small yield increase was reported from Tohoku, Kanto, and Hokuriku regions. Practically the same trends are seen in stabilization. Stabilization and yield-increase effects of group farming operations were, if anything, more pronounced outside the main rice-producing regions.

From the standpoint of labor productivity, 99 percent of the total number of group farming operations recognize the labor-saving merits of joint operation. Savings in labor requirements were estimated at about 20 percent.

What mechanism provides the connecting link between land productivity improvement and labor saving? As stated above, yield increases are ultimately brought about by thorough and meticulous management in group farming operations, incorporating new cultivation technology such as the use of superior varieties, fertilizer application, and adjustment of water depth at different growth stages. And those operations can often be carried out more efficiently when done cooperatively. For instance, the average number of disease and insect control treatments under individual management is ordinarily six per season. Under joint operation these are done about ten times, yet with less labor since they are done jointly. Joint use of a large tractor for land preparation saves much more labor than working individually with small machines.

The decrease in labor requirements is not achieved solely through joint operations. As stated above, the part-time farmer who is dependent on income from an outside job has an insufficient labor supply to perform all the farming operations carried out previously. The remarkable feature of group farming is that relatively large-scale management compensates for this reduced labor force. Wives of part-time farmers have become skilled in performing even the most complicated work. Improvement of labor quality in group farming is one merit which should not be overlooked. Accordingly, elevating the productivity of the remaining farm labor force has contributed significantly to the improvement of land and labor productivity in group farming.

Agricultural Structural Problems: Today in Japan all policies and changes in agriculture are likely to be closely examined and evaluated from the standpoint of agriculture's structural problems. Two salient features of these problems are: (a) a decrease in the agricultural population and the number of farm households, and (b) finding a way to increase the scale of operations and management, and raising the quality of management. The questions being discussed in this connection are: How is the differentiation of farmers into various strata progressing? What measures are being taken to promote it? What are the best ways to promote movement out of farming?

Various positive measures are being carried out, such as expansion of the scale of farm management, simplifying regulations dealing with agricultural land transactions, and amendments to the Agricultural Land Law (a bill to amend this Law is to be presented to the Diet). Increased credit has been provided for the improvement and expansion of cultivable land. In short, competition among farmers is becoming more pronounced, and this should serve to place Japanese agriculture on a sounder and stronger footing. Does group farming contribute to the alleviation of these structural problems?

Some people are skeptical since group farming is frequently carried out by the buraku as a unit comprising all farmers as members, whether they are comparatively large-scale farmers or small-scale, part-time farmers. Thus they see it as an inhibition to the progress of farmer-strata differentiation. But will group farming operations actually inhibit the mobility of the farmer in deciding his future? Although farmers did feel the need to take concerted action against the instability of present agriculture by joining group farming operations, the differentiation in function among farmers is developing through the necessity of high-level knowledge of farming practices, large machinery, managerial abilities of organization, etc. Thus, a differentiation in strata between farm leaders and followers is developing. Group farming operations require a high level of techniques and management. The farm-leader stratum increases its work volume in planning and management, and, in contrast, the work volume of the follower stratum naturally tends to decrease. So far, on the surface at least, it appears that there are practically no transactions in agricultural land between these two strata. Consequently, there is no discernable trend in the expansion of cultivable acreage under control of the more progressive farmers. But in reality, the farm-leader stratum is taking over much of the work of the follower stratum through group farming operations, and it seems important to give due attention to this functional differentiation. There is, of course, a divergence of opinions as to how it may develop in the future.

Typical of the claims of those who hold a negative opinion of group farming is that, because group farming is carried on with the village or sub-community as the unit, it cannot be called functionally effective; instead, it preserves old traditional groupings which serve only to retard the economic differentiation of strata and the reduction in farm numbers. Contrary to these opinions, it has been pointed out above that although the distribution of large- and small-scale farmers may not be greatly changed, functional differentiation into farm-leader and follower strata is progressing.

This is substantiated by polls taken on the satisfaction or dissatisfaction of farmers with group farming. The expressions of discontent can be classified into three categories: (a) management of a group farming organization is difficult and therefore inefficient; (b) there is an inequality of burden or responsibility and benefit among participating farmers; and (c) group farming reduces the potential for developing relatively large farms under individual management. It is safe to assume that such negative evaluations come mostly from large-scale farmers and are evidence that the differences in ability within the group have emerged and differentiation in function is progressing. Because the role of the farm-leader stratum is large and the dependency of many farmers on those leaders is increasing, the imbalance in burdens and benefits is seen as a problem. There is a self-awakening of the farm-leader stratum to the fact that it is making sacrifices for the benefit of the follower stratum, and this leads them to express dissatisfaction with the system, stressing especially the inhibition they feel in development of larger farms for individual management. In recent times, however, districts have recognized the fact that the function of the farm-leader stratum has increased and now consider it necessary and proper to compensate the progressive farmers for the guidance and services that they render.

Although it is true that a group farming operation is an organization comprising many part-time farmers, it is conceivable that many of these small part-time farmers will entrust a few of the larger and more progressive farmers with the actual management of their land. Such commissioning of management may be on an individual-to-individual or an individual-to-leader basis; entire farming enterprises might be commissioned or only certain major operations. At any rate, it is very likely that expansion in the scale of farm management will first develop in such a commissioning form.

Perhaps the greatest merit of group farming operations is the spontaneity of the movement, initiated by farmers for their own sake, and hence unlike many other movements witnessed heretofore which were initiated, guided, and supported by Government. The reason that group farming operations differ in strength and flexibility by districts is that farmers have risen to the challenge of solving their own problems. For instance, in a district where many farmers have off-farm occupations from which they obtain their main source of income (with rice farming a subsidiary occupation), group farming operations are more intensive and inclusive. Group farming permits the farmer to devote most of his energies to his main off-farm occupation while his side enterprise (rice farming) is handled through joint operations. But in a district where rice farming is the main source of income, group farming operations permit a comparatively larger degree of individual freedom and operation. Both respect for farmers' independence and the organization of group activities are indispensable for the development of good management in group farming. Yet these two requirements may create problems and contradictions. Such contradictions must be resolved, and the solutions will vary according to the particular circumstances and the unique ideas and conditions of each joint farming operation. The fact that most group farming operations were independently conceived by farmers themselves and are very flexible is an extremely important and positive factor in their development.

The negative evaluations noted above are not shared by this writer. The path toward progress does not always move in a straight line, but more frequently along a zig-zag course reflecting an elastic and more realistic approach.

Group farming has generally achieved a high level of both land and labor productivity. The formation of a farm-leader stratum which can apply and utilize modern techniques and management in rice cultivation has been progressing step-by-step. The best Japanese farmers have a very high level of scientific knowledge of rice culture. In group farming, the emergence of farmers with the highest level of managerial capacity has been progressing, and group farming has been managed largely on the initiative of this leader class.

Of course, it is difficult to demonstrate, at this time, the formation of such a leader class. Thus some people continue to express anxiety lest all Japanese farmers become part-time farmers, and recommend policies for the encouragement of excellent full-time farmers.

In those situations where part-time farming is increasing, the initiative for group farming or group utilization of machinery should be taken, at least partially, by public organizations--for instance, the cooperative associations. We do have some successful examples of this type. But in the main, for the long-term development of agriculture it will be more desirable that the initiative come from the farmers themselves, and especially from the "progressive farmers" group. One of the most important merits of group farming is, I believe, the formation of a leader class. Through group activities, differences in the abilities among farmers become explicitly recognized, and a class known and trusted by all members for its leadership is the result.

THE VILLAGE COMMUNITY FARM SYSTEM

The Characteristics of This System

I wish now to summarize some of the experiences of the "Village Community Farm System" being discussed by many people in Japan. The idea of the village community farm system emerged several years ago from the Akita Prefectural Government, and has now been widely promoted as the basic agricultural policy of the Prefecture. Some characteristics of this system are given below.

At a glance, the Village Community Farm System looks similar to a group farming operation comprising all farmers in one buraku; it stands on the base of buraku and the principle of participation by all farmers.

Another characteristic of this system is the introduction of one or two enterprises other than rice. This permits some form of agricultural production throughout the year. Today farmers in Akita Prefecture are introducing fruit or livestock production as subsidiary enterprises.

This idea for the Village Community Farm grew out of the following conditions. Akita Prefecture is famous for rice production, and the productivity of labor and land in rice is quite high. So the Prefectural Government continued to emphasize rice and encouraged farmers to increase productivity and to expand rice production following Central Government policy. Labor-saving mechanization was successfully introduced, but since no winter-time jobs were available in agriculture, mechanization stimulated the increase of part-time farmers until nearly 100 percent of all farmers were operating part-time. Many farmers in Akita Prefecture, not only young but also older men, go to the large cities to work in construction industries half of the year as part-time laborers. They are absent from their homes for 100-150 days a year. This has created social problems, and occasionally very serious family problems develop.

Of course the aim of introducing multicropping and enterprise diversification is to build up more rational farm management. But successful multicropping and multienterprise operations require farmers with excellent managerial and technical abilities.

In Japanese agriculture, multicropping has often been strongly promoted by Government. The aim is to expand income sources by fuller use of family labor to compensate for the small scale of Japanese farming. So people tended to think of multicropping and diversification as necessary due to the small scale in farming.

But the present Japanese farming situation is different. Of course, full utilization of family labor remains important, but another, perhaps even more significant, reason for multicropping is the need to use more effectively the fixed capital invested, especially since the amount of fixed capital, publicly and privately invested in rice farming, has become very large.

TABLE 1. FIXED CAPITAL PER FARM IN TOHOKU DISTRICT, 1973
(including only machinery, equipment, and farm buildings)
(1,000 yen)

(Farm Size in Hectares)							
0.1-0.5	0.5-1.0	1.0-1.5	1.5-2.0	2.0-2.5	2.5-3.0	Over 3.0	Average
¥290.5	720.1	1,243.9	1,444.6	1,891.2	2,076.5	2,730.5	1,136.1

One of the major purposes of multicropping and enterprise diversification, therefore, is the fuller utilization of large machinery, facilities, and land. Today in the single-cropping rice areas, over-investment in machinery is usually covered by the income from outside jobs. Land itself has increased in value, at least partially as a result of public and private investments in land improvement. These increases in fixed capital have required greater intensity and multicropping.

But multicropping requires more skillful farmers and frequently larger farms than does single-cropping. This is one of the reasons for our concern with developing a class of progressive farmers--the stimulation of farmer strata differentiation.

Akita Prefecture is also now making efforts to introduce industrial factories in rural areas. With the introduction of these factories, small farmers will not need to go to the large cities for half the year. With income from labor in local factories, we expect that the small and part-time farmers will choose to devote more of their time to factory jobs and thus will want to place more of the responsibility for their farming activities in the hands of excellent and full-time farmers through group farming activities. Sometimes this will include contract farming between small part-time farmers and large full-time farmers within group farming. Thus we can expect the differentiation of strata, with the excellent farmer expanding the size of his farm business even without any transfer of landownership. Roughly speaking, this process of farmer strata differentiation will be as follows:

(a) The excellent and diligent farmer will see possibilities to expand the scale of his farming operations by contract farming with small part-time farmers in group farming activities. This can be done through partial or entire enterprise operation contracts, or sometimes by land-lease contracts covering either the winter months only or the entire year.

(b) The small part-time farmer can enjoy higher land and labor productivity in rice farming by participating in group farming. Of course he will have to pay the excellent farmer for his managerial and other inputs. But he can devote more time to his nonagricultural job without anxiety about his farming.

(c) Thus both the part-time farmer and the full-time farmer will be able to find their way in the village community without necessarily transfering landownership.

The Village Community Farm System and Buraku

Why is this village community farm system coterminous with the buraku? Why is this the most effective unit? Roughly, three reasons can be given as follows:

(1) The idea of the village community farm system emphasizes group farming activities at the initial stage. Buraku is usually a very adequate unit of group farming in rice production because of the convenience of common water utilization and control, as stated earlier.

(2) This system expects to promote the differentiation of farmer strata (leaders and followers) by group farming and multicropping, on the one hand, and by the introduction of factories on the other. Of course, to develop such a leader class requires the enlargement of farms. But as mentioned above, land prices have become very high, and rights of land cultivation have been strengthened by the Agricultural Land Law so that generally farmers do not like to have land-lease contracts with each other. Thus essentially only informal leases based on the reliability and mutual trust of both parties concerned have been carried out. When the persons concerned are well known to each other, an informal land-lease within the buraku may be quite satisfactory. Farmers' Cooperative Associations frequently play the role of mediator in such informal land-lease contracts concerning lease payments, duration of the lease, etc.

(3) Farmers (not only small farmers, but also full-time farmers) have been expressing strong resistance to governmental agricultural policies and a governmental attitude which would promote only a small number of excellent farmers on "viable farms" and eliminate small farmers from agriculture in the name of modernization. Such ideas are more palatable to farmers familiar with the coexistence of full-time and part-time farmers in the buraku.

Evaluation of the Village Community Farm System

In Japanese agriculture, Central Government policy has always been dominant, but important policies adopted by the Central Government have then generally been adapted to the local Prefectural level. Thus there have been very few cases in the past in which a Prefecture Government has had and promoted its own agricultural policy. The idea of the Village Community Farm System was generated by the Akita Prefectural Government, and especially by the Governor of the Prefecture. Today it represents the basic and practical direction of agricultural development in Akita Prefecture. Hitherto in Japan, the attitude of local government has ordinarily been a passive one. So this idea should be appreciated as evidence of a positive and assertive attitude in policy-making by a local government.

The Akita Prefectural Government is encouraging this system by recognition of the village community and by the provision of subsidies. But as stated above, the initial step in implementing this system should be the group farming operation. If people want to promote a Village Community Farm System, a group farming organization should be the starting point. Then multicropping, enterprise diversification, or introduction of industrial factories can follow in the next stage. Frequently a village community seeks recognition by the Prefectural Government because of the anticipated prospects of subsidies even before group farming activities are fully established. In such a case, the introduction of factories will only serve to convert all farmers into part-time farmers.

Ordinarily, farmers, especially in Akita Prefecture, have a very poor understanding of multicropping. They understand it only as a way for the family to get income from several sources. They do not understand the supplementary or complementary relationships in capital and land utilization among enterprises. The important aim of multicropping should be proper utilization of fixed capital, the value of which is increasing year by year. But in many cases at present, multicropping leads to significant increases in fixed capital investments through adding new enterprises resulting in a decrease in the intensity and efficiency of its utilization.

Very often the introduction of factories cannot play much of a role in the differentiation of farmer strata because of the small size of the factories introduced. Generally, small factories employ mainly women, and so their introduction in rural areas has little to contribute to agricultural structural improvement.

But we can appreciate the point that even in Tohoku District, with a cold and snowy winter climate, the significance of multicropping offers a new perspective by its systematic reflection of the weakness of a single-cropping rice system.

In my view, this Village Community Farm System, through the fostering of group activities, is a natural way to assist the spontaneous development of a viable, individual farm management in the current economic and social environment. At the same time, this experience suggests that group activities cannot always grow into cooperative farm management, but that they may be an important process for the growth and development of viable, individually managed farms in Japan.

14. French Experience with Group Farming: The GAEC

Philip M. Raup

Among modern states and developed economies the commitment of French farmers to the principle of private property is legendary. The Code Napoleon provided one of the strongest legal bases for the protection of these private rights that exists in any nation. The overt dedication in France to the support of peasant-type family farming has persisted into the industrial era longer than in any comparably developed country. It is perhaps because of this persistence, and not in spite of it, that France has also been the first country in the western world to develop an explicitly defined legal basis for the promotion of group farming. This is provided by Law No. 62-917 of 8 August 1962 relative to the formation of *Groupements Agricoles d'Exploitation en Commun* or Agricultural Groups for Farming in Common, hereafter abbreviated GAEC.[1]

Efforts to promote group activities in agriculture are not new in France, where they have had a long history. What was new in 1962, in the French context, was the clear-cut determination to give these efforts a distinctive foundation in law, with the intention of creating a new structural form in agriculture. The direct roots of this determination trace from the disorganized condition in which French agriculture emerged from World War II. Scattered and small-scale efforts at group farming and mutual assistance had emerged spontaneously in the immediate post-war years. The first attempt at systematic coordination dates from 1948 with the establishment of a coordinating body which was the predecessor of the organization now known as the *Union des groupements pour l'exploitation agricole*, abbreviated UGEA. The preeminent goal of this union was to promote an agricultural structure that would facilitate the enlargement and modernization of farms and their full participation in a market economy while preserving the social values of the family farm unit, the principle of private property, and at the same time enable farmers to participate in the rewards of communal activity. This approach derived much of its original momentum from leaders of the movement within the Catholic church in France that sought to develop a "social catholicism" or an economy for the service of man. A vigorous exponent of this point of view was René Colson whose book *Motorisation et Avenir Rural* (*Mechanization and the Rural Future*)[2] proposed that small- and medium-sized peasant farmers group themselves together to enable the use of new machines. This proposal was urgently advanced as an alternative to a feared massive take-over of peasant-type farms by large-scale commercial mechanized units that in France were identified with capitalistic agriculture, and were anathema to the Catholic left. This early effort was directed as much to the task of persuading French farmers to accept the tractor and its revolutionizing implications for French agriculture as it was to the task of devising a new form of organization for farm management.

These efforts were strongly supported by the Young Catholic Farmers Association or *Jeunesse agricole chrétienne* (JAC), guided by the conviction that a "third road" was needed between what were regarded as the abuses of capitalism and the excesses of Marxian collectivism. These convictions played a major role in the transformation in 1956-57 of the former youth group of the staid French syndicalist farmer's association (FNSEA) into a reinvigorated *Centre National des jeunes Agriculteurs* or National Confederation of Young Farmers, abbreviated CNJA. It was the political strength of this group, strongly influenced by Catholic social policy, that led to agricultural reform legislation focused on structural policy in 1960 (the *Loi d'Orientation Agricole*) and 1962 (the *Loi Complémentaire*), identified with the name of M. Pisani, the French Minister of Agriculture under whose regime it was developed. The legislation supporting group farming activity is a key part of that body of structural reform law.

Although enacted in August 1962 the law was not effective until adoption of implementing decrees (*décrets d'application*) by the Council of State on 3 December 1964. After six years of legislative drafting and nine months of parlimentary debate, a framework was created for the formal recognition of group farming. It is significant that in the final debate on the law the senators and deputies insisted that its application be restricted to small- and medium-sized peasant farms to insure that it would not become a vehicle for the subsequent creation of *kolkhozes*, or "production cooperatives" of the style that had emerged in eastern Europe following World War II. Group farming was explicitly not designed to be an evolutionary step toward the proletarianization or collectivization of French agriculture.

In broad outlines, the purpose of the law creating the GAEC was to make possible the common management of farms under conditions that are as comparable as possible to those that prevail in conventional family-type farms.

This group effort, it was hoped, would result in a reduced individual labor input, greater productivity, and improved security in both economic and social terms, while retaining the incentive for individual responsibility and the sense of ownership of property and products that had characterized peasant farms in the past.

To implement this purpose, the GAEC was created with the legal status of a corporate body under private law, but of a special type (*société civile particulière*). Its most distinctive feature concerns the manner of its creation and subsequent supervision by the state. Before a GAEC can exist it must be approved by a Committee of Review and Approval (*comité d'Agrément*) to be created in each Department, comprising the prefect and 8 members (4 officials, 3 farmers, and a Notary). These are responsible to a national committee in Paris, comprised of 3 representatives from the Ministry of Agriculture, 1 each from the Ministry of Finance and the Ministry of the Interior, plus farmers and representatives of farm organizations. The national committee acts as a policy-making and appeals body, with primary responsibility for implementation delegated to the Department committees.

A GAEC is largely exempt from the customary fees and charges levied upon the creation of a conventional corporation, with the exception of a single "Recognition Fee" of 50 francs.

Unless there are specific provisions to the contrary in the GAEC agreement, the liability of an individual member is limited to twice the value of his original contribution of capital. In practice, this limitation is frequently modified by the credit requirements of the firm.

Tenants and leaseholders can be members of a GAEC, regardless of whether or not they turn over their leased land in whole or in part to the GAEC. If the tenant elects to turn over his leased land to the GAEC for cultivation in common, he retains his individual liability to his landlord for performance of his rental contract and need only notify the landlord by registered letter of the action he has taken. The permission of the landlord is explicitly not required.

In contrast, if the rental arrangement is a sharecropping agreement (*métayage*), permission of the landlord must first be obtained.

Members of a GAEC must be primarily engaged in farming as an occupation, and must be natural persons. Juridical persons or corporations cannot be members. Spouses or minor children can be members but cannot qualify for the several forms of public subsidy or financial assistance available to GAEC members unless they have been active for at least three years as independent farmers or GAEC members. A GAEC cannot be formed by husband and wife only.

A principle repeatedly stressed in the law authorizing the formation of GAECs is that membership shall in no way impair the rights of an individual member to land, to the products of the land, or to any entitlement he would have had as an independent farmer to participate in governmental programs of agricultural or social assistance. Under French law, for example, the minimum term for the lease of farm land is nine years. A tenant or leaseholder who has held land under lease for a minimum of five years has a right of preemptive purchase, in case the landowner decides to sell the land. If the tenant pools his leased land with other land in the formation of a GAEC, he retains this right of preemption.

If crops or livestock produced by a GAEC are commingled for marketing or processing, the individual members retain a legal title to their proportionate share. As taxpayers, the obligation of members is individual and personal and is not assumed by the GAEC. In computing the entitlement of a GAEC to any governmental subventions or financial aids, the sum is computed by adding together the individual entitlements of the separate members. In the explicit language of Article 7 of the basic law, participation in a GAEC cannot result in an economic, social, or fiscal status for a member or his family that is in any way inferior to the status of other farmers or the families of farmers.

The law is equally explicit with regard to the obligation of each member to participate actively in the work of the GAEC. There can be no silent partners. Each member must be a working member, excepting only the ill, the infirm, or the aged. Article 1 of the law emphasizes that the purpose is to permit work in common "under conditions comparable to those existing in family farms." To this end, the article goes on to specify that the size of a GAEC cannot exceed ten times the size of an individual family-type farm, and cannot include more than ten members.

The lack of precision in this limitation on areal size has never been resolved. The size of a family farm is a relative concept, subject to wide variations, geographically and over time. In approving GAEC applications, the Departmental Committees of Review and Approval in practice have held to the maximum of ten members as the principal limit on size, leaving open the question of maximum permissible area.[3]

Additional evidence of legislative intent that the GAEC shall approach as closely as possible the structure and function of a family farm is the specification that a GAEC can engage in the marketing or processing of agricultural products but this activity cannot become the principal function of the association.

The foundation capital (*apport au capital social*) must be at least 10,000 francs, and cannot fall below this amount. It may be contributed in money or in kind (land, livestock, buildings). A member can also qualify by contributing only labor (*apport en industrie*) although this is relatively rare, or by contributing intangible rights (*biens mis à disposition*), for example, a leasehold. In practice, the foundation capital has been contributed primarily in the form of money or livestock, or rented land. Owned land or buildings have typically not been contributed outright to the GAEC but have been leased to it.

The law recognizes two types of GAEC, partial and total. The partial GAECs are designed to permit the pooling of capital and labor for a specific branch of farming activity, typically some form of animal or poultry husbandry. These partial GAECs are very much in the minority, and in practice the dominant form has been the full GAEC, or *GAEC total*.

It is important to stress the fact that a GAEC is not a cooperative. It involves cooperation, in the fullest sense of the term, and the evolution of the GAEC idea owes much to the historical strength of the cooperative movement, in France and in other countries. But French law regulating conventional agricultural cooperatives is relatively rigid and confining, and was not considered flexible enough to provide a basis for the goal of a "cooperative of labor" that was predominant in the minds of the originators of the GAEC approach.

Although possessing many characteristics of a corporate body, it is equally important to stress the fact that a GAEC is not a corporation in the usual sense of the term. The strongest evidence of this distinction is provided by the degree of supervision by the state. Creation of a GAEC involves approval by the Departmental Review and Approval Committee (Comité d'Agrément) of a charter specifying the following details, among others:

1) The number of members and their contributions and responsibilities to the GAEC.
2) The amount and nature of contributed capital.
3) The liabilities or obligations that members bring with them into the GAEC, or assume as members.
4) The method of payment for labor and any special arrangements regarding the division of labor among members.
5) Provisions regarding the division of profit and loss.
6) Procedures to be followed in any transfer of members' rights by inheritance.
7) Procedures to be followed in case the GAEC is dissolved.

No hard and fast rules are laid down in the GAEC law or implementing decrees for the determination of rates of remuneration for labor or capital. In principle, locally customary rates are used. Any real estate owned by members but leased to the GAEC is paid for at a rental rate that is customary for that region. A "normal" interest is paid on contributed capital. Labor is paid at a wage rate that is in principle to be adjusted annually in accordance with farm wage rates in the community. The basic law of 8 August 1962 specifies (Article 4) that the remuneration of members for their work in the GAEC constitutes a *charge sociale*. This places the labor income of GAEC members on a par with the wages of workers in a conventional business corporation, in case of bankruptcy: payments to members for their labor are a prior claim upon the assets of the group and cannot be appropriated by creditors.

In computing the payment for use of any buildings that are not included in the foundation capital contributed by members but that are used by the GAEC, the use value is taken as the basis for valuation rather than market price. The guiding rule is that work shall determine rewards. The goal is to avoid any situation in which relations among members begin to resemble those among employer and employee, capitalist and worker, or stockholder and manager.

Upon approval by the Departmental Comité d'Agrément, the GAEC charter must be published in two newspapers with a circulation in the region, stating the date of activation of the group. The charter becomes an item of public record, available to all, and a copy must be maintained at the headquarters of the group, available to all members.

Attempts had been made before 1962 to supplement existing French cooperative law in order to encourage communal farming efforts. In terms of its subsequent importance in the evolution of the GAEC idea, the most widespread of these earlier efforts involved the common ownership of farm machinery. With the support of Pierre Tanguy-Prigent, Minister of Agriculture (1944-47) in de Gaulle's first post-war cabinet, peasant groups were created for the collective purchase and use of farm equipment. These farm machinery cooperatives, or *Coopératives d'utilisation du matériel agricole* (abbreviated CUMA), had priority in the purchase of scarce farm equipment in the immediate post-war years. Gordon Wright has succinctly summarized this phase:

> CUMAs sprouted at once like mushrooms after a rain; by 1948, more than 12,000 of them had been organized. During the next two or three years, however, many CUMAs disappeared, and it became clear that the movement had been partly artificial in character. While tractors were severely rationed and CUMAs enjoyed a purchase priority, well-to-do farmers had organized false CUMAs in order to get a machine without delay. The facade fell away as soon as rationing ended. Some genuine CUMAs did survive, however, in spite of the tensions they often generated among the co-owners, and by the mid-1950's, their number began to mount steadily once more.[4]

This brief outline of some of the antecedents and characteristics of GAECs can be brought to life by reference to several case histories.[5] One in question was organized in 1968 in the Burgundy region by two pairs of brothers, with the strong encouragement of the local Catholic priest. The brothers had been working together in various ways for over ten years. Their first communal effort was in 1957, when they purchased a hay-baler with money furnished by their parents. In 1962 they formed a CUMA (see above), primarily because they could secure a price advantage in purchasing equipment. Throughout the 1950s a 10 percent discount or rebate on the dealer price of farm machinery had been available to all farmers, as a form of state subsidy. This was limited, however, by a ceiling of 150,000 old francs, if the buyer was acting as an individual. For a CUMA, there was no upper limit to the subsidy.

The experience with the machine cooperative encouraged them to consider a GAEC. When they completed the organization in 1968, the four members were farming 60, 69, 92, and 55 hectares, respectively, or a total of 276 hectares (682 acres). Of this, 73 hectares were owned land, and 203 were rented. All four had some rented land, but 59 of the 73 hectares of owned land were held by one individual. Among the four, they rented from 46 different landlords.

Coincident with the formation of the GAEC they bought an additional 200 partly forested hectares, giving them a combined area of about 400 hectares (988 acres) of agriculturally usable land. The purchased land was essential to the creation of the GAEC, since much of their 203 hectares of rented land was in pasture. In the Charolais region, laws of the commune typically prohibit the plowing of pastureland. Rental rates for grassland in this area are set in terms of the average price of a given number of kilograms of meat on the Paris market in the preceding year. In 1970, for example, the range in rental rates for pastureland was from the monetary equivalent of 40 to 100 kg. of meat-value per hectare in the commune in question. Assuming an average rental of 60 kg. of meat at a price of 6.15 new francs per kg., the rent for pastureland was approximately 370 new francs or roughly $75.00 per hectare in the early 1970s. In contrast, rents for cropland are set conventionally in terms of the value of a given number of quintals of the crop per hectare, and in this region seldom exceeded 4 to 5 quintal per ha. At 1970 prices, cropland rentals rarely exceeded 200 to 250 new francs or $40 to $50 per hectare. No landlord would willingly permit pastureland to be converted to cropland, because of these rent differentials.

The GAEC provided an escape from this limitation on the conversion of pastureland to cropland. With financial assistance from the *Crédit Agricole* to enable them to purchase additional land and by pooling their rented and owned land in a GAEC, the brothers emerged with approximately 200 hectares of cropland to balance their 200 hectares of pastureland.

It is also probable that an additional incentive for creation of this GAEC was to escape statutory limitations on farm enlargement. One of the first successes of the National Confederation of Young Farmers (CNJA) had been enactment of a law in 1958 (the *Réglementation des Cumuls d'Exploitations*) that set minimum size limits on farm subdivision, and maximum upper limits on farm enlargement. These upper and lower limits vary by Departments and region, ranging from minima of 6 to 20 ha. and averaging 10 ha., to maxima of 30 to 100 ha. with an average of 50 ha.[6] Although exceptions were possible and enforcement was not uniform, the existence of these limits had been a barrier to farm size expansion, especially if financial help was desired from the Crédit Agricole. If a GAEC was formed, it was much easier to secure approval of farm size expansion. In practice, this has been a supporting argument for the establishment of GAECs in some regions.

The principal crops on this GAEC are wheat, barley, and rape, in about equal proportions of 60 to 65 ha. each. The division of labor is by function. One man does all the accounting and operates the combine; one does all the plowing and field cultivation. One does all the seeding, and cares for a sheep herd of some 80 ewes. The fourth is the cattle man, caring for a total of some 300 head, scattered among about 50 pastures. Some pastures have no water supply, and water must be hauled for the cattle, a very laborious task. They employed one full-time worker, paid 800 francs net per month, plus board, the use of a house owned by the GAEC, and payment by the GAEC of the employee's share of social security.

Each of the four families drew 700 francs per month as a provisional salary in 1969, the first full year of operation. The total distribution of profits at the end of the year averaged an additional 700 francs per month, for a total family labor income of about 17,000 francs for the year. Each member also received an income on contributed capital, ranging from 2,200 to 4,800 francs. Total annual family income thus ranged from roughly 19,200 to 21,800 francs or approximately $3,840 to $4,360 at then-current rates of exchange. This was by no means a handsome income, but the members regarded it as satisfactory in view of the heavy start-up costs of the new organization. In effect, organization of the GAEC enabled them to shift a part of their labor from the intensive care of animals (cattle, sheep, some dairying), for which the rates of return per man-hour were low, to field crops with associated heavier use of machinery, and higher rates of return to labor. Capital intensity of the firm increased, labor intensity declined, and returns per man-hour improved.

A second GAEC case study involved an intensive dairy farm in north central France, organized by a father and his two sons. With a relatively large dairy herd of over 50 cows, the motive here was three-fold: to insure a labor supply for the milking task, to qualify for a construction grant from the Ministry of Agriculture for a new dairy barn, and to qualify for a highly subsidized loan at 4 percent interest from the Crédit Agricole, to buy more land. Under rules current in 1972, the Ministry of Agriculture would grant a subsidy of up to 400 francs per head of dairy cows housed ("per stanchion"), toward the construction of a new barn. By contributing as much of the construction labor as possible, the three GAEC members reckoned that this subsidy would cover roughly one-fourth of the cash cost of the barn. If organized as a GAEC, they stood a much better chance of securing approval of a modernization subsidy of this magnitude.

An even more important subsidy was available in the form of a long-term loan from the Crédit Agricole, to buy land. Before organizing the GAEC, the father was engaged in beef cattle production. With the aid of the 4 percent loan which was only available to the family if it organized a GAEC, he sold his beef herd of Charolais cattle, bought a dairy herd, and also bought more land. In effect, the family now has two farms, one that the father considers "his," and one about 2 kilometers away that is in the name of the oldest son. Both the father and the son retain title to the land in their respective names, and "rent it out" to the GAEC.

In this case the GAEC had permitted an expansion in the scale of the business, with more land and better buildings, and had created a financial basis for a shift from beef to dairying. There is a touch of irony in this fact, since French agricultural policy for the past decade has stressed the need to reduce milk output and increase beef production. For the family involved in this GAEC, the relative prices of milk and meat, and the relative capacity of beef production and dairy farming to absorb and reward three man-years of labor input, led to a reverse shift from beef to milk.

Adding more land and shifting to dairying offered the prospect of keeping his sons "on the farm." This was the real stimulus for the formation of the GAEC, from the father's point of view. For the two sons, a very attractive feature of the GAEC was the prospect of occasional vacations and some opportunities for a "day off" in spite of the demanding labor requirements of a 50-cow dairy herd. For the women in the three families, an overriding stimulus for communal organization of a dairy GAEC was the prospect that they could be relieved of much of the heavy milking chore which had traditionally fallen to them.

This GAEC involved no significant shift in land use, or intensity of cultivation. The primary consequence was more land and a better barn, hence an increase in capital intensity, a stabilized labor supply, and a redivision of labor among the family members. For the two sons in the group, the social value of having some help with the milking and of not being "chained to a dairy cow" was clearly a most attractive feature.

The first GAECs were organized during the winter of 1964-65, immediately following the issuance of implementing decrees No. 64-1193 and 64-1194 of 3 December 1964 (see note 1 above). Initial growth was slow, with some 30 GAECs authorized in 1965. The number of approved GAECs total increased to 340 in 1966 and additions ranged from approximately 300 to 450 in each year from 1966 through 1972, as shown in Table 1.

In addition to the 2,753 full GAECs at the end of 1972, approval had also been given for 239 partial GAECs, for a grand total of 2,992 GAECs of all types (Table 2). The proportion of partial GAECs has declined slowly from 12 percent of all GAECs in 1967 to 8 percent in 1972, and 7 percent at the end of 1973. Through 1972, approval had been denied in 260 cases, and 144 GAECs had been dissolved. The rate of attrition is small but increasing. The number of dissolved GAECs was estimated at 12 in 1969, and 50 in 1970.

Summary data for 1973 show a continuation of the growth trend, with a total of 3,500 GAECs of all types as of 10 December 1973. These involved approximately 9,100 members, and a total area of 316,000 hectares, or just under 1 percent of the area of agricultural land in France.[7]

The typical GAEC is predominantly a family affair, as the data in Table 2 make clear. In 1967, GAECs involving parents and children or brothers and sisters were 58 percent of the total; by 1972 this had increased to 68 percent (2,039 out of a total of 2,992). They are also increasingly two-man farms. GAECs comprising two members only were 46 percent of the total in 1967 and 55 percent in 1972. In contrast, GAECs with 5 or more members fell from 10 percent of the total in 1967 to 5.6 percent in 1972.

One of the most interesting trends revealed by the figures in Table 2 relates to size in hectares. The average GAEC reached its

TABLE 1. GROWTH IN THE NUMBER OF GAEC TOTAL BY YEAR OF AUTHORIZATION[a]

Date	Annual Approval	Cumulative
Exact Date Unknown	7	--
1965	29	--
1966	340	376
1967	427	803
1968	317	1,120
1969	456	1,576
1970	404	1,980
1971	342	2,322
1972	431	2,753

[a]SOURCE: Through 1968, H. Nallet, C. Roger, and M. C. Al Hamchari, *Les Groupements Agricoles d'Exploitation en Commun (GAEC), Tome 1: Les caractéristiques structurelles des GAEC* (Paris: Institut National de la Recherche Agronomique, Série Travaux de Recherche, no. 7, janvier 1971), pp. 82-84; for 1969-72, provisional estimates by the Union des Groupements pour l'Exploitation Agricole (UGEA), Paris.

peak of 105 ha. in 1969, and declined 13 percent to 91 ha. (225 acres) by December 1972. With GAECs involving only two persons in 55 percent of the cases, comprising family members in 68 percent of the cases, and averaging 91 ha. in size, the "representative firm" among GAECs looks very much like a consolidation of two medium-sized family farms. As we shall see, this impression will be borne out in a number of significant details.

A second revealing trend concerns the increase in the proportion of GAEC land that is operated under lease. In 1967 leased land was 55 percent of the total area in GAECs; by 1972 this had increased to 62 percent. Almost all of the leased or owned land is held in the name of one of the members of the GAEC. In a detailed study of the land tenure status of 950 GAECs as of 1 June 1968, Nallet et al. found that land owned in the name of the GAEC was less than 4 percent of the total area, and land rented in the name of the GAEC from nonmember landowners (instead of from one of its members) was only about 2 percent of the total area operated.[8] The GAEC device has been used to pool use-rights in land, but it has clearly not been used to create an alternate holder of proprietary rights. The GAEC is definitely the creature of its members, resembling in many ways a personal (or two-person) holding company.

The regional distribution of GAECs is sharply differentiated. As shown in Table 3 and Figure 1, just under half (46 percent) of all GAECs in 1970 were situated in five of the twenty-two regions. These formed two areas of concentration, one comprising Champagne, Bourgogne, and Rhône-Alpes, and the other Bretagne and the Pays de

TABLE 2. TRENDS IN GROWTH OF GAECs IN FRANCE, 1967-1972[a]

Characteristic	Status as of December			
	1967	1969	1970	1972
I. GAEC Numbers				
Approved GAECs	910	1,734	2,190	2,992
Of which: GAEC *total*	800	1,576	1,980	2,753
GAEC *partiel*	110	158	210	239
No. of GAEC dissolved	?	ca. 12	ca. 50	144
No. of GAEC applications rejected	77	178	209	260
II. GAEC Membership				
Among parents and children	250	530	631	944
Among brothers and sisters	280	586	775	1,095
Among nonrelatives	380	618	784	953
2 members only	420	881	1,228	1,645
3 to 4 members	399	727	919	1,178
5 or more members	91	126	143	169
With no hired laborers	415	916	1,235	1,793
With 1-2 hired laborers	330	586	728	923
Over 2 hired laborers	165	232	227	276
III. GAEC Area[b]				
Total area in GAEC farms (ha.)	90,700	181,900	220,656	273,237
Of which: owned land (ha.)	40,300	82,900	88,021	104,710
leased land (ha.)	50,400	99,000	132,635	168,527
Average farm area per GAEC (ha.)	98	105	100	91

[a]Assembled from reports of the Ministry of Agriculture and the Union des Groupements pour l'Exploitation Agricole, Paris, by Erich Schuler, "Steigende Zahl aber abnehmende Betriebsgrössen der französischen Gruppenlandwirtschaften," *Innere Kolonisation* 22, no. 5 (May 1973): 130.
[b]Includes both GAEC total and GAEC partiel.

la Loire. With the exception of hill-farming areas of Rhône-Alpes and parts of Bretagne, these are all regions in which medium-sized peasant-type farms are quite common.

In contrast, the six regions with the fewest GAECs accounted for only 10 percent of the total. These include the Paris region and Provence Côte d'Azur in which large peasant-type and commercial

TABLE 3. REGIONAL DISTRIBUTION OF GAEC IN FRANCE, AS OF 10 DECEMBER 1970[a]

Region	Number	% of Total	Agr. Used Land per GAEC[b] (ha.)	Agr. Used Land per Member[b] (ha.)	Av. No. of GAEC Members per GAEC
Rhône-Alpes	250	11.4	78	28.5	2.7
Champagne	231	10.5	160	66	2.4
Pays de la Loire	196	8.9	71	28	2.5
Bourgogne	173	7.9	140	61	2.3
Bretagne	157	7.1	59	25.5	2.3
Lorraine	130	5.9	123	50.5	2.4
Midi-Pyrénées	127	5.8	113	45	2.5
Aquitaine	112	5.1	94	27.5	3.4
Franche Comté	110	5.0	78	32	2.4
Poitou Charentes	106	4.8	107	44	2.4
Centre	105	4.8	130	55.5	2.3
Auvergne	80	3.6	115	43	2.7
Picardie	74	3.3	148	47	3.1
Basse Normandie	57	2.6	88	27	3.3
Haute Normandie	52	2.3	98	42	2.3
Provence Côte d'Azur	52	2.3	137	41.5	3.3
Limousin	48	2.2	100	27	3.7
Nord	45	2.0	74	24	3.1
Languedoc	37	1.6	98	30	3.3
Alsace	34	1.5	64	24	2.7
Region Parisienne	14	0.6	161	43.5	3.7
Total or Average	2,190	100.0	106	40.1	2.8

[a]Including 1,980 full GAECs (90.5 percent) and 210 partial GAECs (9.5 percent).
[b]As of 31 December 1969.
SOURCE: Ministry of Agriculture, and UGEA, as consolidated in U. Otzen, "Entwicklungen und Auswirkungen von Betriebsfusionen, Dargestellt am Beispiel der französischen Gruppenlandwirtschaft," *Agrarwirtschaft* 21, no. 10 (October 1972): 348.

farms prevail, and Languedoc and Alsace with disproportionately large numbers of small farms.

These regional differences point up a major characteristic of the distribution of GAECs. They tend to be concentrated in areas with a preponderance of farms of medium size. They are relatively rare in areas in which the farm size structure is dominated by farms that are either very large or very small.

FIGURE 1. NUMBER OF GAECs, TOTAL AND PARTIEL, BY REGIONS, FRANCE, DECEMBER 1970

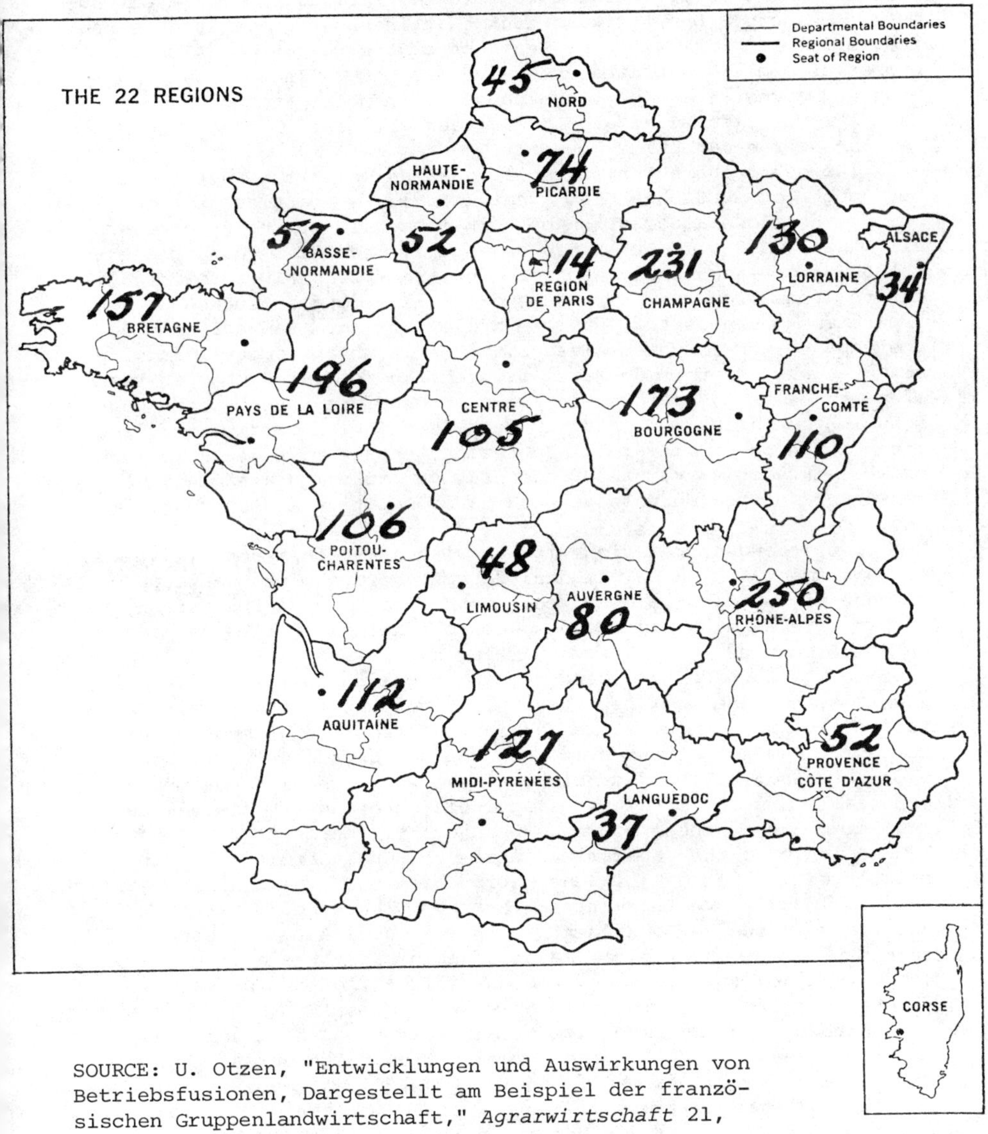

SOURCE: U. Otzen, "Entwicklungen und Auswirkungen von Betriebsfusionen, Dargestellt am Beispiel der französischen Gruppenlandwirtschaft," *Agrarwirtschaft* 21, no. 10 (October 1972): 348-349.

This is in part a reflection of official policy. Supporters of the GAEC legislation saw quite clearly that there would be little value in encouraging farming in common by pooling farms that were too small to provide a tolerable level of living for their numerous members. One striking fact that is shown in Table 3 is that in the five regions with the heaviest concentration of GAECs the number of members averaged under 2.5 per GAEC. In the five regions with the lowest frequency, the number of members averaged 3.3 per GAEC. It is clear that growth has been greatest in areas where two- and three-man GAECs prevail.

Two other characteristics of areas in which the increase in the number of GAECs has been greatest call for particular emphasis. One is the close association between communal farming efforts of all kinds (GAECs, CUMAs, Producer Groups, and CETAs[9]) and the execution of governmentally sponsored programs of strip-farm consolidation (*remembrement*). The second is the strong influence of newspapers and journals that promote group farming. Rambaud grouped the Departments in three classes, those in which communal farming activity was strong, moderate, or weak, and compared them in terms of a series of indicators, including gross farm revenue, the degree to which consolidation of strip parcels had been carried out, security of tenure, professional education, extent of systematic farm bookkeeping activity, use of cooperative marketing agencies, and subscriptions to journals devoted to cooperative and communal farming.

The strongest relationship was with strip-farm consolidation. In Departments in which communal farming activity was weak, only 10 percent of the farm land had been consolidated. In the "strong" cooperative farming Departments, consolidation had been carried out on 40 percent of the land. A similar but less pronounced differentiation was found with regard to subscriptions to farm cooperative journals. In the weak Departments, 17 percent of the farmers subscribed to one or more journals, in the strong Departments, over 25 percent. Rambaud stresses in this connection the pervasive influence exercised by the journals of the Catholic Action movement, and especially the JAC or Jeunesse agricole chrétienne. A study in 1956 showed that these journals reached 19.4 percent of all farm households. In the Departments where communal farming activity is strong today, the rate of penetration of the Catholic Action journals was much above the national average, reaching 36 percent of all farms in twelve Departments of *France de l'Est*. In contrast, in eight Departments of the Mediterranean basin where group farming activity today is very weak, the Catholic Action press reached only 7.3 percent of the farms in 1956.[10]

Although the average size of GAEC farms has declined since 1969-70, they are still quite large by French standards. For all France, the average farm size in 1972 was 20 hectares. The average for all GAECs in 1972 was 91 hectares, and for the GAEC total, 99 hectares, or almost five times the national average. The size distribution of GAEC farms is also highly skewed. In 1969, 10 percent of the number of GAECs total had 40 percent of the total area; 20

percent accounted for two-thirds of all land in GAECs. In terms of the value of foundation capital (*capital social*) the concentration is also great, though less sharply skewed, with 12 percent of the GAECs accounting for 40 percent of the basic capital, and 28 percent controlling 60 percent of the total.[11]

In terms of intellectual capital, the superior position of the GAECs is also noteworthy. GAEC members are better educated than the typical French farmer. In 27 percent of the GAECs all of the members have had some form of professional education. In an additional 37 percent, at least one of the members has had professional schooling. As a result, just under two-thirds of the GAECS have benefited from access to formal professional training in agriculture. For French farmers as a whole, only 8 percent have had this advantage.[12]

If we review the major characteristics of the GAECs that have been established to date in France, we are impressed with several facts. Perhaps most important is the relatively large scale of operation, given the strong role that perpetuation of the traditional French family farm played in the adoption of the original authorizing legislation in 1962. We are impressed as well with the strong role played by ideology in the organization and distribution of GAECs. They have been heavily influenced by the economic action programs of the Catholic church, and can fairly be regarded as an embodiment of Catholic rural policy in France.

On the economic front they have unquestionably created a better base for credit and, perhaps most significantly, have provided a better platform for the procurement of various types of French governmental subsidies available to agriculture. In this regard, they have reduced the total overhead cost of qualifying for various types of agricultural subsidy payments. They can be regarded as a rural institutional response to the development of a "grants economy" in French agriculture. The front-loaded cost of preparing requests for grants and financial aids is a burden on any organizational structure. These preparatory costs of qualifying for the distribution of public funds fall especially heavily on individuals at the lower end of the income scale. There are definite economies of size associated with the creation of the GAECs. One of the most important concerns the greater efficiency that is thus made possible in justifying requests for public subsidies, and satisfying public agencies distributing the funds that expenditure qualifications have been met.

In several important regions of France the GAECs have confronted a dilemma in the internal organization of their activities. Shifts in the domestic terms of trade among agricultural products in France since the formation of the European Common Market have tended to favor grain producers (and sugar beet growers) at the expense of producers of livestock products. The harmonization of internal common market agricultural prices began formally for grains on 1 July 1967, and thus coincides approximately with the beginning of the development period for GAECs in France. The relative increase in grain prices that followed full implementation of the common agricultural policy of the European economic community had the effect of making grain farming more attractive than the conversion of crops through livestock. The one exception concerns milk prices which were kept high on social grounds because of the widespread distribution of dairy cows among small farms. The milk price became a proxy for a welfare program for low-income and small-scale farmers. The resulting periodic milk surpluses have dampened significant rises in the milk prices, whereas the sharp rise in world grain prices following the large purchases by the USSR in the summer of 1972 added a buoyancy to grain prices that has been lacking in the livestock sector. As a result, economic trends over the entire period in which GAECs have developed in France have tended to favor field crops over livestock.

The development of specialized farms concentrating on grain production can result in highly seasonal labor demands, and is successful only if high levels of mechanization can be achieved. The care of livestock has one overriding advantage in that it fills in the valleys between peaks in seasonal labor demand. Livestock provide a wide range of opportunities for the investment of family labor in economically rewarding enterprises on the farm, at a more or less steady rate over time. Returns per hour may be low, but the "security of employment" is high. Livestock enterprises are thus ideally suited to the type of family farms that the GAEC legislation seeks to support. And in fact, the growth of GAECs has been prominent in areas in which livestock play an important part in the mixed enterprises that characterize the medium-sized family farms of the regions.

Economic trends have thus pointed to specialization in field crop production as the most rewarding way to modernize French agriculture in the past ten years. Social considerations and the need for steady employment opportunities point, in contrast, to farms combining a mix of enterprises with heavy emphasis on livestock as most likely to succeed under conditions of group farming in which a major goal is to provide productive work opportunities for all members of the group. This dichotomy has plagued the GAEC movement from the beginning. There have been, in fact, relatively few GAECs in the principal cereal-grain-producing regions of France. The majority of GAECs are focused on a few principal livestock products or are engaged in a variety of enterprises in which animal conversion of pasture and field crops plays an important role.

The picture that emerges is that of GAECs torn between two conflicting goals. One is a desire to increase labor efficiency and labor income by mechanized field crop production. The other is a compulsion to include a high proportion of animal conversion activities in the firm in order to provide stability of employment and opportunity for the utilization of available labor.

Another dilemma concerns the structure of decision-making within the GAEC firm. Perhaps the most important goal in the minds of the originators of the GAEC idea was the achievement of full equality among all cooperating members. We have noted the repeated emphasis placed on this goal in the basic legislation. The ideological leaders of the GAEC movement have stressed the fundamental importance of the development of a communal work ethic, in which no member is dominant. They have seen this as most likely to emerge in full GAECs that involve nonrelatives. Rambaud points out that 73 percent of all articles published in *Agriculture de Group*, the journal of the Union des Groupements pour l'Exploitation Agricole (UGEA), between 1960 and 1972 concerned GAECs total, and nonrelated members.[13]

In spite of this strong promotional effort, we have seen in Table 2 above that 68 percent of all GAECs in 1972 were among relatives. For full GAECs, the percentage was 75. Approximately half of these involve a parent, almost always the father. In the traditional French rural family, as Mendras points out, "the sons of farmers have no independent social existence as long as their fathers are alive."[14] Traditions as deeply rooted as this have been slow to respond to social and economic change. The family tie has been perhaps the most powerful bond that has held GAECs together, yet it was a desire to break out of this rigid mold that led to the revolt of the young farmers in 1958 and the political pressure that resulted (among other things) in the GAEC legislation.

The question of communal decision-making in GAECs involving only brothers is in many cases even more difficult to resolve. When a parent is involved, there is at least a traditional answer to the question: Who is in command? When relatives but no parent are involved, there is a strong tendency for the division of labor to be decided along craft or enterprise lines, as we have seen in one of the case studies above. The problem of communal decision-making tends to be resolved by the creation of several "one-man mini-firms" within the GAEC framework. Over time, it is not difficult to imagine that this could lead to the kinds of problems that have beset craft unions and guilds in the past. A jurisdictional problem in miniature could emerge in the GAECs.

Looking to the future, there are two additional problems that seem likely to increase in importance. One concerns the question of inheritance. Since title to land and buildings is retained in almost all cases by the individual members, difficult problems of valuation and compensation will arise if the children of present members decide they do not want to farm. If they offer to sell out, the remaining members of the GAEC may be unwilling to pay market prices for the real estate. If a concessional price is paid, the heirs of the former GAEC member can easily be persuaded that they have been cheated out of a part of their inheritance. The question of capital gains seems certain to plague the GAECs with the passage of time.

A related problem concerns part-time farming and off-farm work. A total commitment to farming is fundamental to the GAEC idea. This runs counter to one of the major trends of our time in developed economies that combine strong industrial and agricultural sectors. In the larger group farming experiments in Israel, and increasingly in Eastern Europe, a solution has been to bring the "off-farm work" to the farm, in the form of supplemental nonfarming enterprises. These have involved agricultural processing plants, furniture-making, small metal-working or plastic-manufacturing activities, or the organization of teams to perform construction work under contract during periods of low seasonal labor demands on the farm. This has been the communal farming solution to the desire for increased income from nonfarm work.

The GAECs are too small to permit this solution. It seems predictable that the GAECs will come under increasing strain with the realization that the rule that every member must work can be interpreted to mean that every member must work on the farm all the time. A greater flexibility will be required than is apparently provided by the existing GAEC framework.

In one dimension, GAEC members have been participating actively in "off-farm work," through their leadership activities in farm organizations. A certain pioneering spirit has been required in order to establish a GAEC. The members are much better educated than the average French farmer. They have been called an "agricultural elite," and the appellation seems deserved. In a sample study of 179 GAECs with 518 members, Nallet and associates found that 56 percent of the members held at least one position in a farm organization (syndicate, cooperative, CUMA, CETA, etc.). These were often leadership roles, with 19 percent of the GAEC members sitting on governing bodies, and 14 percent serving as President of their organization. In 12 Departments of the west of France, a separate study showed that GAEC members represented 0.3 percent of the number of farms but held 7 percent of the leadership posts in the farm organizations of the region.[15]

Some interesting parallels can be drawn between the GAECs in France and the family-farm corporations that have evolved since the 1950s in the United States. In both cases, the resulting firms resemble large family farms rather than the new forms of farm-business organization they seemed to promise. Both are relatively new institutions, have yet to meet the test of a severe agricultural depression, and have not resolved the problem of the "locked-in" member who wants out.

In economic terms, the financial results of the GAEC have not been better than those achieved on comparably large individual farms. They exhibit a tendency toward over-capitalization and above-average indebtedness for farms of their size and type, but this is a generalization that can also be made about the larger family-type farms in many countries.

In concluding the most thorough available study of GAECs, Nallet and associates observe that the major advantage of the GAEC lies in the social sphere, and particularly in the opportunities they provide for members to have a regular day off, take vacations, and participate more actively in the social, professional, and political life of their communities.[16] If they have not succeeded in creating a new work ethic, if they have not achieved financial results superior to those of individual family farms, these are still encouraging achievements.

NOTES

1. The text of the law, and subsequent implementing regulations have been consolidated in *Groupements agricoles d'exploitation en commun, législation, règlementation, statuts-types*, Journal Officiel de la République Francaise, No. 1288 (Paris, 1966). All references to the law and its implementing regulations will be to this source. For an English translation of the basic GAEC act, see FAO, *Food and Agriculture Legislation*, vol. 11, no. 3, VIII/I, "France: Act No. 62-917, relating to agricultural associations for joint exploitation, 8 August 1962," Rome, 1 March 1963.

2. René Colson, *Motorisation et Avenir Rural* (Paris: Centre National d'Etudes Rurales, 1950). See also J. Chombart de Lauwe, "Aspects Economiques de l'Utilisation en commun du Matériel Agricole," *Bulletin de la Société Francaise d'Economie Rurale*, vol. 4, no. 4 (octobre 1952).

3. H. Nallet, C. Roger, and M. C. Al Hamchari, *Les Groupements Agricoles d'Exploitation en Commun (GAEC), Tome 1: Les caractéristiques structurelles des GAEC* (Paris: Institut National de la Recherche Agronomique, Série Travaux de Recherche, no. 7, janvier 1971), p. 22.

4. Gordon Wright, *Rural Revolution in France, The Peasantry in the Twentieth Century* (Stanford, Cal.: Stanford University Press, 1964), pp. 110-11.

5. Drawn from field studies in 1967, 1969, 1970, and 1972. I am especially indebted to Denis Bergmann, Jacques Brossier, André Brun, and Michel Petit for encouragement and aid in arranging interviews and interpreting results.

6. OECD, *Structural Reform Measures in Agriculture* (Paris, 1972), p. 140.

7. Placide Rambaud, *Les Coopératives de Travail Agraire en France* (Paris: Ecole Pratique des Hautes Etudes VI, Centre de Sociologie Rurale, 1974), p. 160.

8. Nallet, et al., *Les Groupements Agricoles*, p. 42.

9. CETAs are *centres d'étude techniques agricole*, or small groups of farmers organized cooperatively to promote extension-type self-education, under the guidance of a technically trained agriculturist.

10. Rambaud, *Les Coopératives de Travail Agraire*, pp. 84-87.

11. Nallet, et al., *Les Groupements Agricoles*, pp. 40, 44.

12. Rambaud, *Les Coopératives de Travail Agraire*, p. 29.

13. Ibid., p. 36.

14. Henri Mendras, *The Vanishing Peasant, Innovation and Change in French Agriculture* (Cambridge, Mass.: MIT Press, 1970), p. 213. The familial roots of group farming are deeply imbedded in French agrarian history. With reference to the period from roughly the 13th to the 15th century, Marc Bloch points to the practice of family farming in common in groups known as *frereschе*, or an association of brothers. "The children continued to live with their parents even after marriage and on their parents' death frequently remained together, sharing 'hearth and home', working and possessing the land in common." Marc Bloch, *French Rural History, An Essay on Its Basic Characteristics* (London: Routledge and Kegan Paul, 1966), pp. 164-65.

15. Nallet, et al., *Les Groupements Agricoles, Tome II*, novembre 1973, p. 38.

16. Ibid., p. 86.

Part VI: In Retrospect

Editor's Introduction

This final section contains two chapters and a postscript prepared after the conference. The Soviet experience is of special interest and significance because it was the first national system of collective farming--established nearly 50 years ago. Miller, Stuart, and Wädekin agreed to take a close look at this experience (with some East European comparisons) to provide a reasoned judgment on the likely suitability of the Soviet model for countries in the Third World.

Reed reviewed all the conference papers and additional literature on group-collective farming. The focus of his analysis is the special issues that may arise, and the problems which in any case have to be anticipated, when policies are formulated to establish a sub-sector of group farming (while also retaining a sub-sector of private-individual farming), with special reference to the less developed countries.

The brief postscript outlines some of the major issues on which further research is needed. The suggestion is made that field-level empirical investigations (as against macro-level aggregative studies) are now required to extend knowledge and in-depth understanding of the complex experiences which we call group farming.

15. Some Retrospective Thoughts on the Applicability of the Soviet Model of Collectivization

Robert F. Miller, Robert C. Stuart, Karl-Eugen Wädekin

A general conclusion from the conference reports and discussions of group farming throughout the world is the recognition that the preconditions and goals of the various existing forms of group, cooperative, or collective farming differ so much that it hardly seems justified to recommend any one such form for application in other countries or communities. But rejecting the wholesale applicability of group farming models is not to deny the possibility that lessons may be drawn from the various experiences. If nothing else, obvious mistakes, differences in preconditions, goals, and forms can be pointed out, as has been done to a certain degree by Galeski (Chapter 1 in the present volume). And there may also be positive lessons that can be learned from partial successes.

In the following paragraphs we shall attempt to assess the Communist collective farm as a model for less developed countries. Communist collective farming is the only form of group farming which, to the present, has been applied on a national scale. It is considered by some to be applicable, with local modifications, on a world-wide basis. At the same time, Communist collective farming encompasses ideological and socio-political goals which, at least over the long-term future, are placed above economic goals; in this respect, the Hutterite colonies and the *kibbutzim* are similar (Chapters 3 and 2 in the present volume), but these were not applied, nor meant to be applied, on a nation-wide, less still world-wide basis. If for no other reason, Communist experience deserves closer inspection as to its forms, goals, and successes or failures. Among its various forms, the Soviet *kolkhoz* is a paradigmatic case, because it served a pioneering and model function for other Communist countries. Moreover, next to the case of China, the Soviet kolkhoz system is most important numerically, and it has been applied in a country with vast regions spanning a wide range of climatic conditions and nationality groups of greatly differing historical, cultural, and economic backgrounds.

As is generally known, Soviet collectivization, as it was brought about historically, was a relative success in some respects, and a failure in others. It seems worthwhile to try and formulate some conclusions from this case, especially with regard to developing countries where group farming as a possible solution to pressing social and economic problems is of greater importance than in Western industrialized countries. Some similarities and dissimilarities to developing countries of the preconditions prevailing in the Soviet Union, past and present, notably the differing stage of economic development, have been spelled out in the paper on the Soviet kolkhoz, Chapter 4 in the present volume. Here we wish to point out some additional and more general lessons of Soviet and other related experiences with collective farming as they emerged initially and have developed since.

1. Large-scale collective farming on the Soviet model was intended to be, and is today, labor-saving and requires substantial amounts of machinery and other capital inputs. To the degree that these are provided, Soviet collective farming is capital-intensive. In its present-day form it has macro-economic advantages only where labor is a limiting factor, and capital is available--which was not the case in, say, Yugoslavia in the late 1940s or in the less developed countries today.

2. The economic and social costs of Soviet collectivization were tremendous. To a great extent these resulted from the forcible and rapid manner of its application, which in turn was associated with the need felt by the political leadership to overcome the resistance of a non-Communist peasant majority. Political defeat in this venture would have meant a serious, if not fatal, threat to the regime. The collectivizing East European countries after World War II were, with the exception of Yugoslavia, in a different position insofar as they could fall back on Soviet political and military assistance in cases of serious internal trouble, a situation which does not apply to--and is not desired by--most of the developing countries of today.

The above statement does not apply to China, where Soviet backing was of minor importance even before the political split between Moscow and Peking. Chinese collectivization may need closer inspection with an eye to the developing countries than does its Soviet antecedent because of its likely lower social and economic costs and its greater reliance on mobilization of surplus labor rather than on substitution of capital for labor. Unfortunately, in contrast to the Soviet case, much less objective information is available on Chinese collective farming, and almost no significant and truly representative information on details.

3. One of the consequences of the coercive nature of Soviet collectivization was the inhibiting effect it had on increases in labor productivity, manifested both in poor utilization of manual labor and in a lack of competent and careful handling of modern technology. Alienation, in the Marxian sense of the word, made it-

self felt to an extreme degree. This intensified the desire--especially among the younger generation--to leave the farms (a phenomenon certainly not unique to Soviet agriculture). Certain administrative regulations impeded this outflow, but did not prevent it, giving it, rather, a seriously distorted form.

4. The MTS (Machine and Tractor Stations) were a special feature of Soviet and other East European collective farming, instituted not as part of the collective farm but as an independent state-owned servicing station for groups of collective farms. Economies of scale, concentration of scarce machinery, specialized management, and the resulting possibility for maneuver (in addition to their function in providing economic and political control over farm activities) were among the obvious advantages of the MTS during the first three decades after the beginning of Soviet collectivization (they were abolished in 1958-59). As to their undoubted professional-educational role, it should be noted that in the end this benefited the rest of the economy more than agriculture since the majority of newly trained "mechanizers," for the most part young males, continued to leave the farms for nonagricultural employment. Precisely because of their training they had many opportunities to do so, legally or illegally. Apparently, when farming is considered an extremely undesirable occupation in comparison to others, an institution such as the MTS may turn out to be more a channel for depleting agriculture of some of its most talented labor resources than one for supplying it with the skills for modern production techniques.

5. Soviet collective farming increased output rather slowly (if allowance is made for the expansion of the cultivated area, and that mainly by means of state and not of collective farms). Prices to the state (monopsonist) procurement agencies were reduced (but less so to consumers) not by increased labor productivity but by forcibly depressed producer prices. Even today, with higher producer prices, costs of production per unit of output are often higher than in small-scale peasant farming. The reasons are manifold and hard to assess in detail. But there can be little doubt that, together with the alienation of the work force mentioned previously, central command planning of inputs, of allocation, and of the product mix played a major role. These were an intrinsic part of the Soviet system in general and of collective farming in particular. They are difficult to avoid in countries adopting the Soviet model, as East European experience has shown.

6. The development of nonagricultural rural activities, an important potential outlet for a labor surplus, was retarded rather than favored by the way in which Soviet collective farming was organized: exclusive concentration on agricultural production by administrative fiat did not offer encouragement for traditional rural crafts, nor for the creation of more modern nonagricultural employment opportunities in the countryside. Only in recent times has an effort been made to organize nonagricultural activities, mostly for construction and for processing of agricultural produce. On these questions, the Chinese case may be of special relevance and de-

serves close study since there an attempt has been made to bring small- and medium-sized industries and service organizations to the collectivized village.

7. Centralized Soviet command planning of investment had some merit in developing rural educational facilities and in advancing innovation in some of the agricultural production techniques. But the benefits of educational achievements to the rural areas were largely offset by the disproportionate and continuous outflow of the more educated and skilled as a consequence of living and working conditions in collective farms and of the priority accorded to urban industry. Moreover, there is no reason to suppose that educational facilities could not have been extended to the village as a community without the collective farm as its production unit. (There were, on the average, more of these facilities in state-farm villages.) The same holds true for other infrastructural measures, which under the Soviet regime were often, but not always, carried out by the collective farms.

8. The combination of large-scale collective and mini-scale private production, of collective labor organization in brigades and smaller subgroups and of family labor on plots, may turn out to be a promising approach for cooperative farming in less developed countries, provided that a prospering private sector is not treated as a threat to the overall economic and political order. The absorption of an agricultural labor surplus by private plot farming may be as important as the parallel increase of labor productivity on consolidated large fields and livestock operations. The latter more capital-intensive, cooperative farm activities can serve as vehicles for the general improvement of agriculture and the spread of progressive techniques.

9. The main economic disadvantage of a centralized and coercive system is the inflexibility and the resulting potential misallocation of resources and lack of incentives. The problem is worsened by the distortions of information flows from lower levels. Workers, managers, and lower administrative officials consider it more important to report fulfillment of plans and orders than to report the true state of affairs and the need for adaptive changes to meet unique local conditions. In the Soviet context this was balanced, though not completely, by the party organization which was and is an instrument not only of control from above but also of information feedback from lower levels. In the absence of such an organization, the disadvantages of a rigid and centralized system are likely to be even more pronounced than they turned out to be in Soviet collectivized agriculture.

10. The rigidity of the overall Soviet and the collective farm system also extends to the relationship between the farm and its sub-units, mainly for reasons of control and produce extraction, but also because of the lack of management and specialist cadres. Where such constraints are absent, an organization might be considered where the central farm management fulfills the functions of planning and administration, of procurement on behalf of the state organizations, of supplying nonlocal inputs (machinery,

quality seed, processed feed, building material, etc.), and leaves to the sub-units freedom of decision-making with regard to labor organization and incentives, adaptation to local and seasonal conditions, choice of the most advantageous inputs, enterprise combinations, etc. In this way, economies of scale might be combined with the advantages of local, adaptive management. Recent Soviet experimentation with the so-called self-accountancy of sub-units, although not a total success under the rigidities still prevailing, shows that this has been recognized as an urgent problem.

11. Soviet and all other Communist collectivization managed rather successfully to separate land reform from collectivization. They were carried out in two different stages. The first stage, that of land reform, helped to win over the peasants for the new political regime, and thereby to stabilize it. At that time, officials in most East European countries expressly denied even an intention to collectivize in the foreseeable future the newly reinforced peasant small-farm structure. The Communist monopoly of power enabled the political leaders to conceal and postpone their long-term goal of collectivization without fear that at a later stage they would no longer be in a political position to attain it. (In the one case, Soviet Kazakhstan, where collectivization was combined with another transformation of rural life--the settlement or partial settlement of nomads--human losses were immense and surpassed even the huge losses in other parts of the Soviet Union.)

The above strategy has relevance where an existing peasant structure is to be transformed into a collectivized one. The psychological situation may be much different where collective farms are organized on hitherto unused land or on former large estates run with hired labor (not with small tenants).

12. It is very unlikely that, without the political monopoly of the Communist party, Soviet (and other East European and Asian) collectivization could have been successfully concluded. Government interference, its direction, degree, and instrumentalities were largely determined and made possible by this monopoly and its ideological legitimization. There is a notable correlation between ideology, coercion, and centralized decision-making.

13. For the long term, the goal of Soviet collectivization was only in part an economic one. Rather it was one of social policy and ideology directed toward creating the preconditions for the construction of the Communist society of the future. This provided a legitimization for collectivizing, at nearly prohibitive economic and social cost, against the peasants' volition. Such legitimization would be absent from a group farming venture based on predominantly economic considerations and carried out in a more or less democratic setting. Measured by its political and social goals, Soviet collectivization may be considered a success in spite of its obvious disadvantages. With different goals, the yardstick for success must also be different. Absence of Communist goals and of absolute Communist party power precludes an imitation of the Soviet model of collectivization in its Stalinist or similar form; lack of tremendous amounts of capital for agriculture makes imitation of

the post-Stalin modernized kolkhoz unattainable for developing countries, except for isolated cases within a predominantly individual peasant-farm structure.

14. The Yugoslav case (detailed in Chapter 7 in the present volume) provides substantial confirmation of these conclusions and suggests some additional related lessons. Early failures with forced collectivization, which closely replicated the Soviet experience, prompted the Yugoslav leaders to abandon the kolkhoz in favor of less coercive forms of cooperative and group farming. The resulting general farmer cooperatives were successful to the extent that the government was willing to subsidize their popular, but economically marginal, service functions. But the ideologically motivated formal restrictions on genuine autonomy for the cooperatives resulted in a low level of peasant commitment. When economic considerations impelled a shift in governmental resources away from the cooperatives toward other, more highly integrated types of socialist agricultural enterprises (combines and estates), the cooperatives tended to disintegrate, leaving the private and socialized farm sectors in a de facto adversary relationship under which the peasants became increasingly alienated from the system as a whole.

15. The primary lesson of the Yugoslav experience is that genuine cooperatives have an important role in the social and economic transformation of a traditional peasant agriculture, and that this role is worthy of substantial governmental assistance, material as well as moral. However, such assistance must be limited in terms of goals, expectations, and methods of administration. Care must be taken not to impose social and economic objectives that are too widely at variance with peasant aspirations. Above all, officially sponsored cooperatives must be seen by the peasants to be consonant with their own economic and social interests and not primarily a vehicle for their contribution (or sacrifice) to an overriding "higher" interest.

* * *

At the most general level it seems to us that those less developed countries which are able to avoid total ideological commitment to either capitalism or Soviet- (or Chinese-) style Marxism-Leninism will be in a position to choose the successful features while avoiding serious failures of Soviet-type collective farming. The greatest enemy to successful group farming in Soviet-type economies has been the "either/or" mentality engendered by Marxism-Leninism. Useful devices and institutions, such as central planning, state agricultural machinery pools (MTS), state-controlled procurements, etc., were substantially vitiated in practice by enforced rigidities and conformity in application, which precluded meaningful accommodation to local circumstances and traditions--all in the name of an ideologically determined ideal. To be sure, not every institution or device may be borrowed from a specific social, political, or cultural setting--which is why the above fifteen points have tended to emphasize negative features requiring special caution. Nevertheless, there are obviously some features in socialist group farming experience that will be of practical interest for developing countries.

16. Introducing Group Farming in Less Developed Countries: Some Issues

Edward P. Reed

INTRODUCTION

Until recently group farming in the form of agricultural production cooperatives was found on any large scale only in socialist countries operating within a totally directive economic environment, or among largely ideologically or religiously motivated groups, such as the Israeli settlers or the Hutterites. However, in the last decade or so a number of less developed countries have initiated large-scale cooperative production ventures in an attempt to solve their pressing rural and agricultural problems. These countries have either consciously sought direction from the existing socialist or ideologically based models, or attempted to create new, indigenous organizational forms. In either case, the difficulty of organizing and sustaining collective agricultural production in the context of most non-Communist developing countries has become increasingly apparent.

In most developing countries revolutionary change encompassing all political and economic relations is not envisioned by the ruling groups. In most cases the approach to change and development is incremental and uneven, and a total consensus is notably absent. There is no comprehensive, long-range development strategy. Varying official and unofficial approaches are tried in an attempt to cope with the rural and agricultural problems, and with minimal room for maneuver, the less successful approaches soon become apparent and are eventually discarded. Introducing what is in most cases a radically different economic and social organizational form among poor peasants, in the context of a mixed and usually struggling economy without the driving force of either political coercion or ideological fervor, creates a set of problems that requires special analysis.

This paper attempts to identify and examine in a systematic way the key organizational problems that arise on both the macro and the micro levels in introducing and maintaining a collective organizational structure in agriculture in the context of competing approaches, competing political and economic pressures, and competing social demands characteristic of most LDCs. These are some of the critical problems that must be successfully dealt with if the potential benefits of collective agriculture seen by LDC policy-makers are to be realized.

GROUP FARMING IN THE CONTEXT OF DEVELOPING COUNTRIES

The type of group farming referred to here is what is commonly called a production cooperative.[1] This means that the land and major capital items are held in joint ownership by the farm workers themselves, the bulk of the land is collectively cultivated, and any profits from the enterprise are shared by the cooperative members. Ideally, as joint owners members of production cooperatives participate in the decision-making process concerning all aspects of production, distribution, and investment. Thus, this type of group farm is distinguished from the state farm, where workers are wage employees of the state, and forms of cooperation where farmers cultivate their individual plots while carrying out some operations jointly.

In Galeski's typology (Chapter 1 in the present volume) these farms would usually come under Type Two--those founded by peasants to better utilize new land received in agrarian reform programs. However, the state plays an important role in promoting these production cooperatives so that characteristics found in Galeski's Type Three might also be present.

A general classification of the types of group farms found in less developed countries can also be made based on the preexisting tenure arrangements that are the objects of institutional reform:

1) Group production on expropriated large agricultural estates by the former resident workers. This type may include an arrangement whereby landless laborers and small owner-operators near the estate are incorporated into some relationship with the group farm in order to more widely distribute production benefits.

2) Group production on newly acquired or reclaimed lands, either by smallholders who continue to cultivate their original plots privately or by beneficiaries of colonization programs.

3) Group production on lands that are held in a traditional form of communal ownership but were previously cultivated privately (the African case).

4) Group production by former private holders on land consolidated from their owner-operated farms.

Particular problems related to both the external and the internal issues discussed here will vary depending on which of these types of group farming is involved. An attempt will be made to highlight the most important differences as each issue is discussed.

In most developing countries the introduction of production cooperatives represents a new beginning, a clear divergence from former agricultural structures. It is almost always a new experience for the participants themselves, and frequently for the government bureaucracy as well. It is likely, then, that this transition involving sharp changes in political, economic, and social organization and habits will not be smooth. A positive commitment to anticipate and meet problems as they arise is probably the most important element for success. This, of course, presupposes not only a commitment by the government to this transition, but also the existence of the government's capacity to monitor the transition and take proper action when required.

Although most peasant societies have some tradition of limited cooperation, and some of the more successful farmers may expand the range of cooperation on their own initiative, there are very few instances where full-fledged production cooperatives have been formed spontaneously by peasants themselves. Therefore, in the situation we are discussing the move toward establishing group farms is usually made at the initiation of the government. However, since improvement in the lives of the participants and their ultimate acceptance of the change are critical factors for achieving success, the degree to which pressure and overt coercion can be effectively employed is clearly limited.

It is the broad range of rural problems facing the Third World that has led some developing countries to attempt a collective approach to agriculture. It is precisely because group farming involves multidimensional change that policy-makers view it as a means of dealing with their complex political, social, and economic problems in rural areas.

Though abstract ideological commitments are often read into a turn to group farming, it is probably more often a case of searching for a broad enough framework in which a comprehensive approach to these interrelated problems can be made. As one conference participant observed, what is seen as ideology is often just the choice of a long-range development strategy with social and political as well as economic elements, as opposed to short-range economic goals. Ideally, then, there need be no conflict between "practical" and "ideological" considerations. In practice, however, the task is often to convince the non-policy-making participants of the validity of these broad, long-range goals.

Among the social expectations that LDC policy-makers usually hold for collective agricultural organization is greater equality in distribution within the rural sector--through pooling resources and sharing returns at least partially on the basis of labor input. Employment opportunities could be expanded in some cases by more intensive cultivation; but prospects are also improved for utilizing surplus labor in capital-expanding projects and in small agriculture-related industries. Finally, medical, educational, and cultural facilities, as well as farm input and marketing services, could be provided more easily and more uniformly to collectively grouped rural populations.

Government leaders may have political expectations as well. These may be seen as the possibility for an expansion in the degree of direct peasant participation in local and national affairs, and the development of a more effective peasant political movement. Conversely, the motivation might be a desire to align a potential peasant organization to the government party or even to extend government control over emerging peasant political activity. Whichever role is envisioned, the collectivization process may be viewed as a means of politicizing the peasantry or of restraining its independent political action.

Policy-makers evidently have positive economic expectations when they introduce group farming. They are convinced that in certain circumstances collective organization can raise output through exploitation of economies of scale, more rational allocation of labor and material resources, facilitating the introduction of new technology, and more effective channeling of the marketable surpluses. Without going into the many qualifications that a detailed analysis would require, there is evidence that under certain clearly defined circumstances these factors can have a positive economic impact on agricultural production, and that the level of agricultural output attained under a collective organizational structure can equal or surpass that attained in the same physical setting under other organizational structures.

A statement requiring few qualifications, however, is that even a limited realization of these social, political, and economic expectations depends primarily on the approach taken in dealing with the external and internal organizational issues that are inherent in a system of group farming.

MACRO-EXTERNAL ISSUES

Types of Farming

While experiences of the socialist countries have demonstrated that it is possible to establish collective farming in almost any agricultural environment, in the context of the mixed economy of most developing countries it is important to note that some types of farming seem to lend themselves more readily to collective production while others are more easily adapted to production on small plots. Furthermore, it is of central importance that the particular crop or crops around which the collective is formed are sufficiently rewarding financially so as to make collective activity competitive with private production both in the private sub-sector and on the individual plots.

Experience from several LDCs (e.g., Peru, Dahomey) gives some indication of the type of farming system that presents fewer difficulties in introducing collective production. Generally, systems that depend on centralized management, large scale, division of labor, and extensive infrastructure are more easily adapted to group farming. A type of farming that relies on a complex water-supply system, for example, is more logically adapted to large-scale collective production than to small-plot production. Likewise, highly mechanized systems of cultivation have been more easily converted to group farming. Finally, so-called industrial crops--those channeled directly into a particular processing and marketing system--fall into this category.

On the other hand, when crops require more intensive cultivation or a great deal of individual care, or when they are raised mainly for direct consumption, collectivization may be expected to meet with much greater difficulty. These types of farming are probably better suited to small-scale production on individual plots.

These observations are not meant to imply that group farming should only be initiated under the conditions listed above, but are intended to indicate situations in which organizational and financial problems have been found to be more easily dealt with. In fact, other considerations that may be primary in LDCs might mitigate against some of these factors considered favorable for group farming. For example, with labor absorption a major consideration in most LDCs, mechanization on a large scale would not be an appropriate policy even though it might make the transition to collective production smoother.

Therefore, LDC policy-makers should be aware that conditions which predominate in their agricultural sector will usually not coincide with those conditions under which a transition to group farming has been found to be the least difficult. Such considerations might provide a guide as to where to begin if a limited or incremental approach to group farming is contemplated, and can also indicate where organizational problems are likely to be more severe.

Participant Mobilization

Since at least acquiescence on the part of participants, if not enthusiastic acceptance, is a necessity if a shift to group farming is to be successful, both the basis and the style of mobilizing peasants are of critical importance for the introduction of the system and its later internal functioning. There has been some debate over whether or not socio-cultural factors are significant in the successful mobilization of peasants for group farming systems; that is, whether a group of peasant farmers might be more receptive to the introduction of collective organization because of some tradition of cooperation. Though almost all case studies of group farming begin with an examination of the traditional base, the consensus seems to be that while cooperative traditions may play some role (they at least indicate farmer interdependence and may make peasants more psychologically receptive), they are not decisive factors. The qualitative difference between independent peasant farmers cooperating in some tasks and the collective organization of all tasks, the pooling of all assets, and the sharing of all risks and profits is very great. And there are very few cases reported where peasants spontaneously moved from the former to the latter system.

Nevertheless, cooperative traditions have been positively used in at least two ways: first, as the basis for a gradual, though directed, move from simple cooperation to more integrated forms of cooperation, to full collectivization (as in China); and, secondly, as important cultural and institutional symbols on which to base the inauguration of more far-reaching change (as in Tanzania).

The relationship between the economic condition of the peasants and their receptivity to collective production is not clear. On the one hand, peasants barely subsisting in conditions of severe poverty might be more open to radical change if the rewards are fairly certain. However, having only a slim margin for survival, they may be less willing and less able to take real risks. In addition, the poorest of the rural poor are also usually the most suspicious of meddling outsiders. In Tanzania, for example, the "bait" of *Ujamaa* (improved services, public health, schools) has been received best in the poorest sections of the country; however, the transition to totally integrated group farming in such areas has proved very difficult and is dependent on convincing demonstrations of its benefits for the peasants.

The preexisting tenure system is obviously an important factor in mobilizing peasants for group farming. The least difficulty has been encountered with resident laborers on established, market-oriented estates (type one enterprises in the above classification) since the change in this situation may not, at least initially, require radical shifts in work organization or management. Likewise, resettlement of landless workers in areas of newly developed land, or the acquisition and group farming of additional land by private holders (type two) have not met with great barriers. The African case, involving transition from traditional communal ownership to actual group cultivation, is an intermediate situation about which there is little decisive evidence.

For obvious reasons, however, the greatest difficulty has been in consolidating private landholdings for group production (type four) and this approach has been attempted only rarely in LDCs to date. One difficulty is that the private holders most likely to be receptive to some forms of cooperation are those living on sub-marginal plots. However, consolidation of these plots might be difficult if they are widely dispersed, and the result may be a very high man-land ratio. In addition, such organization would be building on the weakest and poorest rural group and would require greater government support. Some attempts have been made to overcome this problem by institutionally linking sub-marginal farms to the large, collectivized estates though these attempts do not seem to have met with great success. In such situations perhaps policymakers might consider a less completely collective approach to cooperation.

A final and extremely important factor is the style of mobilization used by the government. A common approach has been a mixture of exhortation, pressure, and positive and negative material incentives. But the most important key to the success of this mixed approach (and to minimizing negative pressures) is whether the mobilization is carried out on a truly participatory or on a bureaucratic basis. There are clear indications that the more peasants participate directly in the preparation for and the actual implementation of the change, the more effective will be their participation and interaction in the established collective and the greater chance there will be for long-term success. However, since participation requires a decentralized and responsive bureaucracy and a great deal of interaction with and training of local farmers, there is a strong tendency toward a top-down bureaucratic implementation. Experience shows that there is less chance of success with such an approach than with an incentive-based, voluntary one.

Competing Group Objectives

As stated earlier, LDC policy-makers usually turn to collective agricultural production with a view to attaining a complex of broad, long-range economic, political, and social objectives. These objectives for the agricultural sector must be closely linked with overall national goals. But at the other end, in order to be successfully implemented, they must be in harmony with the goals of the communities and individuals that participate in the new organizational system and must also take into consideration the objectives of the nonparticipant groups. Clearly these are areas where serious conflicts can arise.

A broad area of conflict can arise in the allocation of resources within the agricultural sector. Where both a group farm and a private farm sub-sector coexist and where the government is attempting to induce the development of collective units, there may be a tendency to favor the group farms in the allocation of inputs and other government services. If a mixed agricultural sector is envisioned for some time to come, extension services and other support must continue to flow to the private farms as well as to the group farms. In addition, the government's collectivization policy must be clearly spelled out so that both sub-sectors have a reasonable feeling of security with respect to future organization. Otherwise, capital investment and productivity are likely to fall, at least in the private sector, thus offsetting any gains in the socialized sector.

Another area of potential conflict is in the relationship between the group farms and the government. The motivating force behind a group of peasants pooling their assets and efforts in agricultural production could be the promise it offered for improving their families' levels of living and giving them a greater chance of participating in educational and cultural opportunities of the society. The goal of individual gain does not disappear but is linked with the gain of the cooperating group. Such groups will continue to seek their own advancement in competition with other groups and individuals in the society.

Meanwhile, the goals motivating the government's promotion of group farming are usually broader than the goals of an individual collective. Conflict may arise when the government expects to exercise control over the crops to be grown and the channels through which they will be marketed. In Chile, for example, when the government attempted to secure loan repayment by requiring *asentamientos* to channel crops through official markets, members by-passed these channels to take advantage of higher prices in unofficial markets (this was in the context of extremely rapid inflation). Thus, while the group pursued its profit-maximizing goals in a rational way from its point of view, the government's objectives of collection on loans and greater control over the market surplus were frustrated.

Likewise, the government's goal of expanding employment opportunities and equalizing distribution within the whole rural sector may well conflict with the individual collective's desire to exclude new participants since that would result in a smaller share per member. Because of this, the most needy of the rural population--the landless laborers--may be excluded from the benefits of collective systems.

These conflicts are closely related to the question of the degree of autonomy allowed to the group farm as opposed to the degree of direct or indirect control exerted over it by the government. A delicate balance must be achieved between these two elements, since absolute control would surely suffocate and destroy the required peasant acceptance and participation in the system, while absolute autonomy would make it difficult to integrate the collective into the broader economic and social goals envisioned by the government.

The exact mixture of these elements will depend on circumstances in individual cases. However, it can be generalized that peasants, whether as individuals or collectives, will almost always respond rationally for their own benefit to the economic environment. Therefore, the more closely positive economic incentives for the group farm are aligned with the general economic goals of the society, the less need there will be for use of direct control mechanisms. Likewise, the more the government deals with group farms in a way which strengthens their autonomy, the more solid the collective institutional base is likely to become.

Relations with External Institutions

Perhaps the most difficult issue in initiating a group farming enterprise in a developing country is the relationship between the collective and the outside economic and political institutions. Given the coexistence of what is fundamentally a socialist sector with the usually more dominant private farm sector, both within the same largely market-oriented macro-economic environment, some peculiar problems arise in integrating the group farm into the overall system. These issues are critical since the way in which they are resolved affects directly the internal viability of the enterprise.

In most LDCs the government has had an important role in setting up the collectives and it usually provides special assistance and concessions in order to get them established and functioning successfully. Besides being an inducement to peasant participation, these special considerations can be important because the new group farms are often deficient in the necessary managerial and technical skills required to operate a large-scale agricultural enterprise. In addition, especially in the case of new settlements, private supply and marketing arrangements may not exist on the necessary scale.

Experience indicates, however, that in providing these special concessions and in placing a major portion of the risk on the government, serious problems may arise. Instead of promoting a spirit of cooperative responsibility, there is the danger that government assistance may result in peasant dependency and an attitude that positive results are a concern of the government and not the participants. The goal of maximizing returns and minimizing costs in accordance with national social costs and benefits may be seriously jeopardized.

Several steps have been proposed as means of avoiding these pitfalls.[2] First, there is the critical need for support of the socialized sector on the macro level. This assumes the existence of an efficient planning mechanism at the national level with the capacity to collect and interpret accurately reported information from the local operating level. Into this planning framework an efficient and rational tax system, a responsive price policy, and a rational credit and service system can be incorporated. Second, tying all credit extended to group farms directly to the actual and potential productive resources is of key importance, otherwise the possibility of repayment becomes remote and continual dependence is assured. In some cases, as mentioned above, repayment of credit has been tied to marketing requirements. For this to succeed requires careful and efficient bureaucratic operations and perhaps an incentive price policy to prevent selling in parallel markets which are usually available in the private sector. In the short run it might be necessary to accept a transfer of resources into the collectives through incentive pricing in order to place the credit-market relationship on a sound institutional footing.

Finally, some experts have recommended that when special consumption and social subsidies are deemed desirable to overcome initial transformation difficulties or to compensate for special emergencies, these subsidies should be kept strictly separate from the agricultural credit and marketing arrangement in order not to jeopardize its rational basis. The requirement that loans for productive purposes be repaid promptly and in full is a critical requirement if the individual enterprises are to develop the autonomy necessary for efficient and rational economic functioning, and if the government is to have the resources to carry out other investment and welfare aspects of its rural programs.

Group farms may form regional or even national cooperative organizations to carry out some of these economic functions. Ideally, such organizations can deal with both government and private enterprises in funneling inputs to member group farms and marketing their output. These organizations can also become instrumental in setting up agro-related industries which can serve both to process agricultural produce and to absorb surplus labor from member collectives. Finally, such an institution could be an instrument by which inequalities, that almost always arise between group farms based on differing resource bases and opportunity structures, could be ameliorated.

However, such inter-collective organizations have not functioned in a very successful manner. The primary reason seems to be political rather than economic. When a government is more interested in monopolizing the political allegiance of peasant participants, or in strengthening its position vis à vis an opposition political group, than in fostering the development of an independent peasant political movement, it will probably find this more easily accomplished when group farms are isolated and dealt with on an individual basis. In some cases governments, on both national and regional levels, have been somewhat fearful of allowing inter-collective organizations to become well established, or they have taken measures to ensure that, if established, they are tightly controlled. This has diminished the real benefits that could be derived by the member farms.

In Tanzania, for example, when several well-established collective Ujamaa villages in the Ruvuma region spontaneously formed an inter-village Development Association, it was viewed as a threat by the government party and forcibly dissolved. In Peru, on the other hand, the government has attempted to initiate regional cooperative organizations. These have had almost no success because of the suspicion of the primary collectives that they would become instruments of government control, and because of opposition to the idea of redistribution among collectives.

Another potential pitfall is that the political importance of the reformed or socialized sector may grow to such an extent that a rational economic relationship between the primary collectives and the government may be jeopardized. In Mexico during the early *Ejido* period and more recently in Chile, government loans continued to flow into the reform enterprises even though repayment was not forthcoming. The end result in both cases was that the government was supporting economically stagnant collectives in exchange for critical political support.

A group farm, therefore, is in a delicate position vis à vis the external organizations and institutions with which it must relate. Unlike a family farm which has more autonomy in setting its objectives and which can survive long periods of neglect, the group farm ordinarily is in a more dependent position. It is usually established on outside initiative with a complex of goals envisioned by several different groups, and its continued operation as a collective enterprise will depend on the attention and involvement of outside institutions.

The delicate balance required is not one that is likely to be achieved easily or quickly since radical changes in institutional as well as individual methods of operation and interaction are involved. While these considerations do not deny the potential benefits and relevance of group farming in the LDC setting, they should serve to caution policy-makers not to expect group farming to be a rapid and automatic solution to the multitude of rural and agricultural problems facing them.

MICRO-INTERNAL ISSUES

There is obviously an interrelation between the way in which a group farm fits into the macro social, political, and economic setting of a country and the way in which the collective is organized internally. For example, the way in which the autonomy-control dilemma is resolved will have a direct impact on internal issues. However, it is also true that the problems and challenges related to the internal organization of group production--the decision-making process, organization of labor, distribution, incentives--will be relatively similar in almost all circumstances. Therefore, in this area, the experience of others is probably more instructive than in the area of macro integration discussed above.

Size of Operation

It is very difficult to say anything specific about the optimal size of a group farm. Yet size of unit will have an important impact on many internal issues related to the rational and efficient operation of the farm from an economic, administrative as well as social point of view.

Otto Schiller notes that economic/managerial considerations on the one hand and social considerations on the other involve different factors that may not always coincide.[3] With regard to any rational and manageable agricultural operation, Schiller points out that there is a lower limit determined by the nature of the working process itself, and an upper limit determined by managerial requirements. In other words, according to the peculiarities of each kind of operation, the diseconomies of small as well as large scale should be avoided.

In a group farming situation social considerations such as the need for cooperative working arrangements and participatory decision-making will affect the optimum number of families the group can include. The factors determining the optimal number will be such things as the homogeneity of the group and the size of important communal institutions. The size that these considerations indicate as optimal may not coincide with purely economic considerations, so that a trade-off may be necessary in order to arrive at the most desirable and effective size.

Not only can little be said about the specifics of size that would apply to all cases, even in particular situations it is usually not apparent *a priori* what the optimal size might be. In most cases a period of experimentation and adjustment is necessary. China is an instructive case in point. There, over a period of a decade, the size of operation expanded in leaps from small community cooperatives to the giant, almost county-sized communes; and then because of both economic and social difficulties most farming operations were adjusted back down to village-sized groups, while those integrative operations most suitable to larger units were maintained at the higher levels. Thus, a three-tier arrangement has emerged and appears to have attained at least a short-run optimality that secures some of the economies of larger scale without sacrificing the social and economic benefits of small group operations in a basically peasant agricultural setting.

Those planning to establish group farming in other countries may find it instructive, on this issue of size, to examine the Chinese experience more closely. In some cases it may be possible to attain a similar tier arrangement--if so desired--through inter-cooperative organizations, though we have seen some of the difficulties encountered with such institutions.

Management and Decision-Making

Group farming of the type we are discussing presumes that the farmer participants as co-owners share in both labor and management responsibilities. Ideally, then, members should participate to the widest extent possible in the decision-making function. An effective participatory decision-making process is of central importance if a "we-they" gap between members and managers is to be avoided and member morale sustained.

Such a decision-making process assumes, of course, that the group farm has a sufficient degree of autonomy in the areas of production, distribution, and investment to provide a range in which meaningful decisions can be taken at the farm level. However, it is also obvious that making such decisions affecting the overall functioning of a large-scale agricultural operation, and relating the farm to external institutions and economic and political forces, requires a high degree of skill and sophistication, as well as time. As Galeski has shown, while market-oriented farmers might be able meaningfully to participate in such a management process, the more common situation in developing countries with interactive economic systems[4] involves former peasants whose experience has been with farming near the subsistence level. Transforming themselves into collective entrepreneurs when both commercial agriculture and collective life are entirely new to them may be nearly impossible. Horton has described some of the severe managerial problems encountered in highland Peru under these very circumstances.

Therefore, while all group farms have some form of organized representative participation in management decision-making, it has often been necessary for the group to hire, or the government to appoint, an outside professional manager with the necessary skills to handle the complex job. This becomes a very delicate matter which can easily degenerate into a sharp division, marked by mutual suspicion, between the manager and the member-workers.

It is perhaps possible, though by no means easy, to avoid this outcome if there are constant reminders of the overall, long-term goal of maximizing participation in management by the members and the development of internal management and leadership. If an outside manager is deemed desirable in the beginning, it is very important that the members have a prominent role in the selection process and have the right and authority to remove him at any time. From the very start, preparation and training of peasant members for participation in the planning and management of the farm operations should be emphasized. Parallel with this should be a step-by-step scheduled increase in active participation, as skills and experience are gained. Government advice and assistance can be maintained through extension-type services, but within a relationship based on increasing group farm autonomy.

Of critical importance in this process is the attitude of government employees who deal with the peasants. It is quite clear from experience that mobilization and organization with maximum peasant participation at every step is directly related to the development of a viable participatory framework in the long run. While the macro-environmental situation of China is very different from most mixed-economy developing countries, the Chinese experience in participatory mobilization and participatory management at the team level within the framework of strong support and assistance from higher units is very instructive. One especially relevant point is that older peasants, though without modern technical knowledge, have a revered position in the management process due to their reservoir of valuable experience. Peasants usually know their environment and their land and its capabilities much more intimately than any outsiders. Their knowledge in this area can be incorporated into the management process from the very beginning, thus forming a strong base for further developing effective and responsive cooperative decision-making.

Organization of Labor

In a peasant society the collective organization of work may present special problems. Of course, if it is a matter of collectivizing workers on an existing large estate, these problems may be less pronounced since work may be organized essentially as before. But even here work organization may have to be altered in order to promote a greater degree of worker participation and interaction in the decision-making process. However, if peasant families who previously farmed individually are grouped together, work habits will usually have to be changed in more fundamental ways.

In most cases it has been found highly desirable to take into consideration social traditions when collectivizing peasant labor. Usually group farm members are divided into smaller teams to carry out specific tasks. These teams, especially in the earlier stages, may be based on natural social (family, kinship, or neighborhood) divisions, where such divisions do not threaten the overall integrity of the collective. In Tanzania the ten-house cell has served this function, while in China the production teams are usually based on a natural village or neighborhood, while work groups within the teams are often based on kinship structures. Such arrangements give greater stability to the system and assure some forms of social continuity within the context of broader social and economic change. Details with respect to the division of labor and tasks can be worked out by these natural groupings on the basis of traditional and habitual arrangements.

The division of tasks in the overall production process among the smaller work groups, and within them among the individual members, is an area where problems and strong feelings can easily arise. Even when the degree of member participation is high in the overall management process, there is a point where some central though responsible authority will have to allot various tasks which are unequal in their degree of difficulty and desirability, and possibly in remuneration as well. In Tanzania an attempt has been made to minimize friction by rotating this position of authority among the members of the smaller groups. Likewise, it may be desirable to rotate ordinary tasks on an established schedule among the various sub-groups, and within these among individuals. If such rotation is organized so as not to be at the expense of developing necessary expertise, it can be a means of broadening the skills of all members as well as overcoming the boredom that arises with routine tasks.

Another issue that must be considered in the area of work organization is the hiring of outside laborers to supplement the member work force during peak periods. In developing countries where group farms exist in the midst of a private agricultural sector with small farms and landless laborers, there is almost always a supply of laborers available. Besides the larger (macro) issue of involving these laborers in an agrarian reform through group farming, there is also the micro issue of integrating these outside workers into the collective structure in a way that does not threaten its functioning.

Group farm members will usually be willing to hire outside laborers if they are assured that this will not give outsiders a claim to membership or a share in production profits. An obvious danger here is a build-up of animosity between the members and the outsiders which might jeopardize smooth operation of the farm. Another danger, however, is that member farmers might shift more and more of the heavy labor to outside hired workers while supporting themselves on shared profits and/or more remunerative off-farm employment.

The solution to this problem must be sought on both the macro and the micro levels. (The experiments in Peru have already been noted.) Even if the overall problem is not fully resolved it would probably be possible to minimize the dangers involved by establishing some direct legal restraints in this area and by organizing economic incentives in such a way as to maximize member participation in farm labor. In addition, some kind of arbitration mechanism might be established by the government to mediate potential conflicts between group farms and outside laborers.

Remuneration and Incentives

In any group farming situation it is important that the system of remuneration reinforce the collective functioning of the institution, otherwise there is a constant tension between the ideals of the organization and the interests of the individual members. In the context of most LDCs, however, this requirement is of special and critical importance. Though some degree of ideological commitment or simply enthusiasm for the group endeavor may exist among the peasants, experience clearly indicates that this motivation is not sufficient to hold a group together. The system of reward and the distribution of output must be designed to provide rational economic incentives for the collective production of the bulk of farm output. Obviously, then, the system of remuneration must be closely coordinated with the organization and performance of labor.

The concept of a production cooperative usually implies that remuneration is based solely on the quantity, and perhaps quality, of labor performed. Where the group farm is formed by the "socialization" of an expropriated estate, or by resettlement in new areas, this approach may be easier to institute than in the case of pooling private farms in a settled area. In the latter case, the question arises of whether or not remuneration, at least for a time, should be provided for the land contributed by individuals.

Though political factors may be decisive in resolving this question (e.g., larger landholders may be able to "soften" a reform by maintaining the capitalist factor), its resolution will directly affect the economic functioning of the collective. The incentive given to labor is reduced to the extent that the capital element is rewarded, since all remuneration comes from the same limited output. Also, those receiving a relatively large capitalist share may refrain from labor or hire outsiders to work for them, further undermining the objectives of the system.

It is interesting to note, however, that even in the socialization of agriculture in most Communist countries the capital contribution has been taken into account, at least during a period of transition. In the Soviet Union this period passed quickly, while in China it existed for several years and the payments thus made were usually considered complete compensation for the land and capital contributed. Such a transition period may also be necessary in developing countries since the participation of the better-off farmers with greater resources and skills is important. However, in order to avoid the pitfalls mentioned above it is probably desirable legally to limit the percentage of net profits distributed on the basis of land contributed.

The importance of linking reward to labor performed has been universally recognized. The problem has been in implementing such a system given the peculiar difficulties of group farming. The great multitude of tasks associated with large-scale agriculture vary according to the degree of difficulty, desirability, and time required. Likewise, individual workers do not perform the same tasks with the same degree of perfection or in the same length of time. Therefore, the first problem in rewarding labor is to establish an acceptable set of criteria for valuing it. This is usually approached through some kind of point or norm system, where specific tasks are valued according to their difficulty or skill requirements. Points may be allotted on a piecework basis (based on the completion of the assigned task within certain quality standards), or on a time basis (e.g., number of days spent on a certain task). The net surplus at the end of the production year is then distributed according to the points accumulated by each member.

Since the value of a work point depends on the total number of points earned by the group as well as the net distributable profit of the enterprise, the value of labor is not known until the end of the production year. Therefore, in most cases an advance payment is given to each worker on a monthly basis during the year and this amount is subtracted from the member's work-point earnings calculated at the end of the year. The advance, then, looks very much like a salary and is often the same for each worker. In many cases it is set according to the existing wage level for hired workers in the local area.

It is clear that this system, which represents a radical shift in the work and income habits of most peasants, can run into serious difficulties. Problems are often encountered with the quality of work performed and the degree of innovativeness. Quality control can be built into the point system, but this implies close inspection and supervision which is not always desirable or possible. Quality standards can be more easily maintained with piecework, though many tasks cannot be neatly divided and such divisions may run counter to the social goals of participation and interaction envisioned for the enterprise. Finally, if the advance payment is not directly related to collective production members can easily come to view their situation as one of laborers bargaining for wages with an alienated management rather than that of actually sharing in the risk-taking function of management themselves.

Procedures for providing incentives and remuneration will probably always be a source of tension in a group farming situation, and will be a special danger in the early stages. A satisfactory resolution can probably be attained only in the long run by successful institutionalization of economic and social elements in the system. If the great bulk of individual remuneration comes from collective production, the connection between individual effort and reward will be clearer. Likewise, as the decision-making process becomes more participatory and members gain a clearer understanding of the interrelation between the various tasks and functions, individual and collective interests will seem less in conflict.

Especially in the early years, therefore, the existence of a net distributable surplus is of critical importance in the successful institutionalization of the group farm. If there is no surplus the connection between work performed and reward is broken (or never established) and the advance payment becomes simply a wage unconnected with the quality or even quantity of collective labor. This was observed to be one of the most serious problems in the Chilean group farming institutions.

A successful production year can never be guaranteed, and more than the usual problems are likely in a period of institutional transformation. In the early years strong government technical support and the constant assistance of an effective extension service are important. An added advantage would be the introduction of new agricultural technology (not widely available in the private farm sector) at the time of instituting the group farms, thus increasing the chances of their early success and placing them on a stronger competitive basis with the private sector farms. The establishment of an insurance system among a number of group farms might also be considered, though this would require several successful seasons in order to accumulate an emergency fund. It might also be desirable to limit the advance pay to less than the existing wage scale in the early stages. It is clear, then, that the more thoroughly the members understand and commit themselves to the new institutions from the beginning the greater the chance of success. Thus does the whole approach to mobilization become of vital importance to the internal functioning of the system.

Income from labor is usually not the only form of remuneration in a production cooperative. The private usufruct plot has been a part of all group farming systems outside of communes. Even the Chinese have institutionalized the household plot after a brief, unsuccessful experiment with total communization. As long as housing and consumption is on an individual basis it is more efficient for peasants to continue to grow some of the vegetables and foodstuffs of their daily diet. A basic point related to the incentive system in this regard is that if production from the individual plots becomes too large a component of the family's overall livelihood, it will threaten the functioning of the collective enterprise and endanger the achievement of the other social, economic, and political objectives of the system.

In addition, an important contribution to member livelihood can come from collectively provided social services such as education, health care, cultural opportunities, etc. While directly achieving important social objectives, these services can also strongly reinforce the other collective activities of the enterprise.

In this context it is well to call attention to the key importance of a sound and efficient system of records and accounting. In group farming, not only is record-keeping necessary for carrying out the usual tasks common to all commercialized farming--planning for production, investment, and marketing--but there is the additional need of maintaining an accurate record of work points and a distribution system which underlies the entire incentive structure.

Since most group farming participants in LDCs were previously landless or land-poor peasants who are used to a close-to-subsistence way of life, such careful accounting procedures are largely unknown to them. This is an area where outside expertise and assistance will be vital. However, it is also an area of extreme sensitivity. Bookkeeping is bound to be a slightly mysterious exercise in the eyes of most peasants and can come to be viewed as a threatening tool in the hands of an educated outsider. If an outsider is employed in the beginning, extreme care should be taken to ensure that he is aware and sensitive to this situation. At the same time, the training of local members to take over these functions should be initiated as early as possible, and general education of all collective members in the basic principles underlying these procedures should be undertaken. Finally, no matter who is keeping the books, it is of extreme importance that accounts be settled on time in order that the time lapse between labor input and the corresponding remuneration be minimized.

The Relationship between Collective and Private Activities

The potential conflict between the peasant's individual economic interest and the collective economic interest is most clearly seen in the division between private plot and collective cultivation. Unless a fairly stable equilibrium can be institutionalized between the two, this concrete division of land, labor, and resources can undermine the entire effort and lead either to a complete subdivision into individual farms, or to an economically wasteful in-between situation in which the private plots receive most of the resources, while the bulk of the land in the collective sector stagnates.

In the ideal production cooperative it is presumed that the total area under private cultivation is a relatively insignificant amount compared to the total area under collective cultivation, but the actual manner in which this division is carried out varies and the problems associated with differing approaches may also vary.

Probably the fewest problems will arise in the situation of resettlement where a portion of the land in the immediate vicinity of the new village can be allotted in small plots of relatively equal productive capacity. In the case of land mergers the private plots can also be rationally grouped and allotted, but problems may arise if peasants wish to keep part of their personal land as their plot; also the usufruct plots must necessarily be smaller than what many peasants had originally farmed, and these changes may present difficulties.

In the case of transformed estates, peasants' previous experience with usufruct plots, which exist for workers on most such estates, has often created special difficulties. If a landlord had allowed such plots the expectation will surely exist that a change for the better will mean an extension of these plots. Also, in dividing his labor between the landlord's fields and his own plot, the peasant no doubt developed sophisticated methods of deceit in funneling resources and time from the former to the latter, and he will not easily forget these habits. Likewise, if a market previously existed for usufruct plot production, the structure will already exist for channeling not only private but collective production into these markets after reform.

As Barraclough and Affonso show in analyzing the Chilean experience,[5] peasants will react rationally to economic incentives as they perceive them. Therefore, if the opportunity exists for maximizing personal income by spending more time on the individual plot, peasants will do just that. The problem, then, is to create a system of incentives governing labor, resources, and market outlets which achieves the desired balance.

Clearly the most important factor necessary for realizing this rational balance is the overall economic success of the collective enterprise, and the sharing of profits in a manner that relates remuneration to individual effort in the collective sector. However, especially in the early stages, desired behavior may have to precede visible economic results and short-term economic incentives may tend to dominate. Therefore, certain specific restraints and incentives will probably have to be incorporated into the system to channel labor and resources. No specific model can be applicable to every situation, and these institutional guidelines must be worked out with the maximum participation of the members. But some possibilities can be outlined.

Regarding labor, the advance payment, instead of being the same for all workers, could be directly correlated with the number of work days (or points) performed that month in collective cultivation. Also, some minimum number of work days might be required per month in order to qualify for certain collective benefits or even to remain a member of the group farm.

Regarding resource inputs, if no other channel exists for the peasants to obtain the seeds, fertilizer, pesticides, and water for their private plots, it is inevitable that these will be diverted from the collective sector in some way or other. It is probably best to recognize the importance of the private plot in the overall economy, and to make resources (as well as technical assistance) available on a loan, rental, and/or purchase basis. This will clearly delineate the private and collective sectors and provide for a more economically rational division of resources. Even with a shortage of inputs, some amounts should be made available to the private plots while maintaining the relative priority of the collective sector.

In some cases usufruct plots are rotated among families each year in order to discourage over-investment in the private sector. Though this approach may be effective in some cases, it is at the sacrifice of overall productivity.

No matter what efforts are made to structure internal incentives for resource allocation, if the market price structure does not reward production according to the priorities envisioned in the group farm system all such efforts may be in vain. This is a critical area where macro and micro policies interact. If there is one basic market with price differentials reflecting only the accepted variations in time, form, place, and quality of farm products, the problems will probably not be so great. However, if government loans are issued under an arrangement requiring the marketing of products through official government channels and if prices offered in these official channels are below those obtainable on the open market, not only is the incentive provided for bypassing the official market with collective products, but also the economic importance of private plot production, which is usually under no such restrictions, is accordingly enhanced. The result may be both the diversion of collective production to private markets and the diversion of scarce resources to private plots.

As suggested earlier, to avoid this it may be necessary for the government to forego one of its objectives--acquiring collective production at less than existing market prices--and instead offer near-market or even special incentive prices for group production. The other option is not to tie credit to market requirements, and seek recovery of loans by other means. In any case this will probably be a very difficult issue to resolve and will require experimentation and compromise in close coordination with group farm members.

Finally, this complex of issues concerning tensions between collective and private activities suggests that perhaps a group farming system should incorporate a procedure for its own orderly dissolution. The problem arises when the members, by either formal consensus or informal behavior, indicate that their commitment to the venture is no longer sufficient to make it work. Though especially in the early stages there may be times when the experiment runs into temporary difficulties, there can come a time when there is clearly no chance of achieving the major objectives of the project and participant disenchantment is irreversible. If left unattended such a situation could easily deteriorate into some unofficial resolution whereby the peasants work out a living relationship with the system that is economically as well as socially wasteful to themselves and to society as a whole. It will usually be characterized by almost total attention to and even expansion of the private sector unless strong pressures are exerted otherwise.

All that can be said is that there should be a legal procedure for recognizing and settling such cases. Perhaps an organized procedure would allow a phased institutional shift which could result in the retention of many aspects of cooperation within the context of individual operation of the land, along the lines of the Israeli *moshav ovdim*. Such a step might allow the attainment of many more of the original objectives than a complete abandonment of the venture, and could even serve as the basis for the gradual reemergence of more collective production if and when circumstances permitted.

CONCLUSION

Interest is growing in many parts of the Third World in group farming as a means of effecting fundamental changes in economic, social, or even political institutions in rural areas. Group farming seems to offer a framework for making a broad-ranging assault on the many problems related to rural and agricultural development in less developed countries. Nevertheless, as this survey of issues indicates, the very complexity of introducing such multifaceted change in one sector (or part of a sector) while maintaining less socialist-oriented institutional arrangements elsewhere in the society gives rise to many difficult problems.

The issues discussed here indicate that great attention and care must be devoted to a wide range of organizational issues both external and internal to the group farm if it is to be a successful endeavor. While strong but sensitive support by the central authorities seems critical, governments cannot expect that a group farming system will ensure closer and more automatic control over the agricultural sector and its production. Potential conflicts of interest and tensions often exist both between the group farm and groups external to it and also between private and group objectives within the enterprise itself. Overcoming the apparent dilemma of management in group farming--balancing real participation by members in decision-making with the exercise of necessary expertise and guidance--is always a central problem.

As relatively successful cases in several countries indicate, these problems can be overcome. However, while the experience of other countries can be broadly instructive, it is clear that no single formula for successfully establishing group farms exists which can be applied to every case. The solutions must be worked out over time in ways peculiarly suited to each particular situation and with input from those most directly affected by the institutional change.

NOTES

1. Except for an occasional reference, the analysis does not apply to the "commune" or to the collectivized agricultural systems of the socialist states.
2. See, for example, Solon Barraclough and Almino Affonso, "Critical Appraisal of the Chilean Agricultural Reform," Santiago, Chile: ICIRA, 1972, pp. 31-35.
3. Otto M. Schiller, *Cooperation and Integration in Agricultural Production* (New York: Asia Publishing House, 1969), pp. 34-35.
4. Galeski contrasts the "interactive" system in which market forces are dominant with a "directive" system in which economic units function in accordance with a central plan. See Chapter 1 in this volume.
5. Barraclough and Affonso, "Critical Appraisal," p. 21.

Postscript

Classifications of group farming experiences are important for both theoretical and empirical advances in knowledge. It is necessary to sort out common features as well as differences to ensure that when comparisons are made, experiences are indeed reasonably comparable with respect to the key variables used in the classification. The typology presented by Galeski, and the sorting out by Reed of issues that are likely to create the need for special attention when group farming is introduced in mixed systems are a major step forward. And much research remains to be done on the problems and the prospects of group farming.[1]

In virtually all cases reported in this volume, the peasants and other affected groups have traditionally been strongly biased in favor of private ownership and operation of the land. In most instances, those charged with establishing group farming see this as an obstacle, and this "peasant mentality" must be changed--if not eradicated, certainly subdued. I say "most instances" since the form that group farming is taking in such market economies as France and Japan shows that it does not alter fundamentally the individual entrepreneurship in the farming sector or the existing property relationships, but merely provides a mechanism for the pooling of material and technical resources--including labor. Thus much depends on the type of relationships in production that an executing body is trying to foster. At the same time, another feature common to group farming experiences is the need to reconcile, in one way or another, the twin objectives of efficiency in production and equity in distribution. In many cases, although such reconciliation has been rationalized at the ideological level, it has not been translated into practical results. The numerous internal organizational problems, member commitment, motivation, and morale, and the ambiguities in roles of both managers and members in group farming are all closely related to these fundamental requirements.

A better understanding of these issues, and ways and means of resolving or at least alleviating them, is not likely to come from aggregative studies. National or even regional averages and aggregates cover up, and in any event are not designed to reveal, the many organizational differences that surely exist among the collective enterprises in any country. Given the centrality of these issues, and the universality of their occurrence, there are without question many experiments being carried out, many rules being changed, and many diverse consequences being achieved. Systematic analysis of many different cases (different with respect to the rules established to reconcile some of the dilemmas mentioned above) could be most instructive.

The same rationale for more individual case-study type of investigations applies to a host of other questions that need much more attention. How does the decision-making process really work, what is the level of participation and how is it structured, how do these differ by degree of integration (Galeski's five organizational features of joint ownership of land, socialization of work, system of management, organization of consumption, and distribution of output), what differences are traceable to the origin of managerial personnel (and the degree of permanence and/or lack of government involvement in the process), and what is the relation of these structural-organizational variables (singly or in different combinations) to specific and measurable indicators of economic and social performance? There are bound to be wide differences with respect to these and other variables among group farming enterprises in the various countries, and even within the same country. But such differences can be discovered and significant interrelationships tested only through field-level studies.

Such micro, firm-level studies are also required to get additional evidence on accounting systems and the manner in which the accounting/distributional issues are handled in different cases, and the effect such differences may have on performance. Likewise, much more needs to be known, in a comparative setting, about the interrelations between the family household and the cooperative farm, both in production and consumption.

Both macro (aggregative) and micro (firm-level) studies are required to get a better answer to the basic question: how well have the goals set for group farming actually been achieved--not only political-ideological ones, but economic and social goals as well? How successful have these group enterprises been (and again differences between and among such enterprises including underlying factors and reasons for these differences need to be brought out) in expanding employment? This is a crucial issue in most countries. Such employment expansion needs to be measured not only in terms of farming per se, but also as a result of labor mobilization for infrastructural works and the establishment of rural industries.

If group farming systems can be shown to have a much better record than other systems of "industrializing the countryside" and avoiding the concentration of industry and masses of people in the large cities, this would be a major positive factor in favor of such systems. Countries having a dual farm-size structure, and even those with basically a unimodal small farm system, have not been too successful in such decentralization. Another issue that arises in this connection is the size structure of group farming units. Different rural development activities require different sized organizational units for realizing their maximum potential, or for avoiding major diseconomies of scale. The structuring of multi-size units for different functional purposes (such as the three-tier system found in the Chinese commune) is a prospect that should be thoroughly explored. Likewise, much more research is needed to determine the relationship (and the potential) of group farming to resource conservation strategies and needs, especially as related to such factors as management continuity and economies of scale.

Finally, one must not lose sight of the fact that group farming has many variations and levels and degrees of integration. It will continue to be instructive to study certain types of partial group farming experiences such as joint purchase and use of machines, tube wells, etc. These experiences have wide significance for most of the developing countries whether or not they move to more highly integrated forms of group farming.

NOTES

1. I wish to acknowledge the suggestion for research made by the several authors and discussants at the conference. Some of these surfaced during open discussion periods. In some cases authors responded with written suggestions following the conference.

List of Conference Participants

PROGRAM CHAIRPERSONS

Solon Barraclough
FAO Mission in Mexico
Mexico City

Peter Dorner
Professor of Agricultural Economics and the Land Tenure Center
University of Wisconsin-Madison

Judith Graves
International Bank for Reconstruction and Development
Washington, D.C.

Bryant Kearl
Professor of Agricultural Journalism and the Land Tenure Center
University of Wisconsin-Madison

Vernon Ruttan
President
The Agricultural Development Council
Singapore

Abraham Weisblat
Director, Research & Training Network
The Agricultural Development Council
New York

DISCUSSANTS

Martin Abel
Professor of Agricultural and Applied Economics
Director, Economic Development Center
University of Minnesota, St. Paul

Claudio Barriga
Agricultural Economist
Institute for Cooperative Financing
Santiago, Chile

David Baytelman
Head of the Agrarian Reform Corporation under the Allende Government in Chile
Presently with the Institute of Social Studies
The Heyne, The Netherlands

Marion Brown
Professor of Agricultural Journalism and the Land Tenure Center
University of Wisconsin-Madison

Thomas F. Carroll
Senior Advisor, Economic and Social Development Department
Inter-American Development Bank
Washington, D.C.

Lionel Cliffe
Professor of Anthropology
University of Zambia
Lusaka, Zambia

Ira J. Cohen
Graduate Student-Project Assistant in the Department of Sociology and the Land Tenure Center
University of Wisconsin-Madison

Joseph W. Elder
Professor of Sociology and South Asian Studies
University of Wisconsin-Madison

Richard Fraenkel
Professor of Agricultural Economics
Purdue University
West Lafayette, Indiana

Lane Holdcroft
Bureau for East Asia
Agency for International Development
Washington, D.C.

Leonid Hurwicz
Professor of Economics
University of Minnesota, Minneapolis

David J. King
Adjunct Assistant Professor of Agricultural Economics and the Land Tenure Center
University of Wisconsin-Madison

Herbert Lewis
Professor of Anthropology
University of Wisconsin-Madison

Francisco Oliart
Interamerican Institute for Agricultural Sciences
San José, Costa Rica

J. David Stanfield
Sociologist
Terra Institute and the Center for Cooperative Rural Development
Santiago, Chile

William C. Thiesenhusen
Professor of Agricultural Economics, Agricultural Journalism, and the Land Tenure Center
University of Wisconsin-Madison

Dov Weintraub
Professor of Sociology
Hebrew University
Jerusalem, Israel

OTHER PARTICIPANTS

Kurt Anschel
Office of Agriculture
Bureau for Technical Assistance
Agency for International Development
Washington, D.C.

Charles H. Antholt
Sector and Project Planning Staff
Bureau for Near East and South Asia
Agency for International Development
Washington, D.C.

Alphonse Chable
Office of Development Resources
Bureau for Latin America
Agency for International Development
Washington, D.C.

Herman Felstehausen
Departments of Landscape Architecture, Agricultural Journalism, Environmental Studies, and the Land Tenure Center
University of Wisconson-Madison

Frances Foland
Institute of Current World Affairs
New York, New York

Edward Friedman
Department of Political Science
University of Wisconsin-Madison

Frank Groves
Department of Agricultural Economics, University Center for Cooperatives
University of Wisconsin-Madison

Jane B. Knowles
Editor and Executive Assistant
Land Tenure Center
University of Wisconsin-Madison

Mary Jean McGrath
University Center for Cooperatives
University of Wisconsin-Madison

Mangala Moonesinghe
Sri Lanka

Shao-er Ong
Department of Agricultural Economics
Kasetsart University
Bangkok, Thailand

Jiryis Oweis
Office of Development Planning
Bureau for Africa
Agency for International Development
Washington, D.C.

Raymond J. Penn
Emeritus Professor of Agricultural Economics, Environmental Studies, and the Land Tenure Center
University of Wisconsin-Madison

R. Wayne Robinson
University Center for Cooperatives
University of Wisconsin-Madison

Rigoberto Sandoval
Food and Agriculture Organization, United Nations
Rome, Italy

Julia H. Schwenn
Program Coordinator
Land Tenure Center
University of Wisconsin-Madison

James Scott
Department of Political Science and the Land Tenure Center
University of Wisconsin-Madison

William Smith-Hinds
Instituto de Investigaciones Socio-económicas
Tegucigalpa, Honduras

Roger Soles
Inter-American Foundation
Rosslyn, Virginia

Joseph R. Thome
School of Law and the Land Tenure Center
University of Wisconsin-Madison

Index

18/22 Bodoni

MANUFACTURED BY CUSHING-MALLOY, INC.
ANN ARBOR, MICHIGAN

Library of Congress Cataloging in Publication Data
Main entry under title:
Cooperative and commune.
Based on papers presented at the Group Farming Conference held at Madison, June 10-12, 1975.
Includes index.
1. Agriculture, Cooperative--Congresses.
2. Collective farms--Congresses. I. Dorner, Peter, 1925- II. Group Farming Conference, Madison, Wis., 1975.
HD1491.A3C66 334'.683 76-53651
ISBN 0-299-07380-7